Being Jewish in 21ˢᵗ Century Central Europe

Europäisch-jüdische Studien
Beiträge
European-Jewish Studies
Contributions

―

Edited by the Moses Mendelssohn Center
for European-Jewish Studies, Potsdam

Editorial Manager: Werner Treß

Volume 43

Being Jewish in 21st Century Central Europe

Edited by
Haim Fireberg, Olaf Glöckner,
and Marcela Menachem Zoufalá

DE GRUYTER
OLDENBOURG

ISBN 978-3-11-099149-9
e-ISBN (PDF) 978-3-11-058236-9
e-ISBN (EPUB) 978-3-11-057985-7

Library of Congress Control Number: 2020938208

Bibliographic information published by the Deutsche Nationalbibliothek
The Deutsche Nationalbibliothek lists this publication in the Deutsche Nationalbibliografie; detailed bibliographic data are available in the Internet at http://dnb.dnb.de.

© 2022 Walter de Gruyter GmbH, Berlin/Boston
This volume is text- and page-identical with the hardback published in 2020.
Printing and binding: CPI books GmbH, Leck

www.degruyter.com

Table of Contents

Editors' Introduction —— IX

Section I: Contextualizing Jewish Life in the Midst of the "Old Continent"

Raphael Vago
The Quest for the "Authentic" Central Europe —— 3

Sergio DellaPergola
Jewish Demography in the European Union – Virtuous and Vicious Paths —— 17

Julius H. Schoeps
Renewal or Regression? Jewish Self-Assertion and Re-Orientation in Twenty-first Century Central Europe —— 57

Section II: Breaks, Changes, and Continuities in Austria and Hungary

Vladimir (Ze'ev) Khanin
"Russians," "Sephardi", and "Israelis": The Changing Structure of Austrian Jewry —— 73

Ildikó Barna and András Kovács
Jewish Religious-Cultural Traditions and Identity Patterns in Post-Communist Hungary —— 103

Zsófia Kata Vincze
The "Missing" and "Missed" Jews in Hungary —— 115

Avihu Ronen
Memories and Hopes: The Zionist Youth Movements and the Communist Regimes in Central Europe, 1944–1950 —— 145

Section III: Jewish Past and Present in the Czech Republic

Jiří Holý
Jews and Jewishness in Cinema and Literature: The Case of the Czech Republic —— 165

Marcela Menachem Zoufalá
Ethno-religious Othering as a Reason Behind the Central European Jewish Distancing from Israel —— 185

Zbyněk Tarant
Jews and Muslims in the Czech Republic – Demography, Communal Institutions, Mutual Relations —— 207

Michal Schuster
Jewish-Roma Relations in the former Czechoslovakia: An Alliance Against Racism —— 235

Section IV: An Ongoing Struggle with Judeophobia

Dina Porat
Holocaust Denial as a Symptom of Unresolved European History —— 253

Haim Fireberg
The Antisemitic Paradox in Europe: Empirical Evidences and Jewish Perceptions. A Comparative Study Between the West and East —— 269

Appendix: Memories, Reflections, and Prospects

Konstanty Gebert
What is Jewish about Contemporary Central European Jewish Culture? —— 283

Anna Chipczyńska
Preserving Jewish Cemeteries as an Actual Challenge in Contemporary Poland —— 291

Natalia Sineaeva-Pankowska
Holocaust Memorialization in Poland: A Case Study of Polin Museum —— 301

Tomáš Kraus
Thirty Years After. The Yesterday, Today, and Tomorrow of the Czech Jewish Community —— 311

About the Authors —— 323

Index of persons —— 329

Editors' Introduction

The presented collective monograph focuses on the contemporary circumstances of Jewish life, primarily in Central Europe. The book is divided into four thematic sections and an appendix: **Section I** *Contextualizing Jewish Life in the Midst of the "Old Continent,"* **Section II** *Breaks, Changes, and Continuities in Austria and Hungary,* **Section III** *Jewish Past and Present in the Czech Republic,* and **Section IV** *An Ongoing Struggle with Judeophobia.* The **Appendix** carries the title *Memories, reflections, and Prospects.*

What does it signify to be a Jew in the twenty-first century? Academic discussions revolving around this question have comprised a wide range of controversial points of view. Nevertheless, there is one rather rare consensus, expressed by many, referring to massive consequences of the twentieth century historical legacies – the burden which has been borne by both Jews and non-Jews on the old continent.

Zygmunt Bauman famously defined the Jews as "the pioneers of the postmodern condition,"[1] whose most noticeable contribution to European culture was exposing all the contradictoriness and ambiguities that currently shape the lives of all Europeans.[2] The Jewish diaspora is often perceived as archetypal, and the Jews are considered in many aspects as the perennial minority. At the same time, Jews have undoubtedly had a great share in forming the face of Europe as it is today. This unique Jewish experience of having an ambiguous position of insider and outsider may provide valuable views on contemporary reality and the identity crisis in Europe. Thus, the main common characteristic of European Jewish life should comprise an intense confrontation with the heritage of the Holocaust and unrelenting antisemitism on the one hand, and on the other hand, as a seeming paradox, forceful acknowledgement and appreciation of traditional Jewish learning and culture by a growing number of non-Jewish Europeans.

It is needless to emphasise that a common approach to European Jewish existence from the second half of the twentieth century has been radically shaped by the Holocaust. The world-famous British historian and European studies theorist Tony Judt quoted in his book *Postwar: A History of Europe Since 1945* Heinrich Heine, who once stated that "For Jews (…) baptism is their 'European entry

---

[1] Zygmunt Bauman, *Jews and Other Europeans, Old and New* (Institute for Jewish Policy Research, 2008), 1.
[2] Ibid.

ticket'".[3] Almost two centuries later, Judt responded to Heine that our European entry ticket is Holocaust recognition: "Today the pertinent European reference is not baptism. It is extermination."[4]

Presently, when the term of "Holocaust fatigue" broadly resonates in the western world, threatening hard-won Holocaust awareness and ethical commitment, when the number of Shoah survivors and witnesses is constantly decreasing, and European Jews are again encountering antisemitism on a daily basis, the fundamental question emerges once more on whether Europe will remain the "paragon of the international virtues"[5] and whether the twenty-first century can still, after all, "belong to Europe,"[6] as Judt hesitantly supposed.

Prevailing trends, however, expose a much less positive image. As a result of the aforementioned European identity crisis, there is the infiltration of xenophobia and intolerance towards diversity into a public discourse where it gradually turns into a widely accepted norm.

In the second decade of the twenty-first century, Jews across Europe have begun questioning again whether it would improve their living conditions if they would entirely assimilate into the majority society. The public wearing of religious symbols like the kippah or David's star represents a security threat in most of the European countries. In some countries, the parents of Jewish children are confronted with the unsolvable dilemma of whether they should enroll their children in an ordinary school where they often become the target of bullying, or into a Jewish school where they could become victims of a terrorist attack. Emigration, usually to Israel or the U.S., an option chosen by an increasing number of European Jews, is understandably, due to many causes, not an easy decision. Antisemitism is, therefore, still a highly relevant factor in the lives of European Jews. Even though the perpetrators of antisemitism differ between individual countries, the consequences generally lower the quality of life of Jewish communities and are, in the upshot, relatively interchangeable.

A particular challenge of the last decade is the rise of right-wing populism and nativism, which are not limited to the borderline parts of the political spectrum, but shown off in everyday speeches of political leaders with high electoral credit. This trend by itself is alarming for the Jewish minority as such. Many identify a dangerous parallel between this development and European events in the 1930s.

---

3 Tony Judt, *Postwar: A History of Europe Since 1945* (Penguin Books, 2006), 803.
4 Ibid.
5 Ibid., 799.
6 Ibid., 800.

Jews, especially in Central and Eastern Europe, moreover cope with the curious paradox where some local governmental provisions or the public statements of political elites appear bluntly antisemitic, but simultaneously on the level of international relations have indisputably fervent and firm ties to Israel. With the exception of undeniable pragmatism not burdened by moral premises, these relatively contradictory stances could originate from the ambiguous image that Israel represents for the populist governments. On the one hand, Israel is acknowledged as a welcomed and successful ally in the declared nativist battle for European or Western values that are allegedly threatened by the influx of refugees and immigrants. On the other hand, Israel's mere existence inevitably confronts Europe's never thoroughly confessed and processed guilt for the Holocaust and post-war events.

After the fall of the Iron Curtain, these burdensome topics began slowly opening in post-communist countries, and everything attested that there is the will to gradually come to terms with this past. Nonetheless, the last decade persuasively showed that the national identities of single countries of the Central and Eastern bloc are not consistent enough so they can, speaking in Jungian terms, integrate their shadow in the form of accepting their participation in the Holocaust and subsequent anti-Jewish violence. These deep-rooted issues have assured potential to be also discussed within the context of western European countries – however, such a debate would be beyond the main scope of the presented book.

Nevertheless, not every aspect of contemporary Jewish life across Europe is overshadowed by old-new antisemitism and new uncertainties. For instance, the recent developments in Germany intensified by the Russian speaking Jewish influx from the successor states of the former Soviet Union are positively promising. Irrespective of frequent cultural dissonances and conflicts, the local Jewish communities in Germany have enlarged, stabilised, and made up the leeway for a new Jewish pluralism. Also, the Jewish communities and organisations in Hungary reflect persisting dynamics, though only a distinct percentage of Hungarian Jews is organised and "visible." Against this background, it cannot be omitted that even less plentiful numbers of Jews in Poland and the Czech Republic are highly committed and actively working on communal cohesion and continuity. Advanced discussions on Jewish identity and self-conception have also emerged in Austria and other countries of Central Europe. While searching for their own path and *modus vivendi*, it is of great importance for the Jewish communities to contextualize their past, present, and future not exclusively in terms of demography, human geography, and geopolitics, but also in an attempt to decipher a profound sense of belonging – intensely flourishing in Central Europe. On this very point, a seemingly fundamental question then arises, i. e. what is today perceived as the authentic "Central Europe"?

In **Section I** under the title *Contextualizing Jewish Life in the Midst of the "Old Continent"*, **Raphael Vago** touches on the quest of the traditional and also the present-day "real" Central Europe and offers a whole spectrum of geographical, cultural, and political answers provided by writers and intellectuals like Milan Kundera, Alan Palmer, Jacques Rupnik, György Konrád, Anita Shapira, Róbert Kiss-Szemán, and others. The various and distinct, partly contradictory, statements manifest that an "authentic Central Europe" has not yet been materialized, not even in hieratic and deathless forms, although it is often recalled and reverted for new identity drafts, by non-Jews and Jews as well.

Sergio DellaPergola offers a profound overview of the prevailing demographic situation of the Jews in Europe, which is marked by longstanding trends of assimilation, overage, and seclusion. Despite the considerable number of Eastern European Jews who decided for a new beginning in Western Europe after World War II and also at the end of the Cold War, Jewish community life is seriously threatened in the long run. Nonetheless, unexpected trends and developments are not obviated. One of them might be a recently growing birth rate in Jewish families in Great Britain, which contradicts the general negative life balance in European Jewry.

Julius H. Schoeps refers to the high degree of motivation among many Central European Jews for cultural and organisational self-assertion, and in a similar way for re-orientation within the "Jewish scene." Schoeps states that "an authentic Jewish culture, however, ceased to exist in Europe." At the same time, he reflects on the diverse ways of identity searching, especially among the younger generations. As an appearing element of modern Jewish identity, primarily in the middle of Europe, he notices a resilient "involvement in public matters, commitment to civil society, and most of all, exercising criticism wherever freedom and human rights are under attack."

Besides an ongoing debate to what extent authentic, traditional Jewish culture still is – or again is – present in post-cold-War Central Europe, it is an incontrovertible fact that genuineness and persistence of a community journey can hardly be disputed. Central European Jews have indeed undergone a large number of breaks, transformations, and sometimes surprising continuities.

Section II, entitled *Breaks, Changes, and Continuities in Austria and Hungary*, opens with **Vladimir Ze'ev Khanin**, who describes the new situation of Jewish communities in Austria since the end of the Cold War. The fall of the "Iron Curtain" enabled massive Jewish emigration from Eastern to Western Europe accompanied by peculiar intercultural encounters. Besides the small group of originally German-speaking Jews, "Russians," "Sephardi" and "Israelis" are instantly guaranteeing the dynamics and a demographic continuity. Khanin further recounts a permanent identity search among the respective Jewish subgroups in today's Austria.

Based on recent empirical data, **Ildikó Barna** and **András Kovács** are suggesting that Jewish religious and cultural revival has continued, to some degree, within the Hungarian Jewish community for ca. 20 years. According to the authors, this Jewish revival has notably affected the age groups of 25–44 and 45–64-year-olds, and the very young Hungarian Jews at a slower pace. Barna and Kovács also argue that striking trends of intermarriage will not necessarily lead to the disappearance of Jewish tradition in Hungary.

However, the identified revival of Jewish religious and cultural customs in Hungary seems to not be automatically connected with a significant presence in veteran nor recently established Jewish communities. **Zsófia Kata Vincze** estimates that about eighty-five to ninety percent of Hungarians with any Jewish background are not instantly committed or affiliated with any of the functioning Jewish organisations. Nevertheless, Vincze assumes that the constant creation and re-creation of additional institutions indicate that active Hungarian Jews still believe in reaching out to the "greater Jewish population" and eventually attracting "the missing ninety percent."

Avihu Ronen raises the question of hyphenated identities among Jews in Central Europe, albeit going back to the second half of the 1940s. He recalls Jewish communist leadership's attempts shortly after World War II and the Shoah while trying to realise both supporting the project of the Jewish State and also encouraging a new social formation in their countries of origin. The "experiment" continued until 1950, while this idea paradoxically subverted itself. The efforts of the Zionist movements led out of Europe, thus weakening the local Jewish communities. In parallel, the totalitarian regimes aimed to assimilate the remaining Jewish populations.

Further, more recent decline brought many Jewish communities into difficulty at the end of the Cold War when tens of thousands of Jews left, especially Eastern Europe, and headed to Israel or the U.S.

Still, in the early 1990s, remnant Jewish communities with international support became properly re-organized, initiating their own redefinition and opening new chapters with Christians, Muslims, Roma, and other ethnoreligious neighbours.

Such new ways and experiences are exemplarily illustrated in **Section III**, *Jewish Past and Present in the Czech Republic*. Here, **Jiří Holý** depicts how Czech cinema and literature featured the murder of 80,000 Czech and Moravian Jews after World War II. As he writes, Jewish topics and literary debates on the Holocaust sprung up, especially during the 1960s. Holý refers to two different approaches regarding this subject: on the one hand, a "closed narrative" which portrayed the Shoah as a monumental narrative with authentic details, and, on the other, an "open narrative" which presented the persecution of Jews and the war without any heroism, pathos, and sentiment, featuring objective reports, but also farcical and tragicomic elements. After the 1989 Velvet Revolution, this literary tradition witnesses a remarkable continuation.

Marcela Menachem Zoufalá reflects on how Czech Jews relate to the contemporary State of Israel and poses the question of whether Israel appears as a "distant beloved homeland" or rather as an "ambivalent bond." Menachem Zoufalá indeed argues Czech Jews are distancing themselves from Israel, though for quite different reasons than, for example, American Jews. Based on a qualitative anthropological study, the author detects two particular, yet interconnected themes repeatedly emerging. First, the mentioned partial disengagement which seems to occur in the context of the transformation of ethnic setting within the Israeli population. In other words, we are referring to a process often identified as Mizrahization. The second interrelated theme represents nostalgic, idealized views of the Central European Jewish past. This mental imagery, to some extent, weighs into the distancing of Czech Jews from Israel and the Israelis.

Zbyněk Tarant dedicates his contribution to the current situation and the future perspectives of the very small and fragmented communities of Jews and Muslims in the Czech Republic. Both struggle with certain trends of particularism, and they share the reality of being relatively less plentiful (ethno)religious minorities within a quite secularised majority society. Present encounters between the members affiliated with any communities or institutions seem rather rare, and according to Tarant, the relations, while "not ideal," nevertheless can be characterized as being "correct." Nevertheless, escalations of the Middle East conflict(s) unsettle both groups, and the mainstream perception of the consequences of the so-called refugee wave might challenge the relatively peaceful status quo on both sides.

Michal Schuster describes Jewish-Roma relations in the Czech Republic as an essential alliance against present racism. Czech Roma and Jews have both struggled with hostility, persecution, and annihilation during the time of German Nazi occupation, and these common experiences have carved into their collective memories. Contacts between Jews and Roma in the country are solidly cohesive, and the groups support each other in their efforts to establish vivid forms of remembrance. Thus, Roma representatives attend the annual events on the Yom ha-Shoah in Prague, while several Jewish institutions – like the Jewish Museum and the Terezín Initiative Institute in Prague – actively cooperate with the Museum of Romani Culture in Brno.

Like a microcosmos of the whole Central Europe, Jewish life and Jewish-/non-Jewish relations in the Czech Republic reveal striking developments and opportunities that nobody would have had predicted at the turn of the millennium. Though, also in the Middle of Europe – like across the "old continent" – the ghost of Judeophobia seems to have returned. Despite continuing efforts of elucidation, programs of further education, and prevention, antisemitism is also in Central Europe rigorous in the air; or in more latent forms such as Holocaust denial and relativization of the Shoah.

In the opening article of **Section IV**, *An Ongoing Struggle with Judeophobia*, **Dina Porat** interprets Holocaust denial as a symptom of unresolved European history, writing: "Denying the Holocaust might mutate to a 'cultural event,' not taken that seriously by large publics but disdaining the victims and the survivors." Furthermore, Porat claims that it is of utmost gravity to realise that the across-Europe urge to deny or at least to relativize the Holocaust is not a problem of uneducated social strata, on the contrary. A "competition of victimhood" (Bernard-Henri Levy) seems to be taking place behind the scene.

Haim Fireberg analyses an antisemitic paradox, based on empirical data, collected among Jews and non-Jews in Europe. According to him, the level of violent antisemitism does not necessarily indicate the state of antisemitic perceptions. Generally, it could be a necessary condition in defining an antisemitic atmosphere, but undoubtedly not a sufficient one. Moreover, there is growing frustration from the political establishment, from ruling parties, and the deficient solutions they supply to control antisemitism, but even more importantly to provide a common basis for fractioned societies. Fireberg is convinced that without belief in the future of the country, and without confidence that Jews are an essential component of their respective society, Jews feel abandoned, and the problem will continue to exist.

The **Appendix** entitled *Memories, Reflections, and Prospects* mainly turns to present-day challenges for Jews and non-Jews, especially on how to define disparities and similarities in the context of commemorating the past and on how to determine cultural mutualities (or polarities). Most of the contributors demonstrate these issues in the cases of ongoing developments in Poland.

Konstanty Gebert poses the question, "What is, in fact, Jewish about Central European Jewish Culture?" To a certain extent, he builds on Ruth Ellen Gruber's book *Virtually Jewish: Reinventing Jewish Culture in Europe* and shows through examples of today's Poland that non-Jewish interest in former, authentic Jewish (cultural) tradition extends beyond the limit. Gebert argues that "a problem, however, arises when we have non-Jews not only organising or consuming Jewish culture, but producing it: writing novels on Jewish themes, for example, or composing and performing klezmer music."

Anna Chipczyńska, former president of the Jewish Community of Warsaw, raises the challenge of preserving Jewish Cemeteries in Poland. As a point of departure, she refers to the Law of February 1997 on the Relations between the Polish State and the Jewish Religious Communities, which regulates, inter alia, the rules for the return of the pre-WWII Jewish communal property to the contemporary Jewish communities in Poland. The main question remains: how shall a community with just a few thousand organized Jews manage the preserving of more than 1,200 Jewish cemeteries across the land? Chipczyńska also refers to some other very relevant laws that might affect the future of Poland's Jewish cemeteries, such as the 1959 law on Cemeteries and Burying the Deceased, the 2003 law on Spatial Planning, and the 2003 law on the Protection and Preservation of National Heritage. The author argues that "these tools are both a challenge and a chance for the protection of Jewish cemeteries in Poland. Their effective application could theoretically make the Jewish community a very successful watchdog and heritage preservation organisation."[7]

Natalia Sineaeva-Pankowska portrays the "Polish Complexities of Dealing with the Past" from a guide's perspective, specifically from her personal experiences as a guide and educator in the Polin Museum of the History of Polish Jews in Warsaw, which opened in 2014. Sineaeva-Pankowska testifies that the Polin Museum does not only concentrate on multidimensional historical ties of the Polish-Jewish past, but also tackles difficult subjects. For example, it includes quite recent outrageous events, such as the anti-Jewish pogroms in Poland in 1941 and anti-Jewish violence in 1944–1946, in order to reflect the shadow which lies upon the Polish past. This first-hand account is highly relevant, especially in times of constant growth of distinct nationalist trends of revisionism and "whitewashing," which are succeeding not only in Poland, but in other countries of Central Europe as well.

Finally, the focus returns to the Czech Republic. **Tomáš Kraus**, the Executive Director of the local Federation of Jewish Communities since 1991, ruminates on the more than one thousand years of Jewish presence on the territory of today's Czech Republic. He illustrates the significant extent to which Jewish thinkers and artists have influenced Czech culture and identity. Despite that, at the beginning of the twenty-first century, the Czech Jewish communities struggle to find their unique expression. On the other hand, Kraus underlines that the Czech Jews find themselves in a comfortable position: "The Czech Republic – with

---

[7] However, in August 2018, the Polish government had declared that it would map out all the Jewish cemeteries in the country and establish a computerised database with information about the people buried there and the local Jewish history. See Ynet News, August 12, 2018, www.ynetnews.com/articles/0,7340,L-5419925,00.html, accessed March 15, 2019.

all its controversies in the public arena – is a safe-haven for Jews. But is that enough? Can we survive without new programs, without new ideas, without opening-up, instead of closing down because of conservative approaches?"

This is a whole range of crucial questions that might influence Jewish existence in the twenty-first century – not only in Central Europe but across the continent and beyond. This collective monograph represents a rather minor piece in a broader discussion that has been gradually unfolding over the years.

With this, we would like to express our gratitude to all respectable authors for their valuable contributions. Further, our sincere appreciation is extended to the Tel Aviv University, Charles University, and the Moses Mendelssohn Centre for European Jewish Studies at the University of Potsdam, all of which supported the conception and making of this book with remarkable expertise and great enthusiasm. For the important work of proofreading and improving the manuscript we owe a great deal to Anna Hupcejová, Gritt Wehnelt and Vijay Khosa. Lastly, we wish to present our special thanks to De Gruyter Publishing House for including this book in its prestigious "European Jewish Studies" series.

Marcela Menachem Zoufalá, Olaf Glöckner, Haim Fireberg

Section I: Contextualizing Jewish Life in the Midst of the "Old Continent"

Raphael Vago
The Quest for the "Authentic" Central Europe

Is there still a clear meaning to the term "Central Europe," three decades after the demise of communism and the fall of the "Iron Curtain"? This essay presents the many meanings that have been attributed to this term along the twentieth and twenty-first centuries by thinkers and historians. Not only has no consensus been reached on the geographical borders of this territory, but also the efforts to explain what "Central Europe" means from a socio-cultural point of view are truly challenging, especially in terms of the construction of territorial identities and the politics of memory. Jews have had a very important role in participating and creating the Central European space, which, eventually, also became destructive for them during the Holocaust. In the search for Central Europe, which has not yet ceased, we witness not only nostalgia and romanticizing of the past, but also practical efforts to form new-old national and cultural discourses.[1]

In the last decades of the twentieth century, when different signs indicated the imminent collapse of the communist regimes, a long-standing debate resurfaced over the essence and destiny of Central Europe. Part of this debate continues to this day and focuses on the very definition of the term "Central Europe." The discourse on Central Europe deals with the term's implications on national and regional levels, cultural identities as well as on economic and political processes. Although part of this discourse is reflected in the intellectual sphere, it also has important repercussions on many other realms.

The discussion focuses on several questions: does Central Europe exist? How can it be defined? Who needs it and for what purpose? Moreover, given the unification processes in Europe, is there still significance to regional demarcations and identities? This subject undoubtedly requires an interdisciplinary approach involving both the humanities and social sciences. Are the definitions based on a classical historical approach regarding the divisions of Europe, in different eras, into political entities which underwent changes because of wars and international agreements? Is the definition based, *inter alia*, on geographic and geopolitical grounds? Can we define this region along ethnic lines, as a battleground for conflicting national movements? Can we identify Central Europe as

---

[1] The first version of this essay was published under the title "In Search of Central Europe," in Hebrew in *Zmanin* 114, Spring 2011, Tel-Aviv University.

the largest killing ground of the Jewish and Romani people, wherever from its killers arose? Can we define it as a cultural sphere according to "mentality," according to "Central European" qualifiers such as *politesse?* Can we also attribute to this region a culinary history and characteristics such as the introduction of "Turkish coffee" – a heritage of the Ottoman Empire – into the culture of the Viennese coffee houses? Can we at all offer just one practical definition to the term "Central Europe?" I would claim that we cannot reach one single definition and that, for research purposes, we need to embrace *ad hoc* several of the different definitions and present the historiographic aspects and implications of the discourse which has been taking place mainly since half of eighties of the twentieth century.

Europe – a Continent with Many Centers

An essay by Milan Kundera, the Czech writer in French exile where he has lived since 1975, revived public interest in Europe – initially only in Western Europe – and more specifically the discourse on the essence and destiny of Central Europe. The essay was published in several languages under the title "The Tragedy of Central Europe," and stimulated many responses after its publication in the *New York Review of Books* on April 26, 1984 in its English version. Kundera defined Central Europe, which was then under a communist-Russian regime (he stressed the word "Russian" rather than Soviet), though it always was attached to Western culture and remained an integral part of Europe. He wrote, "Central Europe is not a state: it is a culture or a fate. Its borders are imaginary and must be drawn and redrawn with each new historical situation."[2] Alongside this statement, he claimed that what characterizes this region is "the greatest variety within the smallest space."[3]

Although Kundera generally accepts the definition of Central Europe as a region, which in different periods had been under the influence of German culture and under the rule of the Habsburgs, he regards it mainly as an imaginary territory – as an idea or a state of mind, and less in terms of changing political borders. That means he regards Central Europe as a mental space inside a changing geopolitical region.

One should clarify there are numerous, complex problems regarding the definition of "Central Europe" as a region that was shaped by the cultural legacy of

---

2 Kundera 1984, 35.
3 Ibid., 33.

the vast Austro-Hungarian Empire and influenced by German culture. This specific concept of "Central Europe" is common mainly in the intellectual discourse in post-communist countries in that region. Those states that were in the past somehow integrated into the empire and later embraced by the Austro-Hungarian monarchy dissolved after World War I. It seems that the definition of Central Europe as a region that has spiritual rather than political roots suited the aspirations of these countries that were liberated from communist rule in 1989 rather well.

The pursuit of a center of the continent had different significances since the beginning of the nineteenth century and was connected primarily to the request for a united Germany, German nationality, and the definition of the identity of multinational and multicultural Habsburg Austria. The book of the liberal-democrat German theologian, Friedrich Naumann, *Mitteleuropa*, published in 1915, emphasized Germany's location as "the country of the middle" ("Land der Mitte," commonly used in German discourse) and the pivotal role of Germany in an economic union of the whole German space together with other parts of Europe except Russia, France and Britain. The Nazis' use of the term "Mitteleuropa", meaning middle Europe, expressed Germany's expansionist imperialist aspirations. This term was cautiously reintroduced in the German lexicon after World War II by Chancellor Willy Brandt, father of West Germany's "Ostpolitik", while giving a speech in 1973 in the Bundestag mentioning Germany's role in Mitteleuropa. Two years later, the succeeding Chancellor Helmut Schmidt claimed that "East Germany, being a Central European country (mitteleuropäischer Staat), should contribute to détente in Europe." Is the common use today of the German term Zentraleuropa an attempt, among others, to disconnect from the term used by the Nazis? And is there a geographical difference between Mitteleuropa and Zentraleuropa? Probably, though these questions will remain unanswered.

Where is "the center of Europe" situated? If we could define it geographically maybe we could then progress towards a geopolitical, historical and even mental definition. From the point of view of the Germans, it is their country which is situated in the center of Europe; however Switzerland, Austria and some other countries also beseech this attribute to themselves. If we assume, as broadly accepted, that Europe ends at the Ural Mountains, then according to various measurements (that are themselves in dispute) the center of Europe might be situated in Dilove – a forlorn village in Ukraine (or many other places claiming the right to be the "center"). If indeed the center of Europe is located in western Ukraine, or south Poland, close to the Slovak border, where then is Eastern Europe situated? These problems arise from our imaginary maps. We generally regard shining Prague as situated in Eastern Europe and Vienna in Western

Europe, although geographically Prague is situated west of Vienna and Vienna is therefore a more "eastern" country, at least from a geographical point of view. On the other hand, the Ukraine is conceived by us as "East European" although its western part was included in the Habsburg monarchy, and according to one of the common criteria, the term "Central Europe" is synonymous with the late Austria-Hungarian Empire. Odessa with its pluralistic cultural atmosphere, situated on the Black Sea coast, was sometimes considered as a "Central European" city, to a large degree because of its Jewish residents.

However, at the same time, it could be regarded as a thrilling laboratory of an encounter between Russian and Western cultures, including the realized Jewish potentials emerging from within. The founder and first president of Czechoslovakia, Tomáš Garrigue Masaryk, mentioned in this context "the zone of small and smaller nations extending between West and East, specifically between Germans and Russians,"[4] in which he included peoples from Lapland in north Scandinavia to Turks and Albanians in the Balkan, but by no means Germans nor Austrians.

The difficulty of defining the term plays a very prominent role in the discourse on Central Europe, and many publications begin with explanations and justifications for their particular approaches. The multiple attempts attest to the confusion, and difficulties and differences of opinions are inseparably bound to the search for a Central Europe. If we could define where Western and Eastern Europe is, we could then define where the center is. Perhaps it would be better to call this region "The Lands Between", a name given by the British historian Alan Palmer to his book, referring to the countries between the German world in the West and the Russian world in the East.

In 2003 in London, at an event organized for the inauguration of an academic journal entitled *Central Europe*, which started its publication in the school for Slavic and East European Studies at University College, Robert John Evans from Oxford University, one of the senior scholars of the history of the Habsburg world, gave the opening speech. He reviewed several of the delineations of "Central Europe" and mentioned among others the often-cited 1996 research by Lonnie Johnson, which incorporated in Central Europe today's Germany, Poland, the Czech Republic, Slovakia, Austria, Hungary, Slovenia, Croatia, and during earlier periods also the more eastern areas. Evans concluded his speech by saying that "Central Europe" has always been more of an aspiration rather than a reality. He thereby joined the opinion of other thinkers, according to which there exists a

---

4 Masaryk 1920; 2014, 35.

spiritual Central Europe – a "supreme" one beyond an "earthly" or "physical" Central Europe.

This was Kundera's tone as mentioned above, and this was also the opinion of the Hungarian writer György Konrád, who belongs among the eminent "seekers of Central Europe." His essay "Does the Dream of Central Europe Still Exist?", completed in 1985 (the amended version of his acceptance speech upon receiving the Herder Prize in 1984), was firstly published in Vienna because at that time it was not possible to reveal any strong criticism against the obfuscation efforts of the communist regime against Hungary's Central European identity. Here he expressed the view that being Central European means to adopt a certain worldview – a sensitivity to complexities: "Being Central European means, to consider variety as a value."[5]

If one accepts the common definition according to which Central Europe encompasses the areas of the Austro-Hungarian monarchy, Germany's place then becomes a central issue. Among the Germans, also after the unification of the two Germanys in 1990, there is a consensus that Germany should be included in this definition. It is probably not accidental that in former East German universities there are departments and centers that promote the study of Central Europe in order to emphasize the affiliation of the Eastern part of Germany to Central Europe.

But what about those countries that in the past partly belonged to Austria-Hungary, but were expanded by the addition of other territories that were not part of the German-Austrian world? Are these nation-states also part of Central Europe? Is all of Poland, as defined in the journal *Central Europe*, a Central European state, although it includes areas that after WWI were under Russian rule? Are there cultural characteristics, an identity, a mentality, which make all parts of Poland worthy of the term "Central Europe"? What makes Vilnius, which between the two world wars was part of Poland and today is the capital of Lithuania, a city that deserves to be called "Central European?"

It seems that there should be no doubt that the term Central European fits the Polish city Wrocław, in Silesia, known before as Breslau, which came previously under the rule of Bohemia, Austria, Prussia, and Germany. Two British historians, Norman Davies and Roger Moorhouse, described in their book *Microcosm: A Portrait of a Central European City* Wrocław and its rich past as epitomizing all the aspects of Central Europe, including the Jewish presence and influence until the rise of the Nazis to power.

---

[5] Konrád 1986, 109–21.

The literature on the city of Lviv, or Lwow, or Lemberg, today in the Ukraine, affirms that it is a Central European city par excellence – and nobody questions this; the authorities of the city advertise it as such and renovate it after years of neglect in the manner of a city that returns to its grandeur during the Austrian days. Chernowitz, called "Jerusalem on the river Prut," the eastern stronghold of Vienna, was between the two world wars the Romanian Cernăuți, and is today the Ukrainian Chernivtsi. If so, is Ukraine, or at least its western section, part of Central Europe? And if so, where does eastern Ukraine belong? Is this great country divided into west – which is more pro-western and tries to push towards the European Union, and east – in which an ethnic Russian minority of millions of people lives and has a stronger attachment to Russia? This question is even more burning since the Russian military war against Ukraine, which started in 2014 after the Euromaidan revolution, and which increases the danger of breaking the country into western and eastern parts.

Different definitions describe modern Romania as a crossroad between Central and Eastern Europe, between the Balkan and the regions situated north of it. Most of the Romanian population is Christian Orthodox, whereas Central Europe is identified with Catholicism (next to Protestantism, Judaism, and atheism); therefore only Transylvania, which is today part of Romania, lived under the rule of Austria-Hungary and inherited its legacy. Is Romania therefore a Central European country? In post-communist discourse the answer is positive, although the emphasis is placed on its being a bridgehead of cultural encounters. Given its cultural tradition, its historical past, its multinational character, the German ethnic long-term presence, and the Hungarian population, isn't Romania part of Central Europe?

In his book, *Balkan Ghosts: A Journey Through History*, Robert D. Kaplan includes Romania in his voyage to the Balkans. However, when he writes about the city of Cluj, the central city of Transylvania (and the place where I was born, R.V.), he says: "For me, no city in the whole of the Balkans is quite as intoxicating as Cluj, with its steep, gabled roofs and its yellow baroque facades lining narrow streets of cobblestones [...] there is an Indian-summer quality to this provincial outpost of Central Europe,"[6] as if the city found itself in the wrong neighborhood of a Balkan state, Romania, because of historical circumstances.

More examples of the importance of regional identities and the difficulties of definition can be found in the former Yugoslavia. Slovenia and Croatia define themselves as countries belonging to the Central European sphere, because of their historical affiliation with Catholic Western Christianity as part of the Habs-

---

6 Kaplan 2005, 153.

burg Empire. Since the beginning of the 1990s, their discourse on this issue has broadened. Yugoslavia was an artificial entity, a "Balkan" structure that did not accord with the historical legacy of these two entities. In Croatia, emphasis is put on the affiliation with the sphere of the Adriatic Sea and its diversified traditions, whereas in Slovenia the intellectuals speak with arrogant "orientalism" about their former partners in communist Yugoslavia. From Ljubljana, the Slovenian capital, the eyes are directed beyond the mountains, towards Vienna. The Bosnians, the Albanians, the Macedonians, and the inhabitants of Montenegro are sometimes in overstatement perceived as an underprivileged population, which has a Balkan legacy of violence, and will never be able to recuperate from hundreds of years of Ottoman occupation. At the same time, Croatia and Slovenia nurture the Austria-Hungarian legacy as a site of memory, which they regard as preferable to the Balkan partnership. If so, were Croatia and Slovenia incorporated in the "Balkan" just because of the creation of Yugoslavia after World War I and its renewal after World War II? Were these two republics taken out from Central Europe and engulfed in a world of intrigues, identified with the Byzantine Balkan legacy? In this context the Austrian legacy in Bosnia-Herzegovina at the end of the nineteenth century and the beginning of the twentieth should not be omitted. Sarajevo, the capital of this region, in this time also underwent rapid modernization processes, freed itself from the burden of the Ottoman legacy, and became temporarily a "Central European" city. This makes the picture even more complicated.

During the communist era, it was not possible to speak about regional differences and about the diverse cultural legacies. In fact, the very use of the term "Central Europe" was forbidden in the countries of "Eastern Europe," because it hinted at a possibility that the unity of the socialist countries might be severed. According to the communist narrative, there existed an integrity and continuity from the coasts of the Baltic Sea in Poland to the Bulgarian Black Sea Coast, represented by unity, solidarity, cooperation, and a strong kinship, with deep historic roots to the Russian people, the Russian culture, and the Soviet Union. The Cold War and the division of Germany created a map according to which the "German Democratic Republic" was part of communist Eastern Europe. It is obvious that the eastern part of Germany does not belong today to Eastern Europe. It is evident that with the end of the Cold War the division of Europe also changed – therefore in our imaginary map it is as if the "German Democratic Republic" moved to the West and became part of Western Europe, or, according to some definitions, became again part of Central Europe.

So What is There Anyway?

The interest and discussions that grew after the publication of Kundera's paper in 1984[7] accompanied the gradual decline of the communist bloc – a process that was not yet clearly evident when Kundera wrote his essay, but he predicted the end of the communist regimes even before Mikhail Gorbachev came to power. His essay was translated to Czech and was circulated in his country by the *samizdat* system under the title *The Stolen West*.[8] Kundera's main argument was that the Soviet Union, which inherited from Czarist Russia an "Asian" and non-European mentality and manners of conduct, hijacked Central Europe from its natural position as an inseparable part of the Western world, and subjugated it to communist rule and to Marxist ideology, which was nothing but a camouflage of traditional Russian imperialism. Central Europe, which was identified as part of the Western world and culture, was abandoned mainly in the Yalta conference; the West gave it up and it fell into the hands of the Russian kidnappers. The culture and patrimony of Central Europe are those of Western Christianity, whereas, according to Kundera, Orthodox Christianity is Eastern and Russian. In other words, according to Kundera, Asia begins on the border between the Western Christian world and the traditional "dark" world of Orthodox Russia. He thus expressed the same opinion as that of Klemens Wenzel von Metternich, the Austrian State Chancellor, who already in the beginning of the nineteenth century regarded the edge of Asia somewhat east of Vienna. In Kundera's essay, he indeed presented Central Europe, similarly as Masaryk, as "an uncertain zone of small nations between Russia and Germany",[9] as an eternal victim, and as a boxing ring between the Powers from the West and from the East. Like other thinkers, Kundera recalled the historical image of Central Europe as the front defender – a Western Christian entity which protects Europe's gates against its enemies from the East – and now this loyal "gatekeeper" has been betrayed and given up by Europe. He stressed the organic association of Central Europe with Western Europe and spoke in detail about the significance of the detachment between the two parts of the continent because of the abduction by the "Russian Bear."

Kundera's critics maintained that he was nostalgic and longed for the spirit of Central Europe, which in the past was indeed multinational and multicultural.

---

[7] First published in French under the title "Un occident kidnappé ou la tragédie de l'Europe centrale" in 1983.
[8] *Únos Západu*, 1985.
[9] Kundera 1984, 35.

However, it also brought the birth of Nazism, fascist movements, antisemitism, and racism that caused the largest genocide in history – aspects that Kundera did not ignore. The critics reminded him that the decline of Central Europe did not begin with the Soviet-communist expansion from east to west, but rather with Hitler's march in the opposite direction. The writer György Konrád stated in blunt terms that, "After all, we Central Europeans began the first two world wars" (Garton Ash 1999).[10]

With the disintegration of the communist bloc, there have been intense debates in many journals and books in the quest for "Central Europe." By the end of the Cold War, the map of the continent was re-drawn and the European Union began a campaign to have the former Eastern bloc join it. Many claimed that Eastern Europe and its predominant part, Central Europe, are coming back to Europe – the very Europe that abandoned them in the twentieth century.

Distinguished intellectuals from the former communist countries, among whom were György Konrád in Hungary, Václav Havel in Czechoslovakia, and Adam Michnik in Poland, watched closely the comeback of Central Europe into Europe and the return to the historical roots of the peoples in the region. Intellectuals and researchers from the West joined them, such as François Bondy and Jacques Rupnik from France, as well as Tony Judt and Timothy Garton Ash from Great Britain. Garton Ash published several editions of his collection of essays on the destiny of Central Europe, which he updated according to the changes taking place.[11] Not only did the discourse on Central Europe recognize the division, enmity, and national conflicts, but it actually rather emphasized them. Undoubtedly, Central Europe is a unique place for its rich and diversified culture, but it is also a region characterized by internal divisions that were not only caused by great power intrigue, but by internal forces as well. Central Europe was at the forefront of the pursuit of the "other" – the history of antisemitism in the region is proof of this – and attempts to change and blur national identities. The policy of "Magyarization" (the attempt of the Hungarian regime to impose Hungarian culture on other ethnic groups) may serve as an example. This aspect of Central Europe was after the fall of the Iron Curtain seemingly abating. However, recent developments have indicated a forceful return of the nationalistic rhetoric.

The nostalgic weight placed on the glorious and sometimes imaginary days in a world that has disappeared, but also on the disastrous consequences of na-

---

10 Garton Ash 1999.
11 E.g. Timothy Garton Ash, "Does Central Europe exist?" *The New York Review of Books* 33, no. 15 (1986): 45–52.

tionalism, was rapidly adapted to the needs of the era after 1989. The debate on the essence and location of Central Europe is meant to shape identities for the future, by the means of identities from the past. The insistence on the affiliation to Europe is aimed at strengthening the rapprochement to, and the inclusion in, a unifying Europe. Many of the political leaders of the new era were intellectuals who were once opposed to the communist regimes, cultivated the Central European identity, and eventually made it a basis for cooperation between the states in the region. Central Europe became a "club for candidate members" at the gates of the European Union; in the name of the historical partnership, real and imaginary, they wished to shorten the preparatory steps and time for their access. In parallel, it was often claimed in mainstream discourse, the states situated east and south of the "center" were not ripe for joining "Europe;" the first wave of countries to enter the Union should include members of the "Central European club" – the senior and lost sons of Europe – and so indeed it was.

Were the Jews the Real Central Europeans?

There was also a Jewish aspect in the search for Central Europe, and Kundera's contribution to this theme was cardinal. He described the Jews in emotional words as a central factor in paving the way to Central Europe:

> Indeed, no other part of the world has been so deeply marked by the influence of Jewish genius. Aliens everywhere and everywhere at home, lifted above national quarrels, the Jews in the twentieth century were the principal cosmopolitan, integrating element in Central Europe: they were its intellectual cement, a condensed version of its spirit, creators of its spiritual unity. That's why I love the Jewish heritage and cling to it with as much passion and nostalgia as though it were my own.[12]

Kundera refers to prominent personalities of Jewish origin who shaped the Central European spirit and expressed it: Sigmund Freud, Franz Kafka, Gustav Mahler, Josef Roth, Kundera's close friend the Yugoslav writer Danilo Kis, and others. Kundera pointed out that these personalities, who became symbols of Central European culture, originated from different parts of Austria-Hungary and especially from small marginal towns, like Brody in Galicia (today's Ukraine) – like for example Josef Roth who laments in his works the twilight period in his motherland, Austria-Hungary.

---

12 Kundera 1984, 35.

The Jews were indeed the true original Central Europeans, and maybe the only ones, as they were often said to be with a grain of cynicism. According to Kundera, their cosmopolitan cultural contribution undoubtedly served as cement for the Central European existence. However, Central Europe killed its Jews, and large portions of its inhabitants became their hangmen out of their own volition. Karl Kraus, the Austrian Jewish satirist of Czech origin, perceived this part of Europe as "an experimental station in the destruction of the world."[13] This expression, recently, inspired a title of Robert Wistrich's book: *Laboratory for World Destruction: Germans and Jews in Central Europe*.

With the most recent generation of young Europeans, a whole industry has developed to promote the remembrance of Jewish sites, which have ceased to exist, but have left their signs until today. As time passed, "the missing Jews" became the phantoms of Central Europe. The beautiful cities and sites of Central Europe – Budapest, Prague, Bratislava, Krakow, and gradually also Lviv, Chernivtsi, and many others – try, some more and some less, to revive and replenish the void the Jews have left (see for example Omer Bartov's Hebrew article called "To obliterate and to forget: last remnants of Jewish Galicia in today's Ukraine"). In his book *My Czernowitz*, Zvi Yavets wrote about this Central European world before World War II and the Holocaust, describing how the multi-faceted Jewish German heritage had to face Romanian rule, political radicalization, and the rise of antisemitic elements. It was also in Czernowitz where the tragic poet Paul Celan was born and grew up.

Can Central Europe return to its rich cultural past without the Jews? Is there a Central European humor without the Jews? Is Central European literature the same without Jewish themes? Such questions are rather rhetorical in nature. It is evident today that the Jews were not only the cement or glue that solidified Central Europe, as Kundera puts it, but that the history of the Jews – their contribution and their annihilation – has become cardinal in the Central European myth. It is therefore no accident that in the last generation there are also Jewish thinkers among the many intellectuals who cherish the idea of the Central European spirit.

Central Europe also plays a crucial role in the research of the history of Zionism. Scholars like Shlomo Avineri, Anita Shapira, the recently deceased Robert Wistrich, and others regard Zionism as a product of the Central European experience. Can one, for example, understand Herzl's or Nordau's ideas without recognizing the national tensions and the national revival movements in Austria-Hungary? To what extent is Central Europe the heart of that "Glorious, Accursed

---

13 Wistrich 2007.

Europe," as Jehuda Reinharz and Yaacov Shavit entitled their book? One of the major paradoxes of the phenomenon called "Central Europe" is that the more the Jews contributed to modernization, culture, and pluralism in a certain region, the more that narrow-minded local nationalism tried to obliterate them.

Has the Idea of Central Europe Come to an End?

It seems that in recent years, the debate on the concept of Central Europe has somehow lost its impetus. It may be that the first generation of intellectuals, who fought for this cause, have witnessed the realization of their ideas in the return of Central Europe not only to European consciousness, but also on the ground, and have felt that their task had been accomplished. Moreover, the aforementioned generation of thinkers and politicians from the first years after 1989 have been replaced by a new one, more occupied with examining the influence of their economic inclusion in the European Union than with developing a Central European heritage from Slovenia to the borders of Belarus and Ukraine.

However, some Central European frameworks continue to exist. Quite noticeable among them is an alliance of the Czech Republic, Slovakia, Hungary, and Poland known as the Visegrad Group (V4). An origin of this alliance's deliberate interactions dates back to the year 1335 when Hungarian, Polish, and Bohemian kings met in the Visegrad Castle to discuss their mutual relationships, political goals, and cooperation. The Visegrad Group may serve as an excellent source for understanding the importance of cherishing historical identity for current political purposes. The group has gradually become a Central European voice representing specific regional interests within the overall European framework.

Successively, Austrian observers joined the discussions of the group, and the framework was expanded gradually to include the participation of representatives from Italy, Germany, the Baltic States, and Slovenia. In 2016, the Czech president Miloš Zeman, together with the representatives of the right-wing populist political Freedom Party of Austria, even proposed a stable extension of V4 for Austria. These efforts of building "a state within a state," in order to oppose the EU, were, so far, jointly rejected by Czech and Austrian politicians. However, the V4 might certainly be seen as a newly emerging center of power within the Union, promoting opinions and policies, often diverging from the views of their western neighbors. In recent years, the Visegrad Group has become the main "bulwark" against the acceptance of greater numbers of refugees and immigrants from North African and Middle Eastern states – a position that differs at least from a few other central and western members of the EU, with Germany ahead. Unsurprisingly, the "rebellion" has gained wider attention with the ongoing Brexit negotiations.

The continuation of the European Union enlargement process into the regions of Eastern Europe is a reality that largely symbolizes the end of the above-mentioned political advantage of the Central European idea, and sends the discourse back to its proper place – i.e. historical and cultural research. The intellectuals point to the fact that the cycle has been completed. If the classical Central Europe was an open multicultural space, where ideas and people flowed freely – at least looking at it with some nostalgic retrospect – today after the end of communism and the Cold War, and with the open borders of the Schengen Area, the glorious image of Central Europe seemingly materialized itself.

However, as indicated above, treacherous pitfalls consist in the fragmentation of a common European idea into petty local nationalisms, xenophobic proclamations, and blaming the "others" in search for a scapegoat.

Róbert Kiss-Szemán, a Hungarian professor of Slavic Studies at the University of Budapest, published an essay entitled "Homo Visegradicus." In this essay the author describes how member countries of the group deal with their histories, and how they have learned to surmount their past faults and have adopted cooperation as a major means of conduct for the present and for building the future. The author finishes in an encouraging tone, nevertheless, observing that "age of 'isms'" should receive an independent chapter in a textbook on the Central European mosaic, particularly warning against "Visegradism":

> Ecce homo visegradicus!
> Here is the Visegrad man, explosive but generous with hospitality, cautious and careful but fresh and capable of winning, because he looks to the future in an ingenious and optimistic fashion. Who would not want to belong to this breed? Let's all cheer him on, that he might settle this Visegrad land as soon as possible. There is but one danger he faces: The danger of "isms", which could distort the Visegrad idea into "Visegradism."[14]

Bibliography

Bartov, Omer. "To obliterate and to forget: last remnants of Jewish Galicia in today's Ukraine." (in Hebrew). *Zemanim* 13 (Summer 2008): 14–27.

Davies, Norman, and Roger Moorhouse. *Microcosm: A Portrait of a Central European City.* London: Pimlico, 2003.

---

14 Kiss-Szemán 2006, 170.

Garton Ash, Timothy. "Does Central Europe exist?" *New York Review of Books* 33, no. 15 (1986): 45–52.

Garton Ash, Timothy. "The Puzzle of Central Europe." *New York Review of Books*, March 18, 1999.

Kaplan, Robert D. *Balkan Ghosts: A Journey Through History*. New York: Picador, 2005.

Kiss-Szemán, Róbert. "Homo Visegradicus." In *The Visegrad Group – Central European Constellation*, edited by Jagodziński, 169–70. Bratislava: International Visegrad Fund, 2006.

Konrád, György. "Is the Dream of Central Europe Still Alive?" *Cross currents: A Yearbook of Central European Culture 5*, 109–121. Ann Arbor, MI: University of Michigan, 1986.

Kundera, Milan. ""Un occident kidnappé" ou la tragédie de l'"Europe centrale." *Le Débat*, no. 27 (1983/5): 3–23.

Kundera, Milan. "The Tragedy of Central Europe." *New York Review of Books*, April 26, 1984, 33–8.

Kundera, Milan. "Únos západu" ["Un occident kidnappé"]. In *150 000 slov* IV, no. 10 (1985): 112–119.

Masaryk, Tomáš Garrigue. *Nová Evropa* (first published in 1920) [online]. [*The New Europe: The Slav Standpoint*, 1918]. Prague: Městská knihovna v Praze, 2014. http://web2.mlp.cz/koweb/00/03/99/99/13/nova_evropa.pdf. Accessed August 23, 2018.

Naumann, Friedrich. *Mitteleuropa*. Berlin, 1915.

Palmer, Alan. *Lands Between: A History of East-Central Europe Since the Congress of Vienna*. New York: MacMillan, 1970.

Reinharz, Jehuda, and Yaacov Shavit. *Glorious, Accursed Europe*. Boston: Brandeis, 2010.

Wistrich, Robert C. "Laboratory for World Destruction: Germans and Jews." *Central Europe*. Lincoln, NE, and London: University of Nebraska Press for the Vidal Sassoon International Center for the Study of Antisemitism (SICSA), The Hebrew University Jerusalem, 2007.

Yavets, Zvi. *My Czernowitz*. (in Hebrew) Dvir, Israel, 2007.

Sergio DellaPergola
Jewish Demography in the European Union – Virtuous and Vicious Paths

Introduction: Where and what is Europe?

The horrific images of the sites of bloody acts of terrorism in the main European capital cities are matched by the dramatic scenes of hundreds of thousands of displaced persons trying to reach the shores and the land borders of Europe from many countries in Africa and the Middle East – often at the price of their own lives. These troubling views and captions not only portray a huge political and humanitarian issue, but also propose again the dilemma of the meaning of what for one multitude of people is supposedly a promised land, for another multitude is a religious and cultural enemy to be erased from the earth, and for many more is simply home. Under these circumstances, a longstanding question needs to be asked again: where is Europe? And what is Europe? (Seton-Watson 1985).

The extant European Union geopolitical framework was created over and beyond past conflicts, with an eye to common interests in spite of many ideological disagreements and deep regional socio-economic gaps. The preamble of the Treaty establishing a Constitution for Europe states the following fundamental principles (European Communities 2005):

> Drawing inspiration from the cultural, religious and humanist inheritance of Europe, from which have developed the universal values of the inviolable and inalienable rights of the human person, freedom, democracy, equality and the rule of law;
> Believing that Europe, reunited after bitter experiences, intends to continue along the path of civilisation, progress and prosperity, for the good of all its inhabitants, including the weakest and most deprived; that it wishes to remain a continent open to culture, learning and social progress; and that it wishes to deepen the democratic and transparent nature of its public life, and to strive for peace, justice and solidarity throughout the world.
> Convinced that, while remaining proud of their own national identities and history, the peoples of Europe are determined to transcend their former divisions and, united ever more closely, to forge a common destiny.

According to Article I-2, the Union is founded on values that include:

> [...] respect for human dignity, freedom, democracy, equality, the rule of law and respect for human rights, including the rights of persons belonging to minorities. These values are

common to the Member States in a society in which pluralism, non-discrimination, tolerance, justice, solidarity and equality between women and men prevail.

Facing the ongoing dramas, on the one hand European governments are displaying quite an appalling lack of uniform vision and strategy, each trying to shift the security and migration emergency from one's own shoulders to those of others, raising walls and barriers toward the outside and even inside the Union, and questioning existing policy agreements that were aimed at easing the circulation of people within Europe. On the other hand, in one of the oddest intellectual alliances of our time, during 2015 the Roman Apostolic Catholic Pope, Franciscus Jorge Mario Bergoglio, and the recently deceased Polish/Jewish/British sociologist, Zygmunt Bauman, were on record calling for a joint effort of greater human solidarity that would transcend country frontiers and social classes. Both men vigorously criticized the selfishness of the rich versus the poor, and implicitly questioned the historical and political logic that has brought about the current geo-political order.

One of the most intriguing asides of these formal and more personal calls for enhanced universal values and lowered political boundaries is that they openly challenge, if not deny, particularistic cultural, religious, and ethnic identities that constitute the cornerstone of the definition of modern nation-states. The same particular identities also animate the long-term cultural and communal survival of religious or ethnic minorities within states, and of national or transnational diasporas spread across multiple countries. One burning issue – sometimes overtly stated, sometimes concealed under a veil of political correctness – is whether Europe and the EU within it are supposed to preserve a Christian identity or can give way to religious diversity if not pluralism under the impact of Islamization inherent in the current migration wave and security infringements.

The truth is that since the onset, the European Union followed quite a rigid concept of economic correctness and – in spite of its long and detailed charter of principles – it lacked a real project for a shared cultural and identificational destiny that would truly transcend national boundaries and sovereignties. Facing the pressure of continentally widespread economic crisis and under the massive dislocation of the current immigration wave, the diverging perceptions and national interests between different member countries seem to allow for growing destabilization of the EU's institutional order. The United Kingdom's Brexit constitutes only a symptom of a widespread broader malaise. Another symptom of destabilization is the growing emergence of regional autonomism recently demonstrated in Scotland and Catalonia that puts into question the whole concept of belongingness in a European Union of growingly diverse and fractious internal identities.

Under these complex and contradictory circumstances, it is relevant to review the position of the Jews in the EU (for earlier reviews see DellaPergola 1993, 2011, 2014). Jews in Europe are a case in point of a population with very ancient migration roots and an uninterrupted presence in the continent for over 2,000 years, as well as a more recent past of large-scale immigration over the nineteenth and twentieth centuries (DellaPergola 2006). Despite historical persecutions since antiquity and of the Shoah more recently, the Jews have successfully integrated into the complex multi-national mosaic of contemporary Europe and are currently unavoidably associated in the dilemmas and predicaments that Europe must face at a time of crisis and transition. One of the facts that stimulated the present study is the strong revival in Jewish emigration from Europe, primarily to Israel, but also to other destinations. This is but a symptom of a broader set of issues that needs to be elucidated. Some of the lessons to be drawn from the Jewish case can probably be applied to the broader European case, but in addition the Jews' particular history and societal set-up is sufficiently interesting to allow for an investigation on its own merits.

Micro and Macro Research Approaches

In this paper I will present a picture of the trends currently at work in the demography of Jews in the European Union (EU), and along the way trying to disprove some of the myths that circulate in the field. This study largely relies on hard data, and the logic followed is that of statistical inference. There is no pretension here to affirm the superiority of quantitative over qualitative research when assessing the reality of contemporary European Jewry. Different disciplinary approaches in history, literature, and the social sciences, as well as the respective different methodologies, are all legitimate and useful when tackling the issues, provided each is conducted systematically and with its own appropriate disciplinary tools.

One important difference across disciplines is that some focus on the specific experiences of individual Jewish actors, while others focus on the aggregate of collective Jewish communities, or even on the non-Jewish societal environment at large. The micro-social research approach often infers the broader reality from the experiences of relatively small groups, such as intellectuals, writers, and their work, who can provide the lead to other broader ones. The macro-social approach assesses the picture based on the collective performances of the largest possible number of anonymous informants, within which the elites are included, but are not the dominant factor. Each approach has its advantages and disadvantages, the main trade-off being between depth and representativeness.

In this study we follow a broad macro-social approach focusing on the aggregate Jewish communities of different countries as the main actors. Most of the analyses and conclusions shown here are based on comparisons between Jewish and total populations in different countries. The general subject is Jews in Europe, and more particularly in the European Union, within the context of the broader non-Jewish population. It is the difference between country profiles and experiences that produces the variation necessary to articulate a judgment capable to take into account at the same time internal diversity and overall trends. The countries included in this study are the twenty out of the twenty-eight EU members with the largest Jewish populations, plus Switzerland (CH) – not an EU member, but located at its midst. A further distinction sometimes used will be between Western and Eastern European countries. For our purposes the latter include all those countries that in the past were part of the Soviet or pro-Soviet bloc. In this paper, however, we do not deal with the former Soviet Union (FSU), with the exception of the three Baltic republics that actually came to be part of the EU. The FSU needs to be dealt with as a different complex of countries, although in some respects today it may share certain issues and trends with other European countries.

Analytic Configuration

The basic assumptions of this study do not pretend to be new or original: (1) Jewish experiences cannot be understood in isolation from the broader societal framework of which they are a part; (2) the Jews are normally rational people, and as such, if allowed, they will tend to improve their standard of living and quality of life using among other instruments the faculty to migrate from one country to another. The central question under examination here in a demographic perspective is: what is the future for European Jews? Growth or decline?

Numerous countrywide and local variables can be considered to build an analytic framework that may help to verify these general hypotheses and queries. Here we rely on a very parsimonious model based on only four elements in each of the countries examined: (1) the level of socio-economic development; (2) the level of antisemitism; (3) the level of Jewish emigration; and (4) the Jewish population size. The basic linkages and directionality of expected effects are illustrated in Figure 1.

Positing Jewish population size as the main dependent variable, the model logically assumes that a reverse relationship must exist between the volume of

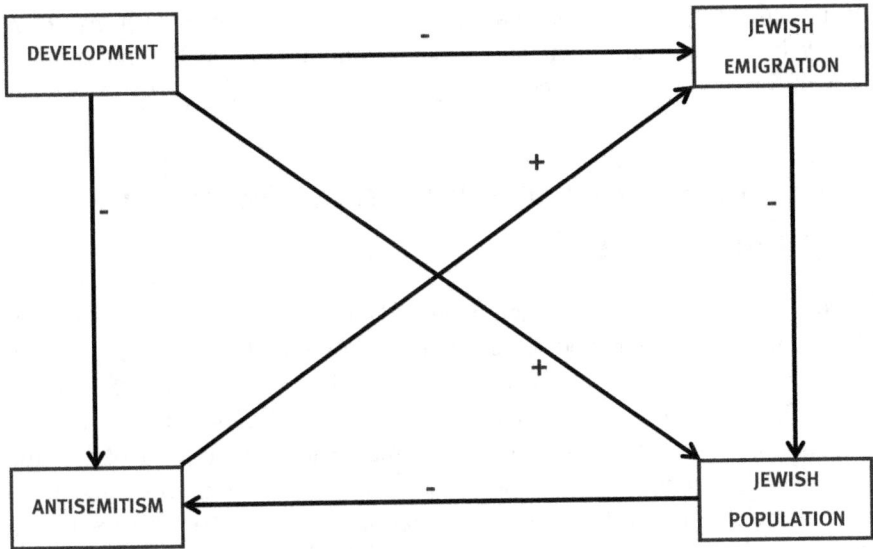

Figure 1: Hypotheses about mutual relations of socio-economic development, antisemitism, Jewish emigration, and Jewish population size in contemporary European countries.

emigration and population size: the more the emigrants, the smaller the population – and the fewer the emigrants, the larger the Jewish population. Jewish emigration in turn is supposed to be prevented or reduced by a higher level of socio-economic development in a country. Development generates more employment opportunities, especially so in view of the generally higher than average education attained by Jews and the occupational specializations that ensue. Development also attracts immigrants, thus strengthening the positive relationship with Jewish population size.

On the negative side, the amount of antisemitism measured in a country is supposed to stand in direct relation to stronger Jewish emigration, and consequently operates to reducing the Jewish population in a country. Here, besides this more obvious effect, we also wish to verify what the effect of a larger or increasing Jewish population may be on antisemitism. The relationship may be complex, and a negative attitude towards Jews would possibly lead one to believe that more Jews are a factor generating more antisemitism. We shall instead assume a negative relationship, meaning that under contemporary conditions a larger Jewish or growing population is associated with lower antisemitism. One of the reasons is that abundant research (see e.g. Wistrich 2010) shows that antisemitism often does not operate out of real contact with and knowledge of Jews, but rather reflects irrational prejudice with no empirical grounding. Another reason is that a larger Jewish population can develop more efficient Jewish community institutions, and therefore is better equipped to combat antisemitism through the media, the local political system, or otherwise.

The European Jewish Population: Descriptive Data and Trends

In order to verify the hypotheses outlined in Figure 1, we rely on a large database of relevant demographic and social indicators. The total Jewish population of the twenty-one countries studied here consisted of 1,106,700 in 2015 by the core definition (see below) (DellaPergola 2015). Not included here among EU members are eight countries: Croatia, Cyprus, Ireland, Finland, Luxembourg, Malta, Portugal, and Slovenia, each ranging between one hundred and 1,700 Jews, with an estimated total Jewish population of 6,100 in 2015. The main reason for excluding these smaller Jewish communities is the limited and sporadic amount of documented emigration, including many years without a single case, which would diminish the efficiency of data processing.

Sources and Estimates

Sources of data for estimating the size and characteristics of Jewish communities in Europe are numerous and different, though often of poor quality and reliability. The tradition of including a question on religion or ethnicity in population censuses has diminished over time. Some countries like France, Belgium, and Spain never asked the question – at least not since the nineteenth century. Several countries, such as Germany, the Netherlands, and Italy, used to ask the question, but later discontinued it. Other countries, especially in Eastern Europe, such as Hungary, Poland, Slovakia, or the former Yugoslav republics, do ask the census question on religion, but only tiny shares of the total Jewish population tend to answer. This raises intriguing queries not only about the reasons for such reluctance to unveil a Jewish identity in those countries, but also about the actual Jewish population size there (Table 1). Part of the problem is that a census question on religion often obtains very high frequencies of respondents, including Jews, who prefer to declare no religion. In some cases, a census question about Jewish ethnicity or national group results in more reliable data.

Table 1: Selected national census figures on Jewish population versus our estimates.

Country	Year	Census	Ours, 2015
Bulgaria	2011	706	2,000
Croatia	2001	495	1,700
Czech Republic	2011	345	3,900
Estonia	2000	257	2,100
Hungary	2011	10,965	47,700
Lithuania	2001	1,272	2,800
Macedonia	1991	26	100
Poland	2002	1,100	3,200
Romania	2002	6,179	9,000
Serbia	2011	578	1,400
Slovakia	2011	631	2,600
Slovenia	2002	99	100

Source: DellaPergola 2015.

Countries that were part of the FSU, such as Lithuania, Latvia, and Estonia, besides the census, keep a comprehensive and more accurate national population register including information about ethnic groups. Finally, as a recent positive development, in 2001 the national census of the United Kingdom reinstated a question on religion, thus providing detailed information about one of the largest Jewish populations in Europe.

Another possible source of information comes from membership registers kept by Jewish communities, like in Germany, Italy, Switzerland, and Austria. With the limit that they do not cover non-members, these registers provide useful data on population numbers and composition.

An alternative source of data is through population surveys independently carried out, usually at the initiative of Jewish community organizations in collaboration with universities or other public research organizations. These surveys cover relatively small representative samples of the total population and, besides being very costly, are subject to sampling and other biases. Examples include Italy, France, the Netherlands, Hungary, and the UK.

There also were a few significant transnational projects, such as the American Joint's survey in five East European countries in 2008–2009 (Kovacs and Barna 2010), and the EU Fundamental Rights Agency/JPR survey of Jewish attitudes to antisemitism in nine EU countries conducted in 2012 (FRA 2013).

Such large databases, with all their limitations and inconsistencies, in addition to evidence from past historical or socio-demographic research, allows for drawing an empirical picture of European Jewry, its trends, and prospects.

Table 2: Selected population indicators for 20 EU countries plus CH, 1990–2015.

Country	Jewish pop. 1990	Jewish pop. 2015	Total pop. (000) 1990	Total pop. (000) 2015	Jews p. 1000 1990	Jews p. 1000 2015	Jew. pop. % change 1990–2015	Tot. pop. % change 1990–2015
TOTAL	1,153,700	1,106,700	454,791	486,340	2.54	2.28	-4.59	6.94
Austria	7,000	9,000	7,583	8,500	0.92	1.06	28.57	12.09
Belgium	31,800	29,800	9,845	11,200	3.23	2.66	-6.29	13.76
Bulgaria	5,500	2,000	9,010	7,200	0.34	0.28	-63.64	-20.09
Czech Rep.	4,000	3,900	10,370	10,500	0.39	0.37	-2.50	1.25
Denmark	6,400	6,400	5,143	5,600	1.24	1.14	0.00	8.89
Estonia	3,500	2,100	1,582	1,300	2.21	1.62	-40.00	-17.83
France	530,000	467,500	56,138	64,140	9.44	7.29	-11.79	14.25
Germany	40,000	117,500	77,573	80,900	0.52	1.45	193.75	4.29
Greece	4,800	4,400	10,047	11,000	0.48	0.40	-8.33	9.49
Hungary	57,000	47,700	10,552	9,900	5.40	4.82	-16.32	-6.18
Italy	31,200	27,600	57,061	61,300	0.55	0.45	-11.54	7.43
Latvia	18,000	5,200	2,679	2,000	5.90	2.60	-71.11	-25.35
Lithuania	9,000	2,800	3,755	2,900	1.94	0.97	-68.89	-22.77
Netherlands	25,700	29,900	14,951	16,900	1.72	1.77	16.34	13.04
Poland	3,800	3,200	38,423	38,500	0.10	0.08	-15.79	0.20
Romania	17,500	9,300	23,272	20,000	0.75	0.47	-46.86	-14.06
Slovakia	3,800	2,600	5,330	5,400	0.71	0.48	-31.58	1.31
Spain	12,000	11,900	39,187	46,500	0.31	0.26	-0.83	18.66
Sweden	15,000	15,000	8,444	9,700	1.78	1.55	0.00	14.87
Switzerland	19,000	18,900	6,609	8,200	2.87	2.30	-0.53	24.07
United Kingdom	315,000	290,000	57,237	64,700	5.50	4.48	-7.94	13.04

Sources: Jewish population: Schmelz and DellaPergola (1992), DellaPergola (2015); total population: PRB – Population Research Bureau (1990 and 2015).

Table 2 presents the population data used in this study. Estimates for total populations in Europe were taken from the yearly compilations by the Population Research Bureau. Regarding Jewish population size, the respective estimates are based on a critical reading and adjustment of all available sources of data.

At the beginning of 2015, the total Jewish population of the twenty-one European countries included in this study was 1,106,700. This represented a decrease of 53,300 as compared to 1990 – a reduction of 4.6%. The total population of the same 21 countries was 486,340,000, an increase of over thirty million, or 6.9% above 1990. Consequently, the proportion of Jews per 1,000 population in the total of the twenty-one countries diminished from 2.55 in 1990 to 2.28 in 2015. The most extreme case of Jewish population increase was in Germany, from 40,000 to 117,500 (+193.8%), while the most extreme case of decline was in Latvia from 18,000 to 5,200 (-71.1%). The two figures are interrelated through the impact of emigration from the FSU, and to Germany in particular.

Definitions

Jewish population size is largely a matter of the underlying definitional assumptions. Table 3 demonstrates several possible alternative approaches in the study of Jewish population, and their implications for quantitative estimates.

The primary definitional concept suggested is the core Jewish population, i.e. the total number of individuals ready to recognize themselves as Jewish or are recognized as such by others in the same family, or who do not wish to declare any personal group identity, but have Jewish parents, and in all cases do not hold another religion. Under these criteria the total core population of our twenty-one countries was 1,106,700 in 2015. This is an empirical, not a normative concept. It does vastly overlap with the Halakhic definition, but does not correspond exactly with it. It does not require any form of membership or standard of knowledge, belief or participation, besides the readiness to self-identify or at least not to deny one's own Jewish identity or origin.

Three alternative definitions are also presented in Table 3. The population with Jewish parents adds to the core all those who currently hold an explicit non-Jewish identity, or affirm to be partly Jewish, and are the children of at least one Jewish parent. The enlarged Jewish population adds to the previous all the additional non-Jewish members who live in the same household. The Law of Return, the legal instrument that in Israel determines eligibility to immigration and automatic citizenship, extends such rights to all Jews, children of Jews, grandchildren of Jews, and the respective spouses – whether themselves Jewish or not.

Table 3: Core and enlarged Jewish population estimates in 20 EU countries plus CH, 2015.

Country	Core Jewish population	Population with Jewish parents	Enlarged Jewish population[a]	Law of Return population[b]
TOTAL	1,106,700	1,326,400	1,607,900	1,878,500
Austria	9,000	14,000	17,000	20,000
Belgium	29,800	35,000	40,000	45,000
Bulgaria	2,000	4,000	6,000	7,500
Czech Republic	3,900	5,000	6,500	8,000
Denmark	6,400	7,500	8,500	9,500
Estonia	2,100	2,600	3,400	4,500
France	467,500	530,000	600,000	700,000
Germany	117,500	150,000	250,000	275,000
Greece	4,400	5,500	6,000	7,000
Hungary	47,700	75,000	95,000	150,000
Italy	27,600	33,000	40,000	45,000
Latvia	5,200	8,000	12,000	16,000
Lithuania	2,800	4,700	6,500	10,000
Netherlands	29,900	43,000	50,000	57,000
Poland	3,200	5,000	7,500	10,000
Romania	9,300	13,500	17,000	20,000
Slovakia	2,600	3,600	4,500	6,000
Spain	11,900	15,000	18,000	20,000
Sweden	15,000	20,000	25,000	30,000
Switzerland	18,900	22,000	25,000	28,000
United Kingdom	290,000	330,000	370,000	410,000

a. Including all non-Jewish members of households.
b. Including all children and grandchildren of Jews and respective spouses.
Source: DellaPergola (2015).

It follows that the catchment area of each of these definitions expands significantly, up to a ratio of over twice in some countries. In 2015 the overall addition of non-Jews with non-Jewish parents was estimated at over 200,000; the further addition of non-Jewish household members was another 300,000; and the additional expansion to the Law of Return criteria involved another 270,000 people, for a total gap between the core and the Law of Return of about 770,000.

These estimates do not cover one additional potential constituency of growing interest: the descendants of families who ever were Jewish in past history, such as the descendants of *Conversos* of the Inquisition period or even earlier times. In many parts of Europe that once were under the influence of the Spanish or Portuguese empires, as well as all across Latin America, there is growing interest in recovering notions about, contacts with, and even formally belonging in a Jewish community. This renewed interest perhaps also reflects disappointment with the current religious identification frameworks of these persons and constitutes a potential for future change on the Jewish demographic scene.

Probably many of the debates about the actual consistency of Jewish population size in Europe reflect inconsistent use of the various definitional frameworks suggested here. Clearly if the same definitional criteria are not used across countries and over time, much of the validity of comparisons risks being lost – whichever criterion one may prefer for the purpose of argument.

Jewish Emigration

The intensity of Jewish emigration from each country is only partly known though it is effectively reflected by annual data on immigration to Israel (see Appendix). No other data source exists with as good a consistency and detail. Jewish migration to Israel is of course underestimated because it does not consider migration to other countries whose incidence – absolute and relative to the number of migrants to Israel – may vary according to the year and countries of origin. Immigration data also do not reflect return migration to the countries of origin. The data available from Israel, however, provide a good sense of inter-country differences and of fluctuations over time. The number of immigrants refer to the Law of Return definition and hence include all persons – Jewish and non-Jewish – who are eligible under the Law's comprehensive criteria (Gavison 2009).

Between 1990 and 2014, a total of 1,315,093 immigrants came to Israel in the framework of the Law of Return, the vast majority from the FSU (CBS, annual). Of this total, about 110,000 came from the twenty-one European countries examined in this study. Because of the way the data were tabulated at Israel's Central Bureau of Statistics, data for the Czech Republic could not be separated from those for Slovakia and they are jointly displayed here. The country studied here with the highest number of immigrants was France with over 47,000, followed by the United Kingdom with over 12,000, as well as over 11,000 from Latvia, over 8,000 from Lithuania, about 5,800 from Romania, and about 4,600 from Bulgaria. Among other European countries, Hungary, Belgium, Germany, and Switzerland each sent between 2,300 and 3,500 migrants to Israel. Estonia, Italy, the Netherlands, and Poland each sent between 1,000 and 2,000. Among the twenty-one countries of origin studied here, the highest number of emigrants in a single year came from France with 6,544 in 2014, and the smallest number was from Greece with one in 2008.

The annual unfolding of emigration to Israel (aliyah) from the twenty-one countries studied here is shown in Figure 2, distinguishing between those located in Western and in Eastern Europe.

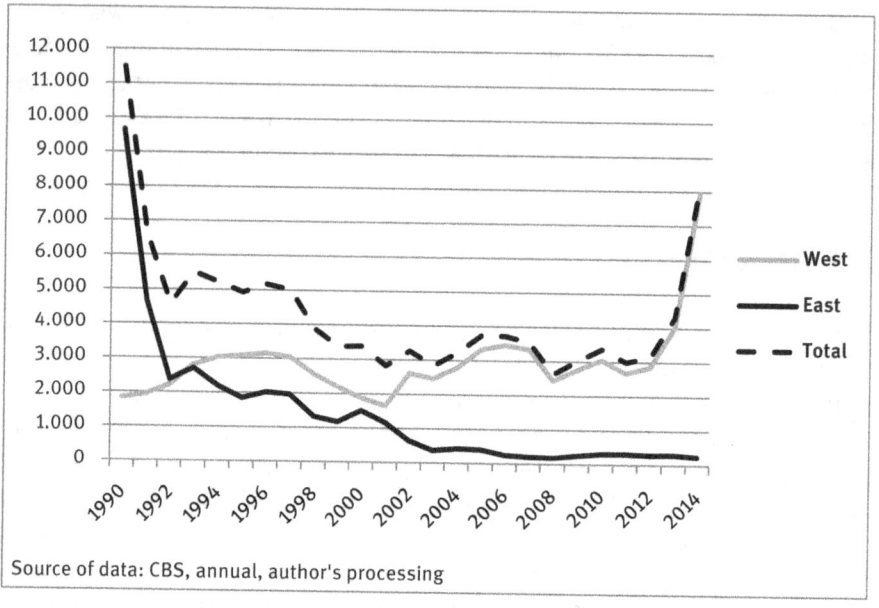

Figure 2: Annual number of immigrants to Israel, by major areas in Europe, 1990–2015.

The two regional trends are markedly different, and in fact specular. With the fall of the Berlin Wall at the end of 1989, changes in migration policies immediately followed in the FSU and other Eastern European countries, and new opportunities emerged for residents seeking a better standard of living and greater civil liberties elsewhere. Jewish emigration from Eastern Europe reached massive levels and brought about a significant decline in Jewish population in some of these countries, therefore eventually reducing their potential for further emigration. The peak of migration appears at the beginning and the trough is at the end of the twenty-five years' time-series. In Western Europe exactly the opposite occurred: the number of emigrants to Israel steadily grew over time. Temporary increases were observed in the mid-1990s and in the mid-2000s. The latest years, since 2012, display unprecedented levels and for some countries the highest numbers ever since the founding of Israel in 1948. A growing unease emerges among Jewish communities in Western Europe, primarily in France, but also in Italy, Belgium, and Spain. For the total of our twenty-one countries, 2014 was the second highest after 1990. It can be added that data for 2015 released after the completion of this paper indicate further increases in migration from Western Europe.

Another way of looking at Israel immigration data is to try to discover regional patterns based on differences in yearly variation in the number of immigrants. Patterns of co-variation between countries hinting at shared causal mechanisms are examined here using Structural Similarity Analysis (SSA), a technique of statistical analysis based on Facet Theory where latent dimensions of series of data are displayed on a map. Each point on the map represents the whole time-series of migration from a country. Closeness and distance between countries reflects similarity or dissimilarity in the evolution of the year by year number of migrants. The whole map can be partitioned into regions that catch and outline the main patterns of continental variation. Figure 3 shows the overall typology of aliyah between 1990 and 2014 for twenty-one countries.

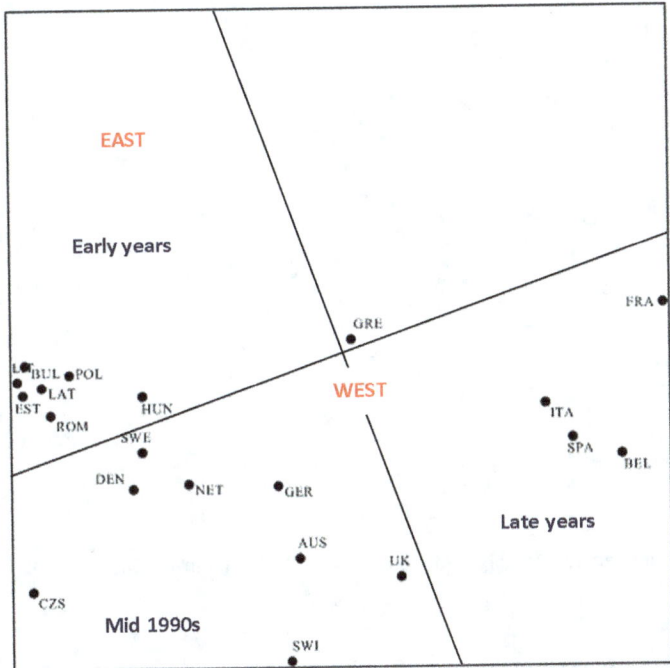

Figure 3: Structural similarity analysis of migration to Israel from 20 EU countries plus CH, 1990–2014.

In Figure 3, two orthogonal lines neatly partition the whole space into four regions. One includes France, Belgium, Italy, and Spain, four countries where the most recent surge in aliyah has been more evident. The next area includes all west-northern European countries plus the merged data for the Czech Republic and Slovakia which for this purpose behave like a Western and not an Eastern country. The next space includes all Eastern European countries (besides Czech Republic/Slovakia). The last space includes Greece, possibly reflecting that country's peculiar location between East and West, and its socio-economic and political features that have drawn so much attention in recent years. The causal mechanisms of these patterns are discussed later in greater detail.

Figures 4 and 5 display the same migration data separately for Western and Eastern Europe for the more recent period 2002–2014.

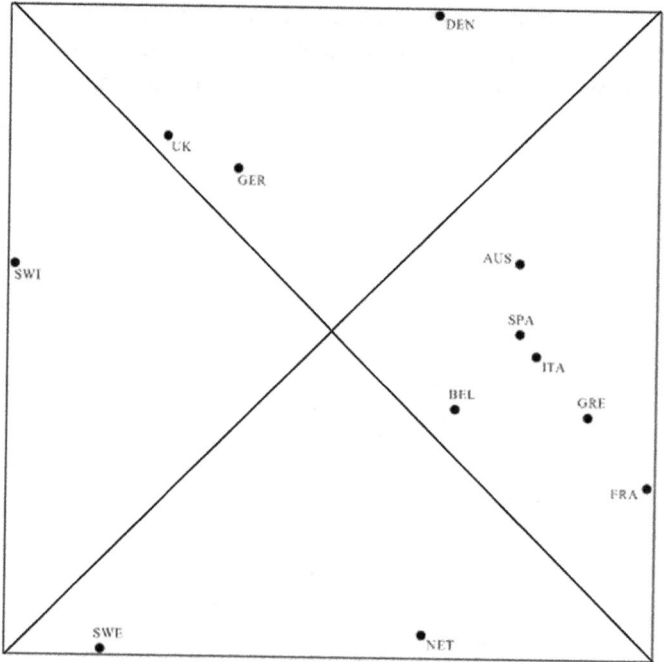

Figure 4: Structural similarity analysis of migration to Israel from 11 EU countries plus CH, 2002–2014.

Again, simple orthogonal partitions of the whole space unveil regional differences and country peculiarities, behind which stand different combinations of the mechanisms that explain migration. In the West, France, Belgium, Italy, and Spain again, with the addition of Austria and Greece, form a cluster of countries with increased migration in recent years. Other western-northern countries display distinctly different time-related migration intensities. Similarly, in Eastern Europe a simple orthogonal partition unveils a cluster including Estonia, Latvia, Lithuania, Romania, and Bulgaria, with stronger migration intensities in earlier years, while the other countries are each distinctive on their own. There is no evidence of one single Central European pattern of migration to Israel, and this raises the question of the contemporary meaning of a concept of Central Europe. This is further discussed below.

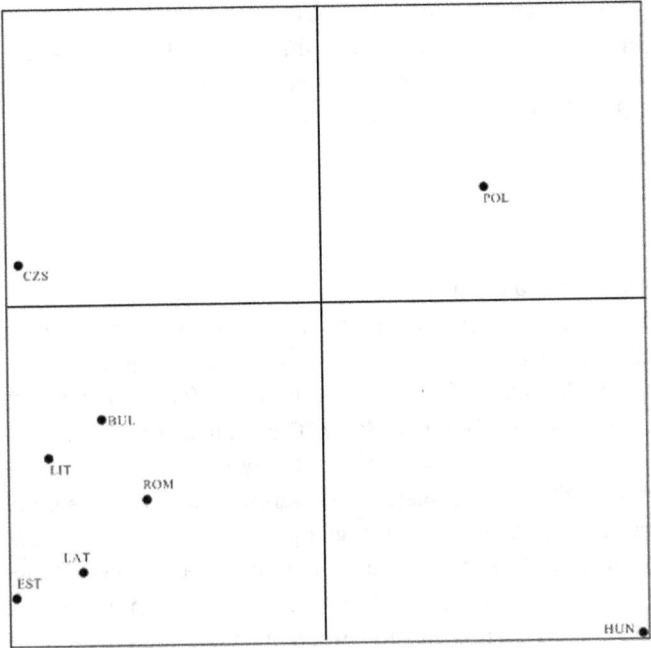

Figure 5: Structural similarity analysis of migration to Israel from 9 EU countries, 2002–2014.

Socio-economic Development

Turning now to an assessment of the level of development of each country, in this study we used the Human Development Index (HDI), a simple and effective measure suggested by the United Nations Human Development Programme. HDI is a synthesis of three basic indicators of the nature of a national society: the level of health, measured through infant mortality and life expectancy; the level of education, measured through current school enrollment and educational attainment; and the level of income, in terms of real purchasing power of the US Dollar in each country. The HDI reflects the general societal context within which Jewish communities live and operate. Of course, it often occurs that Jewish populations may have better average indicators for themselves than those observed among the total population of their country of residence. A country's general development is relevant to understand Jewish populations in that it provides a measure of the quality of the framework of constraints and opportunities that can allow a better or worse Jewish existence.

The respective country HDI values from 1990 to 2014 are reported in Table 4. They show consistent improvements for all countries though with different speeds. Both in 1990 and in 2014, the country with the highest HDI was Switzerland and the one with the lowest HDI was Bulgaria. The relative gap between the two countries remained basically the same over time.

Antisemitism

The intensity of antisemitism in a country can be investigated from different perspectives and with different tools. One relies on the recording of antisemitic events directed against Jewish persons, or Jewish institutions, or Jewish sites. For example, a systematic database of such events was developed at the Roth Institute and Kantor Center at Tel Aviv University. The limit of this method is that it covers only a tiny fraction of all those who actually hold antisemitic attitudes, namely those many who display aggressive discriminatory or contemptuous positions. A second way can be the measuring of perceptions within the Jewish population of antisemitic attitudes and behaviors that prevail in the broader society. One example is the survey of a sample of over 6,000 Jews in nine countries undertaken in 2012 at the initiative of the European Union and coordinated by the Jewish Policy Research Institute in London (FRA 2013). The limit of this method is that it does not grasp the objective diffusion of antisemitism in society, but only the subjective perceptions of the Jews. A third method would be an attentive scanning and filtering of all texts circulated in the printed and electronic media, in academy, politics, culture, and the arts that might include antisemitic contents. The limit of such detailed content analysis is that unfortunately it has not yet been systematically performed as of now. Finally, a fourth method consists of asking a representative sample of non-Jews in a given country about their attitudes toward Jews. One example is the global survey conducted in 2013–14 at the initiative of ADL in 102 countries among 53,100 adults. The limit is that such enquiry can detect certain aspects of antisemitic prejudice, but is less tuned to uncover actions of actual anti-Jewish discrimination.

Here, as a proxy for the levels of antisemitism, we will use the ADL – Anti-Defamation League global survey, keeping in mind that it only illustrates some attitudes about Jews and Judaism among the total population in each country

and does not refer to other possible dimensions of antisemitism. Antisemitism was operationally defined as the percentage of individuals who in a country responded "probably true" to at least six out of the following eleven questions that suggested anti-Jewish prejudices:
- Jews are more loyal to Israel than to [this country/the country they live in].
- Jews have too much power in the business world.
- Jews have too much power in international financial markets.
- Jews don't care about what happens to anyone but their own kind.
- Jews have too much control over global affairs.
- People hate Jews because of the way Jews behave.
- Jews think they are better than other people.
- Jews have too much control over the United States government.
- Jews have too much control over the global media.
- Jews still talk too much about what happened to them in the Holocaust.
- Jews are responsible for most of the world's wars.

This, as noted, is admittedly not the only way for investigating the incidence of antisemitism. In spite of possible limits about the accuracy and reliability of fieldwork on some of the countries covered, the ADL survey offers the unusual advantage of providing comparable – or at least similarly biased – data at the same point of time for a large number of countries on five continents. The countries covered comprise all those included in this study, with the exception of Slovakia.

According to the ADL survey, in 2013–14 of the twenty-one countries covered here, the one with the highest diffusion of antisemitism was Greece, with sixty-nine percent of the population reporting significant anti-Jewish prejudice. The second highest was Poland (forty-five percent) followed by Bulgaria (forty-four percent), Hungary (forty-one percent) and France (thirty-seven percent) – the highest in a Western European country. The lowest incidence of antisemitic prejudice was found in Sweden (four percent), followed by the Netherlands (five percent), the United Kingdom (eight percent) and Denmark (nine percent). Once again, the East-West divide shows up quite clearly here, but again there is no evidence of a coherent Central European pattern. Incidentally, the ADL survey of antisemitic attitudes was replicated in 2015 in nineteen countries, and the results were quite consistent with those of 2013–14, besides a marked reduction in France (ADL 2015).

Table 4 presents the data on migration, socio-economic development, and antisemitism used in the following of this study in addition to the population figures shown in Table 2.

Table 4: Selected demographic and social indicators for 20 EU countries plus CH, 1990–2014.

Country	Total Aliyah 1990–2014	Yearly aliyah 1990–94	Yearly aliyah 2010–14	Aliyah per 1000 1990–94	Aliyah per 1000 2010–14	HDI 1990	HDI 2014	ADL % Antisemite 2013–14
TOTAL	109,949	6,681	4,401	5.79	3.98			
Austria	469	23	19	3.3	2.2	786	881	28
Belgium	2,792	86	189	2.7	6.3	805	881	27
Bulgaria	4,569	499	24	90.8	11.9	696	777	44
Czech Rep.	526	31a	9a	4.0a	1.5a	762	861	13
Denmark	511	48	10	7.6	1.5	806	900	9
Estonia	1,875	247	11	70.6	5.4	730	840	22
France	47,248	1,179	2,899	2.2	6.2	779	884	37
Germany	2,482	124	97	3.1	0.8	782	911	27
Greece	209	10	9	2.2	2.1	749	853	69
Hungary	3,389	220	119	3.9	2.5	701	818	41
Italy	1,905	70	157	2.2	5.7	763	872	20
Latvia	11,330	1,441	53	80.1	10.2	710	810	28
Lithuania	8,327	1,163	19	129.2	6.9	737	834	36
Netherlands	1,556	106	43	4.1	1.4	826	915	5
Poland	1,054	96	20	25.3	6.4	714	834	45
Romania	5,791	619	42	35.4	4.5	703	785	35
Slovakia	b	b	b	b	b	747	830	n.a.
Spain	837	26	62	2.2	5.2	755	869	29
Sweden	767	51	21	3.4	1.4	807	898	4

Table 4: Selected demographic and social indicators for 20 EU countries plus CH, 1990–2014. (Continued)

Country	Total Aliyah	Yearly aliyah	Yearly aliyah	Aliyah per 1000	Aliyah per 1000	HDI	HDI	ADL % Antisemite
	1990–2014	1990–94	2010–14	1990–94	2010–14	1990	2014	2013–14
Switzerland	2,217	102	81	5.3	4.3	829	917	26
United Kgd.	12,025	538	515	1.7	1.8	768	892	8

a. Total for Czech Republic and Slovakia.
b. Included in Czech Republic.
Sources: Jewish population: DellaPergola (1992 and 2015); total population: PRB – Population Research Bureau (1990 and 2015); HDI – Human Development Index: United Nations Development Programme (2014); Aliyah – emigration to Israel: CBS – Israel Central Bureau of Statistics; Antisemitism: ADL – Anti Defamation League (2015).

A synthesis of the relevant variables on population, migration, socio-economic development, and antisemitism, with the respective highest and lowest values for the countries studied here is presented in Table 5. Notable gaps emerge between the minimum and maximum values of the different variables, and in the distribution of these differentials across countries.

Demographic Processes Among Jews in Europe

Before turning to validating the hypotheses formulated above, we add a few notes about the nature of ongoing demographic transformations among European Jewry. All populations evolve as the product of vital events (births and deaths), of migration (arrivals and departures), and in the case of minorities defined by some cultural peculiarity, also of changes in the willingness of individuals to pertain to a given group (accessions and secessions). Demography of the Jews in Europe has long been characterized by low fertility with the consequence of aging of the extant population and a tendency towards higher death rates than birth rates. Social and normative openness of society allowed for growing numbers of intermarriages, whose children were more often than not socialized in a non-Jewish environment. The balance of conversions to and from Judaism, including frequent cases not sanctioned by a formal rite of passage, tended to be on the negative for the Jewish community also reflecting a traditionally restrictive approach toward neophytes. In turn, migration into and outside of Europe, as well as between European countries, strongly affected the overall size and internal demographic balance of European Jewry (DellaPergola 2011).

One illustration of the patterns that prevail in many Jewish communities all across Europe is provided for Germany, where a systematic database on population composition and movements is kept by the *Zentralwohlfhartstelle* – the central organization responsible for Jewish community services in the country (see Table 6). Germany offers perhaps a rather extreme case of the demographic trends at work. It was indeed the country in Europe whose Jewish population grew the most between 1990 and 2015, by a factor of nearly three. It is also the country where the HDI grew the most in absolute value over the same years, and the country from which the rate of aliyah per 1,000 Jews in 2010–2014 was the lowest. But nonetheless Germany has the third largest Jewish population in the EU and well exemplifies its broader trends.

Table 5: Variables used in this study with highest and lowest country values, 1990–2015.

Variable	Year	Total	Lowest value		Highest value	
			Country	N.	Country	N.
Total population (thousands)	1990	454,791	Estonia	1,582	Germany	77,573
Total population (thousands)	2015	486,340	Estonia	1,300	Germany	80,900
Total population change, n. (th.)	1990–2015	31,549	Romania	-3,272	France	8,002
Total population % change	1990–2015	6.9	Latvia	-25.3	Switzerland	24.1
Jewish population	1990	1,160,000	Estonia	3,500	France	530,000
Jewish population	2015	1,106,700	Bulgaria	2,000	France	467,500
Law of Return population	2015	1,878,500	Estonia	4,500	France	700,000
Jewish population change, n.	1990–2015	-53,300	France	-62,500	Germany	77,500
Jewish population % change	1990–2015	-4.6	Latvia	-71.1	Germany	193.8
Jews per 1000 population	1990	2.55	Poland	0.10	France	9.44
Jews per 1000 population	2015	2.28	Poland	0.08	France	7.29
Aliyah total	1990–2014	109,949	Greece	209	France	47,248
Aliyah yearly average	1990–1994	6,681	Greece	10	Latvia	1,441
Aliyah yearly average	2010–2014	4,401	Greece	9	France	2,899
Aliyah per 1000 Jews	1990–1994	5.76	United Kg.	1.7	Lithuania	129.2
Aliyah per 1000 Jews	2010–2014	3.98	Germany	0.8	Bulgaria	11.9
HDI	1990		Bulgaria	696	Switzerland	829
HDI	2014		Bulgaria	777	Switzerland	917
HDI increase	1990–2014		Belgium	76	Germany	129
Antisemitism %	2013–2014		Sweden	4	Greece	69

Table 6: Jewish population changes and size in Germany, 1990–2014.

	1990	2014	Difference
Total Jewish population	27,711	100,437	72,726
Thereof by age:			
0–16	3,868	9,075	5,207
17–40	8,526	20,718	12,462
41–60	6,696	24,956	18,260
61+	8,891	45,688	36,797
International migration	**Incoming**	**Outgoing**	**Difference**
1990–2014	110,352	5,931	104,421
Natural balance	**Births**	**Deaths**	**Difference**
1990–2014	3,869	22,876	-19,007
Other changes	**Incoming**	**Outgoing**	**Difference**
1990–2014	29,532	42,220	-12,688

During the period from January 1, 1990 to December 31, 2014 the Jewish population, as recorded in the community, grew from 27,711 to 100,437 – a total increase of 72,726 (*Zentralwohlfhartstelle*, annual). The peak of such growth largely due to immigration from the FSU ended in 2006, with a total of 107,794 community members. Between 1990 and 2014, over 110,000 Jews immigrated to Germany and came to be part of the recognized community, of which nearly 105,000 came from the FSU and 5,000 from other countries. These immigrants were part of an enlarged migrant population of about 250,000 Jews and their non-Jewish family members that were allowed into Germany under the prevailing legal provisions (Erlanger 2006). During the same period, nearly 6,000 registered Jews emigrated from Germany to another country. Of these, about 2,500 went to Israel and the other 3,500 went to other countries. The natural balance featured nearly 4,000 births versus nearly 23,000 deaths, with a negative balance of 19,000. Other changes included passages from the community from one city to another, the incorporation of new groups (especially Liberal and Reform) into the official community framework, passages into and out of to Judaism, and withdrawals from the organized community. This resulted in 29,532 acquisitions and 42,220 losses, with a negative balance of 12,688. All in all, the large increase in Jewish population was very unequally distributed across age groups, with children below seventeen receiving an additional input of slightly over 5,000, and elders over 60 receiving an additional input of nearly 37,000 people. Jewish population growth was then accompanied by significant aging of the community.

In addition, there surely were Jews, namely immigrants, that never registered with an organized community in Germany (some of these are accounted for in our current population estimate for Germany) and there were high numbers of non-Jewish members of Jewish households who were not eligible to be members of the Jewish community although they were often counted in the reported estimates of total Jewish immigrants from the FSU to Germany. The latter probably constituted an amount equal to or greater than that of registered Jews. But the main lesson here is that the impact of very large immigration upon the long-term size and vitality of the community was far more limited that could be imagined because of the elderly age and weak Jewish commitment of many of the newcomers.

The consequences of these and similar demographic trends for age composition are illustrated in Table 7 for the Jewish populations of several European countries, some of which are observed at repeated points of time. Whereas immigration can produce aging of an existing community, reflecting previous aging in the countries of origin of migrants, emigration is surely related to aging of the sending community because of the tendency of younger people to be more mobile that the older. Consequently, the same Jewish population at different dates often testifies to a marked process of becoming older, namely a smaller proportion of children below fifteen and a higher proportion of elders sixty-five and above. Median ages of national Jewish populations grew over time from around thirty to well over forty, and in some extreme cases to over fifty or even close to sixty. However, there are a number of recent exceptions, like in the United Kingdom or in Vienna, where the influence or the influx of more religious families has caused an increase in the Jewish birth rate and some limited rejuvenation in age composition. But overall, the Jews in Europe are much older than their non-Jewish peers in the same countries. The Jews in Europe are also somewhat older than Jews in the US, and definitely older than in Israel.

Explanatory Power of Causal Relationships

Some Generalizations

Having formulated the basic set of hypotheses and expectations to be verified, the first important question concerns the amount of co-variation that exists between the different social, demographic, and antisemitism indicators chosen for this analysis. Table 8 presents the correlation coefficients between the four main types of variables outlined in Figure 1 and used in the following analysis of Jewish population change in the twenty-one countries examined in this study.

The correlation matrix shows generally high values, close to or above 0.5, with the exception of the link between Jewish population size/change, and the

Table 7: Selected Jewish populations, by age composition, 1975–2014.

Country[a]	Year	Total	0–14	15–29	30–44	45–64	65+	Median Age
Antwerp	1987	100	<u>25</u>	<u>25</u>	17	17	16	30.0
Greater Paris	1975	100	21	<u>25</u>	18	<u>25</u>	11	34.1
France non-Paris	1978	100	17	<u>28</u>	18	23	14	34.2
United Kingdom	1977	100	20	<u>21</u>	18	23	18	36.9
Italy	1986	100	14	23	18	<u>26</u>	19	40.8
Switzerland	1980	100	18	18	19	22	<u>23</u>	41.1
United Kingdom	1986	100	17	19	19	21	<u>24</u>	41.1
United Kingdom	2011	100	19	17	18	<u>25</u>	21	41.3
Vienna	2008	100	15	19	19	<u>26</u>	21	42.0
United Kingdom	2001	100	16	17	19	<u>26</u>	22	44.2
Hungary	1995	100	14	18	19	23	<u>26</u>	44.4
Italy	2009	100	15	16	19	<u>27</u>	23	44.9
Germany	1989	100	12	15	22	25	<u>26</u>	45.8
Salonika	1981	100	12	19	17	25	<u>27</u>	46.5
Lithuania	1993	100	14	15	18	<u>27</u>	26	47.7
Turkey	2002	100	10	16	22	<u>34</u>	18	47.2
Germany	2001	100	10	15	18	<u>30</u>	27	49.9
Ukraine	2001	100	6	10	14	35	<u>35</u>	56.4
Russian Federation	2002	100	5	11	14	33	<u>37</u>	57.5
Germany	2014	100	7	13	14	28	<u>38</u>	57.8
Romania	1979	100	5	11	10	34	<u>40</u>	59.1
Israel	2013	100	<u>27</u>	21	20	20	12	31.5
United States	2013	100	16	19	17	<u>31</u>	17	43.2

a. Countries sorted by rising median age. Largest age group underlined in each population.
Source: DellaPergola 2015.

intensity of antisemitism. In some cases, we present the correlations after excluding one or two countries whose extreme values highly affect the total country distribution. This is the case for the overwhelmingly rapid Jewish population growth in Germany since 1990, and for the outstandingly high value of expressed antisemitic prejudice in Greece. If these extremes are excluded, the level of correlation between the respective variables with other variables increases substantially. As to the directionality of the relations, they all stand in conformity with the expectations expressed in Figure 1.

Table 8: Correlation coefficients between variables, 20 EU countries plus CH, 1990–2015.

Variable type	Development (HDI)	Antisemitism (ADL)	Emigration to Israel (aliyah)	Jewish population (Core)
Development	1.000			
Antisemitism	-0.553	1.000		
Emigration to Israel	-0.597	0.377; 0.625 without Greece	1.000	
Jewish population	0.584; 0.754 without Germany	-0.164; -0.466 without Greece, Germany	-0.649	1.000

The next question concerns the strength of the relationship between basic trends in the size of the total and the Jewish population in the countries studied here. Figures 6a and 6b show significant correlations between total and Jewish percent of change over the years 1990–2015. The simple correlation for twenty-one countries is 42.3%, but this is greatly affected by the rather anomalous growth of the Jewish population in Germany following immigration from the FSU after the fall of the Berlin Wall. When removing Germany from the computation, the correlation between total and Jewish population change reaches 86.2%. Assuming the directionality of explanatory effects goes from general to Jewish, this translates into coefficients of determination (R^2) of 0.220 with Germany, and 0.785 without Germany. Simply stated, nearly three-quarters of the variation in Jewish population change across twenty countries (excluding Germany) between 1990 and 2015 are statistically explained by the overall demographic trends in the respective societies. The remaining one quarter of unexplained variation in Jewish population change reflects other still unmentioned determinants, but that residual is relatively small. This is a dominant finding that must be kept in mind in assessing all that follows.

The powerful influence of total over Jewish population trends calls for explanation. Conventional logic that assumes a positive relationship between the quality of life in a country and the propensity of people to live in that country can be empirically tested. Country HDI indexes in 2014 compared with total population change between 1990 and 2015 indeed produce a high R^2 of 0.669 (see Figure 7). This means that the total population of more developed countries grew more and that of less developed countries grew less or declined. In other words, country development exerts tremendous explanatory power upon total population trends, with the consequence of a strong expected effect on the Jewish population as well.

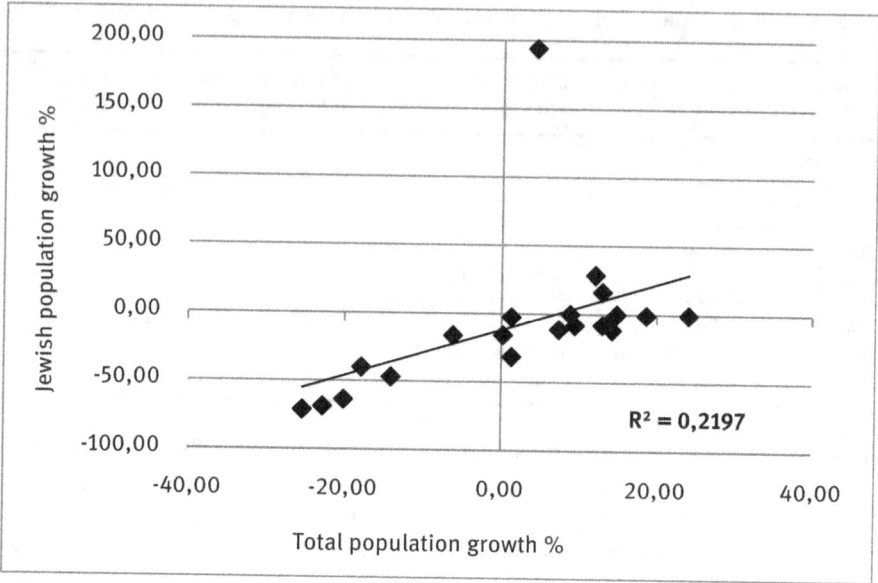

Figure 6a: Relationship between total and Jewish population change in 20 EU countries plus CH, 1990–2015.

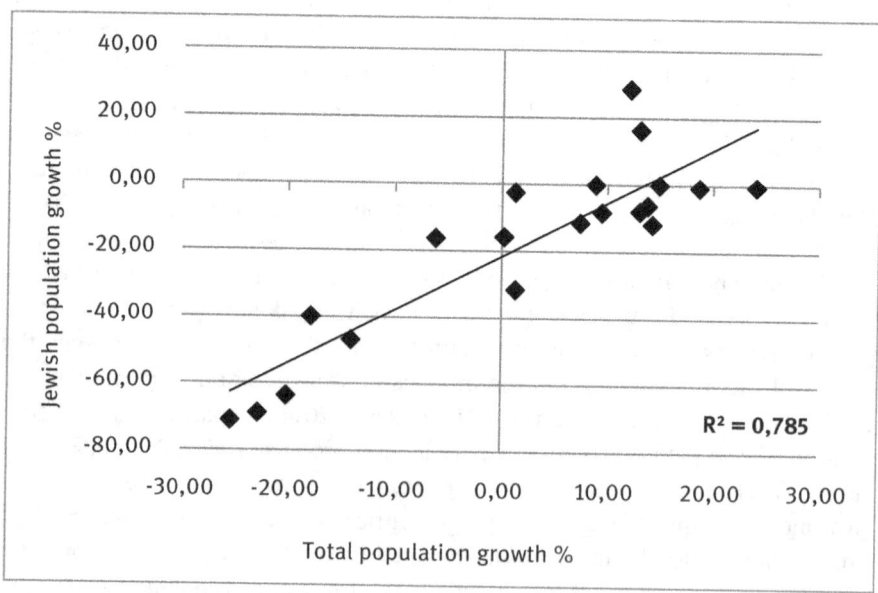

Figure 6b: Relationship between total and Jewish population change in 19 EU countries without Germany plus CH, 1990–2015.

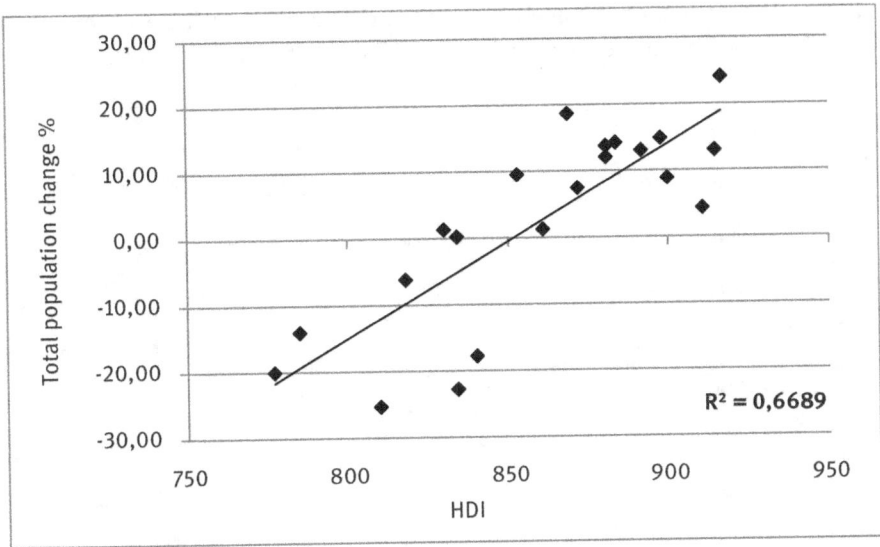

Figure 7: Relationship between country HDI, 2014, and total population change, 1990–2015, for 20 EU countries plus CH.

Jewish Emigration and Population Size

Turning now to a more detailed examination of the trends at work, the most obvious consideration is that when the level of Jewish emigration from a country grows higher then its Jewish population must shrink the most. The classic case in Europe was demonstrated by the exodus of Jews from the FSU, beginning with the 1970s and then decisively growing in the early 1990s and continuing at lower but still significant levels in subsequent years. As expected, there exists a strong correlation between the total amount of Jewish emigration from a country, here measured with the Israeli data on aliyah, the incompleteness of which we already noted, and Jewish population change in that country.

The simple correlation between total aliyah and total Jewish population change in 1990–2014 was 64.9% and the coefficient of determination (R^2) that expresses the explanatory power of aliyah in total Jewish population change was 0.421 (Figure 8). One might perhaps expect an even higher effect. The actual finding means that there are other significant factors at work in determining Jewish population change in Europe, primarily immigration from other countries and emigration to other countries, but also differential birth and death rates, differential rates of intermarriage and of Jewish socialization of the respective children, and differential Jewish accession versus secession balances. Nonetheless, the effect of international migration, much of which is directed to Israel, clearly produced quite massive reductive effects on European Jewry between 1990 and 2015.

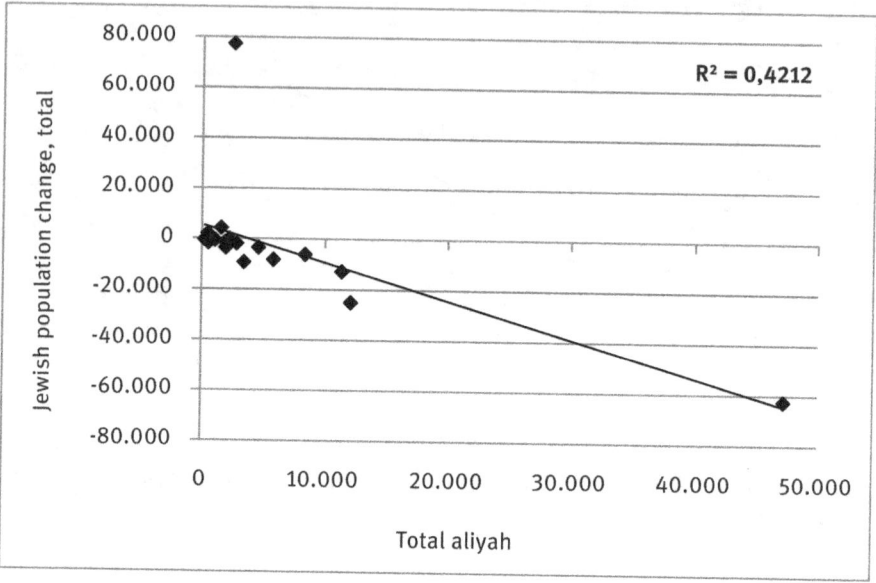

Figure 8: Relationship between total emigration to Israel and total Jewish population change, 1990–2015, 20 EU countries plus CH.

Indeed, migration from Israel to European countries to some extent moderates the effects just demonstrated and in fact explains at least part of the unexplained variance in total Jewish population change. But it must also be stated that journalist reports about Israeli emigration are overtly exaggerated and do not find confirmation in official sources of data on Israeli citizens in Europe. For example, at the end of 2013 the number of officially recorded Israeli citizens in Berlin was 3,578 versus 3,065 in 2011, clearly on the increase, but very far from the high figures often mentioned (up to 50,000) in popular discourse (Amt für Statistik Berlin-Brandenburg 2014).

Socio-economic Development and Jewish Population

In light of the relation just outlined between country HDI and total population growth, as expected socio-economic development has a strong and positive effect on Jewish population size (Figure 9). This effect operates both directly through a variety of attraction mechanisms and indirectly by reducing the amount of emigration from more developed countries. Indeed, the strength of the relationship is demonstrated by an R^2 of 0.392 for the twenty-one countries. However, one notes the quite extreme value of Jewish population growth in Germany. Once repeating the calculation after omitting Germany, the R^2 coefficient of determination rises to 0.568.

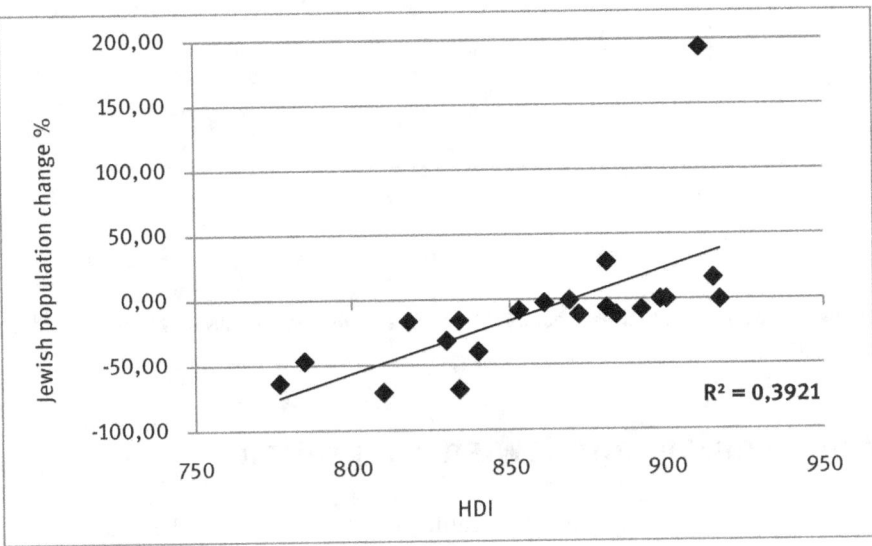

Figure 9: Relationship between country HDI, 2014, and Jewish population change, 1990–2015, 20 EU countries plus CH.

Socio-economic Development and Jewish Emigration

In conformity with what we have just seen, socio-economic development exerts a powerful hold effect on the Jewish populations of the more developed countries and a strong pull effect on Jewish populations in less developed countries. In terms of Jewish emigration, a strong negative relationship therefore emerges between country HDI in 2014 and the rate of aliyah per 1,000 Jews in the same country in 2010–2014 (see Figure 10). The respective R^2 is 0.357. Comparing the HDI for 1990 and the aliyah rate for 1990–1994 results in a similar R^2 of 0.365.

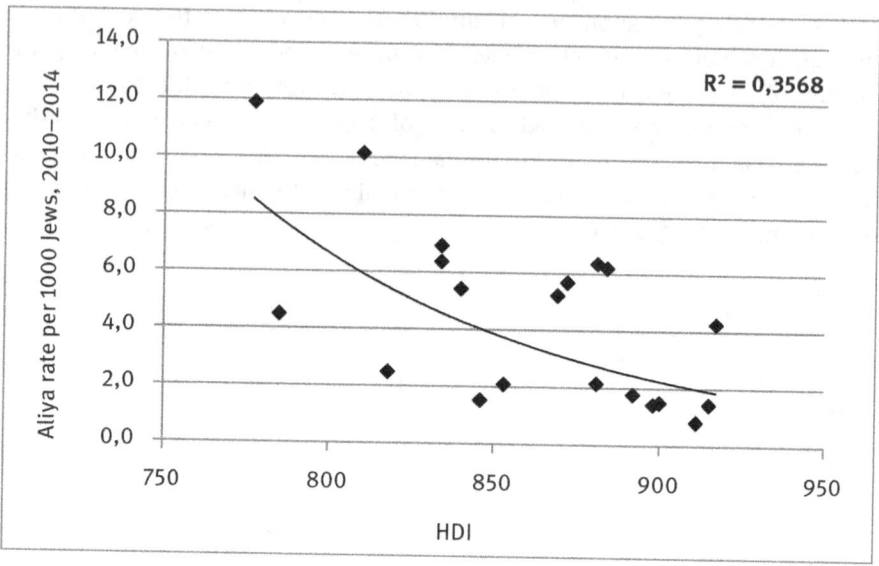

Figure 10: Relationship between country HDI, 2014, and frequency of Aliyah, 2010–2014, 20 EU Countries plus CH.

Antisemitism and Jewish Emigration

Turning now to antisemitism, and recalling the doubts expressed about the quality of measurement of antisemitism according to various methods and sources, one first validation is presented in Figure 11. We compare here the results of two studies, the ADL 2013–14 and FRA 2012 surveys, limited of course to the eight countries covered in the latter. The correspondence of findings between the two studies is quite high and persuasive, with an R^2 coefficient of determination equal to 0.542 – assuming the directionality of the effect goes from the amount of antisemitic prejudice expressed in society at large towards the perception of antisemitism among the local Jews. France and Hungary are the highest in both cases, with the UK lowest in both cases. Such coherence strengthens our confidence in the ADL data displayed below.

The relationship between the intensity of antisemitic prejudice (based on ADL data) and the frequency of Jewish emigration is expectedly a direct one, and indeed Figure 12 confirms such an expectation. The strength of the relationship between the percent of antisemitism in a country and the rate of aliyah from that country for 2010–2014, though, is significantly weaker than that for socioeconomic development, with an R^2 of 0.142. Here again, one country, Greece, af-

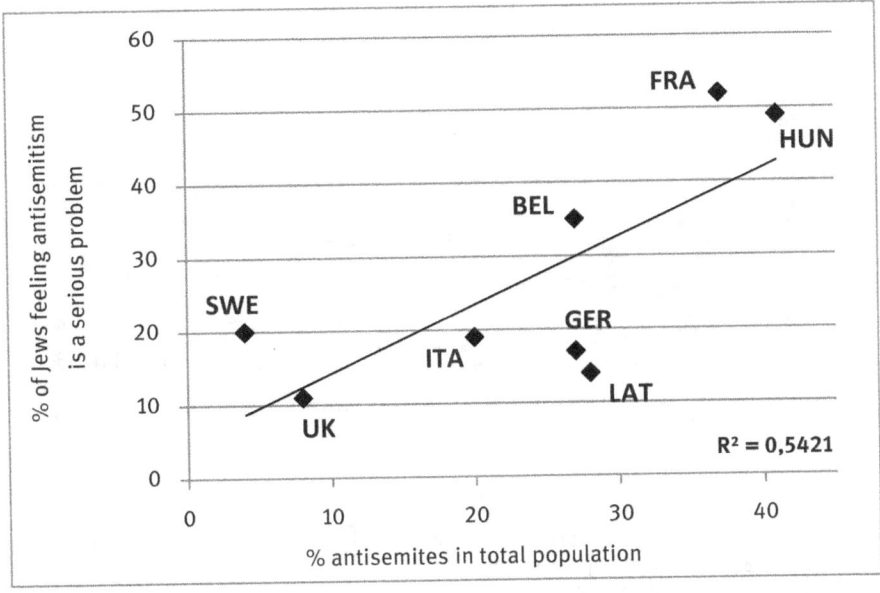

Figure 11: Comparing measures of antisemitism in 8 EU countries: FRA 2012 vs. ADL 2013–14.

fects the whole series with its quite extreme position on the scale of antisemitism. If we remove Greece from the calculation, the R^2 rises significantly to 0.391. In other words, Greece has an unusual combination of a high rate of antisemitism and a very low rate of aliyah, and this seems to reflect the peculiar institutional and economic arrangements of that society and political system. We might postulate that in Greece the achieved advantages overcame the manifest disadvantages of living in a country with lesser economic development and high anti-Jewish hostility. The recent internal and international troubles of Greece indicate the end of such quite an unusual situation and might foreshadow more Jewish emigration from that little community.

Socio-economic Development and Antisemitism

Having outlined the relationships between development and emigration, and between antisemitism and emigration, it is interesting to verify the nature of the relationship between the two motivating variables themselves. We postulated above a negative relationship, assuming socio-economic development is associated with institutional arrangements that promote democracy and civil liberties and in various ways tend to moderate the influence of totalitarianism and prejudice.

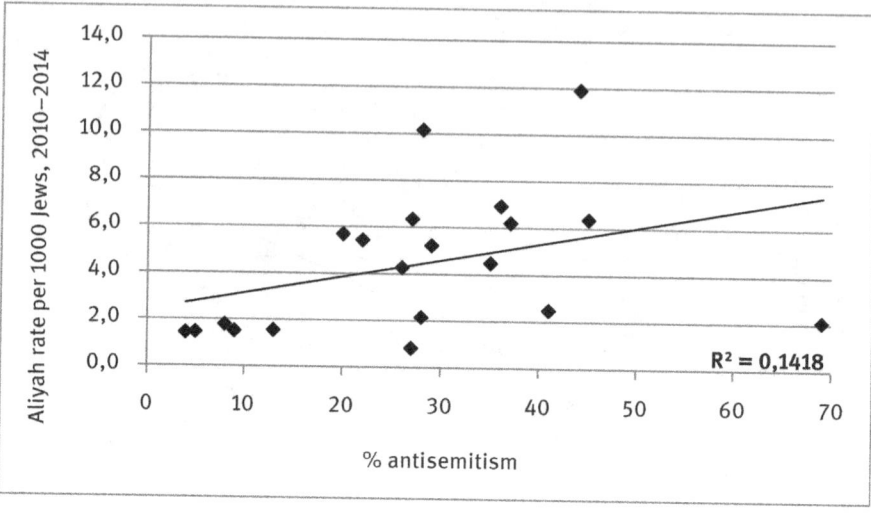

Figure 12: Relationship between percentage of antisemitism, 2013–2014, and frequency of Aliyah, 20 EU countries plus CH, 2010–2014

Figure 13 confirms these assumptions: the relationship is negative with an R^2 of 0.306. Incidentally, if instead of using the actual values of both HDI and antisemitism one uses the respective ranks of each country within the twenty-one studied here, the R^2 increases slightly to 0.320.

Jewish Population Size and Antisemitism

The final question to be examined concerns the relationship between Jewish population size and its trend to grow or decline, as well as the incidence of antisemitism. Here the view is highly affected by assumptions about whether antisemitic attitudes and behaviors do or do not have any relationship with the real and observable presence and characteristics of Jews. If the latter assumption is taken in a negative connotation, the presence of more Jews would cause an increase in antisemitism, while by the former assumption there would be no such relationship, or the relationship might be reversed into a positive one.

Figures 14a and 14b test some of these assumptions. Looking first at the relationship between Jewish population size in 2015 and the percent of antisemitism in 2013–2014 (Figure 14a), one finds no relationship at all, with an R^2 of 0.003. The very fact of no relationship may be interesting, but certainly not conclusive.

If instead of Jewish population in 2015 we measure the relationship of antisemitism to Jewish population increase in 1990–2015, the relationship is

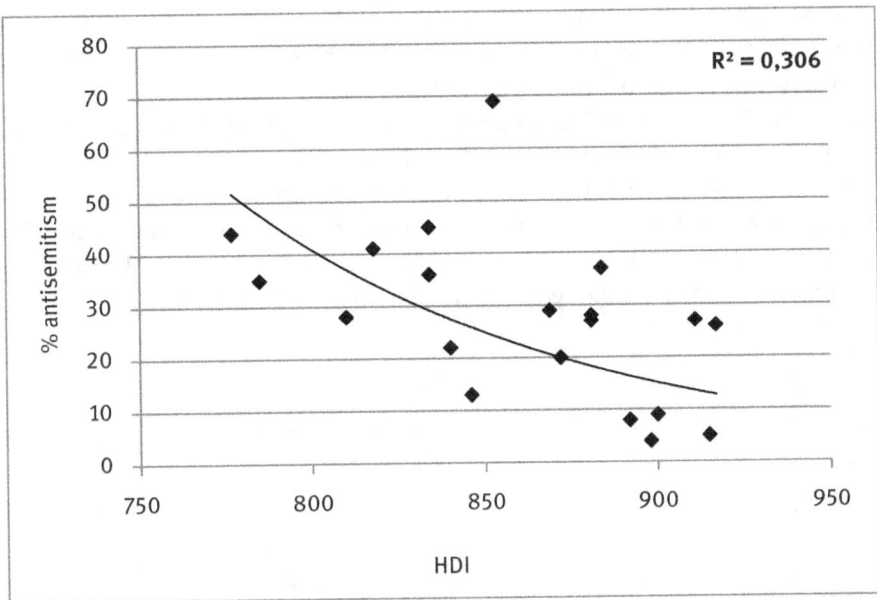

Figure 13: Relationship between country HDI, 2014, and percentage of antisemitism, 20 EU countries plus CH.

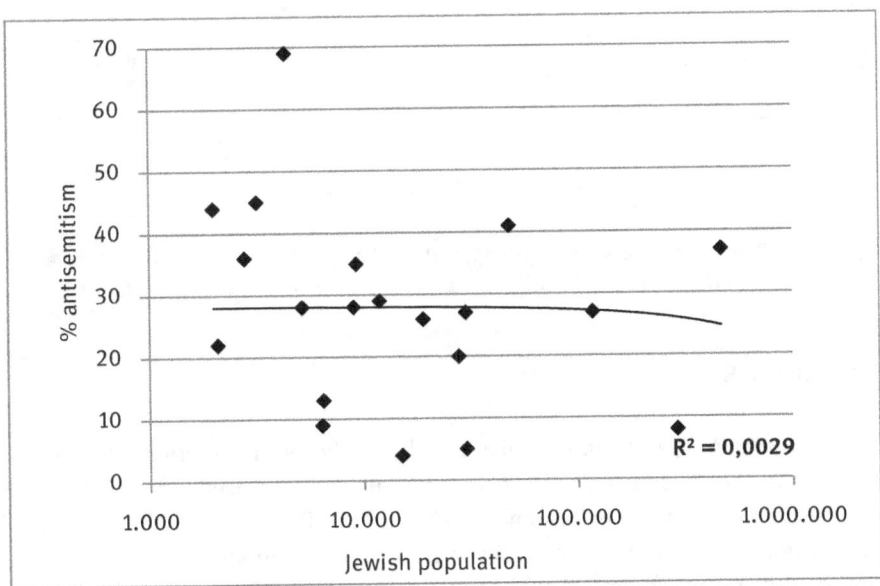

Figure 14a: Relationship between Jewish population, 2015, and percentage of antisemitism, 2013–2014, 20 EU countries plus CH.

still extremely weak and not statistically significant, with an R^2 of 0.027 (not shown here). In Figure 14b the same analysis is repeated after the exclusion of Germany and Greece, the two countries with extreme values respectively on Jewish population growth and on antisemitism. The relationship here becomes more visibly and significantly negative with an R^2 of 0.217.

The main conclusion is that in spite of the weaker nature of these relationships, the amount of antisemitism in a country cannot be related to a negative influence of the actual Jewish presence. If anything, the effect of a stronger Jewish presence appears to be reducing the impact of antisemitism.

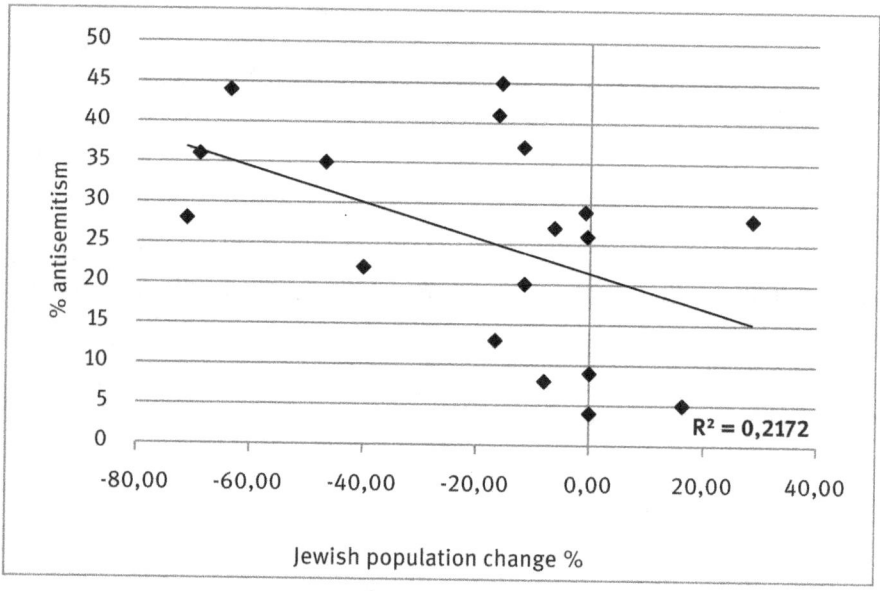

Figure 14b: Relationship between Jewish population change, 1990–2015, and percentage of antisemitism, 2013–2014, 18 EU countries without Germany and Greece plus CH.

Synopsis

As a synopsis of the previous analyses, Table 9 shows the determination coefficients in the various causal relationships initially hypothesized in Figure 1. In the sociological literature it is customary to consider a R^2 of 0.3 and above as proof of a strong causal relationship. In seven out of eight relationships considered the observed causal effect is higher than 0.3.

Table 9: Determination coefficients in causal relationships, 21 EU countries plus CH.

Explanatory variable	Dependent variable	Direction	R^2
Total population change %, 1990–2015	Jewish population change %, 1990–2015	Positive	0.220; **0.785** without Germany
Human Development Index, 2014	Total population change %, 1990–2015	Positive	**0.669**
Total aliyah, 1990–2014	Jewish population change, 1990–2015	Negative	**0.421**
Human Development Index, 2014	Jewish population change %, 1990–2015	Positive	**0.392**; **0.568** without Germany
Human Development Index, 2014	Aliyah rate per 1000, 2010–2014	Negative	**0.357**
% antisemitism, 2013–2014	Aliyah rate per 1000, 2010–2014	Positive	0.142; **0.391** without Greece
Human Development Index, 2014	% antisemitism, 2013–2014	Negative	**0.306**
Jewish population change %, 1990–2015	% antisemitism, 2013–2014	Negative	0.027; 0.217 without Greece and Germany

The main findings can be summarized as follows:
1) There exists a very strong dependency of the dynamics of Jewish population change on general population trends.
2) This strong relationship is significantly explained by the effect of general socio-economic development on total population growth across European countries.
3) Within the Jewish population, the strongest and most obvious relationship is the negative effect of emigration on population size.
4) Socio-economic development exerts a strong and positive effect on Jewish population size, both directly by attracting people to stay and indirectly by reducing emigration.
5) Emigration to Israel is strongly and negatively related to a country's socio-economic development.
6) Emigration to Israel is directly related to antisemitism, but the latter relationship is weaker than the one with socio-economic development.
7) A country's socio-economic development is strongly and negatively related to antisemitism.
8) Jewish population growth is weakly related to antisemitism, however, in any case the basic relationship found is reverse and not direct as might have been postulated.

In some cases, one or two countries significantly depart from the general distribution of a given variable. If that or those countries are omitted from the analysis, the explanatory power of the relationship between the concerned variables increases substantially.

Conclusions: Virtuous and Vicious Paths in the Future of European Jews

One fundamental assumption that cuts across and seems to be supported by this study is that Jewish population trends can be understood primarily in the light of a general principle of rationality and through the apt operationalization of a relatively small number of variables. This assumption contradicts the often-heard skepticism about the predictability of Jewish history and society, or even the failure of conventional logics to understand the world and destiny of the Jews, which is assumedly operating in force of its own unique laws.

A second important assumption supported by this study is that important insights on Jewish society can be gained by referring to the macro-societal level, namely by looking at entire countries rather than at smaller, local Jewish communities or even at selected individuals as the main unit of analysis. This further assumes the existence of widely spread processes whose logics cuts across national boundaries and expresses representation of broad continental or global patterns.

This study shows that a number of simple relationships do withstand the criteria of statistical inference. Indeed, a number of basic relationships involving Jewish and total populations in twenty EU countries plus Switzerland indicated strong and persistent linkages between explanatory and dependent variables under the conditions of the early twenty-first century. The model can be made more cogent and efficient by adding a few more variables and letting them operate in one single multivariate data processing to evaluate the overall effect and relative power of all factors together. This is left to further research. What should essentially be stressed here is that the mutually reinforcing effects of these factors may have far-reaching implications for the future population size and well-being of Jewish communities in Europe. Indeed, a virtuous and a vicious path can be outlined based on the data presented here.

Going back to the model introduced in Figure 1, we learn that more development in a country attracts a larger Jewish population and decreases Jewish emigration, thus keeping the Jewish population high. It also depresses antisemitism which therefore does not operate to increment Jewish emigration, hence maintaining the Jewish population higher. This in turn may have beneficial effects in further depressing antisemitism. This virtuous loop links a country's higher development and lesser antisemitism to a larger and growing Jewish population.

But there may be another scenario as well. Diminishing or stagnating development may produce more Jewish emigration and greater antisemitism which also may enhance Jewish emigration, with reducing effects on the Jewish population. A smaller, diminishing Jewish population may be related to increasing antisemitism. This vicious loop links a country's diminishing or doubtful development and more antisemitism to a declining Jewish population.

The singular circumstances of the most recent years in the European Union have featured successive falls in the economic cycle of many countries and growing internal cleavages under the impact of unprecedented immigration. Increasing Jewish emigration, declining Jewish populations, economic difficulties widespread among the Jewish no less than among the general population, and Jewish perceptions of growing antisemitism altogether unfortunately point more to the second, vicious, than to the first, virtuous, scenario.

The reader may be surprised that in this paper not a single word was said about Jewish culture and identity. The original contents of Jewish community life are of course the very *raison d'être* of Jewish corporate existence and should be at the center of any sensitive analysis of the meaning of being Jewish – in general and in Europe in particular. The modest goal of this study was to show that even putting aside the fundamentals of the issue of the Jewish future, relevant insights and conclusions can be reached based on a very minimalist set of assumptions. From here, further analysis and the evaluation of policy implications can hopefully become stronger and more cogent.

Bibliography

ADL – Anti Defamation League. *ADL Global 100: An Index of Anti-semitism*. New York: ADL, 2014.

ADL – Anti Defamation League. *ADL Global 100: An Index of Anti-semitism. 2015 Update in 19 Countries*. New York: ADL, 2015.

Amt für Statistik Berlin-Brandenburg. *Statistisches Jahrbuch Berlin 2014*. Berlin: Amt für Statistik Berlin-Brandenburg, 2014.

CBS – Israel Central Bureau of Statistics (annual). *Statistical Abstract of Israel*. Jerusalem: Central Bureau of Statistics.

DellaPergola, Sergio. "Jews in the European Community: Sociodemographic Trends and Challenges." *American Jewish Year Book* 93, 25–82. New York-Philadelphia: American Jewish Committee, 1993.

DellaPergola, Sergio. "Jewish Communities in Europe." In *Handbook of Global Religions*, edited by Mark Jurgensmeyer, 215–221. Oxford: Oxford University Press, 2006.

DellaPergola, Sergio. "Jews in Europe: Demographic Trends, Contexts, Outlooks." In *A Road to Nowhere? Jewish Experiences in Unifying Europe*, edited by J. Schoeps, O. Glöckner, and A. Kreienbrink, 3–34. Leiden/Boston: Brill, 2011.

DellaPergola, Sergio. "Reflections on the Multinational Geography of Jews after World War II." In *Displacement, Migration and Integration: A Comparative Approach to Jewish Migrants and Refugees in the Post-War period*, edited by F. Ouzan and M. Garstenfeld, 13–33. Leiden and Boston: Brill, 2014.

DellaPergola, Sergio. "World Jewish Population 2015." In *American Jewish Year Book*, edited by A. Dashefsky and I. Sheskin, 115, 269–360. Dordrecht: Springer, 2015.

Erlanger, Simon. "Changes in the German Jewish Community." In *The Jewish People Policy Planning Institute Annual Assessment 2006, Major Shifts – Threats and Opportunities*, 83–88. Jerusalem: JPPPI, Executive Report 3, 2006.

European Communities. *Treaty establishing a Constitution for Europe*. Luxembourg: Office for Official Publications of the European Communities, 2005.

FRA – European Union Fundamental Rights Agency. *Discrimination and hate crime against Jews in EU Member States: Experiences and perceptions of antisemitism*. Vienna: European Union Agency for Fundamental Rights, 2013.

Gavison, Ruth. *Sixty years to the law of return: History, ideology, justification*. Jerusalem: Metzilah Center for Zionist, Jewish, Liberal and Humanistic Thought, 2009.

Kovács, Andras, and Ildiko Barna. *Identity à la carte: Research on Jewish identities, participation and affiliation in five European countries. Analysis of survey data*. Budapest: The American Joint Distribution Committee, 2010.

PRB – Population Research Bureau. *1990 world population data sheet*. Washington, DC: PRB, 1990.

PRB – Population Research Bureau. *2015 world population data sheet*. Washington, DC: PRB, 2015.

Schmelz, U. O., and Sergio DellaPergola. "World Jewish Population 1990." *American Jewish Year Book* 92, 484–511. New York-Philadelphia: American Jewish Committee, 1992.

Seton-Watson, Hugh. "What Is Europe, Where Is Europe? From Mystique to Politique." *Encounter*, July-August 1985, 9–17.

United Nations Development Programme. *Human development report – Sustaining Human Progress: Reducing Vulnerabilities and Building Resilience*. New York: United Nations Development Programme, 2014.

Wistrich, Robert. *A Lethal Obsession: Anti-semitism from Antiquity to the Global Jihad*. New York: Random House, 2010.

Zentralwohlfahrtsstelle der Juden in Deutschland. Annual. *Mitgliederstatistik; Der Einzelnen Jüdischen Gemeinden und Landesverbände in Deutschland*. Frankfurt a. M.: ZWJD.

Appendix

Number of new Immigrants to Israel, 20 EU countries plus CH, 1990–2014.

Year	FRA	UK	BEL	NET	DEN	SWE	GER	AUS	SWI	ITA	SPA
1990	_864_	488	_46_	78	42	53	118	15	_51_	45	25
1991	966	472	65	100	44	40	91	14	77	58	13
1992	1,182	459	93	77	_49_	53	91	21	102	49	30
1993	1,372	647	106	_123_	31	_61_	151	29	132	103	28
1994	1,512	626	121	151	76	49	167	_37_	146	95	36
1995	1,635	**669**	111	108	39	42	129	35	146	101	59
1996	1,869	544	127	107	30	_64_	109	_39_	132	78	37
1997	1,934	484	108	85	25	55	114	30	89	75	28
1998	1,667	393	93	56	16	36	102	9	84	62	23
1999	1,366	383	83	41	22	22	100	17	92	44	21
2000	1,152	326	100	_29_	7	20	78	20	85	42	15
2001	1,002	307	62	46	10	17	62	20	90	_22_	10
2002	2,035	_277_	57	31	5	18	_55_	8	83	29	_8_
2003	1,789	330	90	46	5	19	58	_5_	75	21	20
2004	2,003	363	120	46	12	11	85	8	68	42	26
2005	2,545	383	80	42	7	30	96	17	54	41	30
2006	2,408	594	91	50	13	17	112	12	85	42	33
2007	2,335	562	84	43	8	22	96	8	84	58	33
2008	1,562	505	84	30	12	_7_	86	17	60	52	22
2009	1,556	**708**	125	51	10	24	96	11	79	62	29
2010	1,775	632	185	36	8	28	119	18	110	97	35
2011	1,619	485	175	40	_3_	22	97	19	59	94	53
2012	1,653	569	140	36	14	15	100	18	81	_137_	_76_
2013	_2,903_	403	_222_	55	13	29	79	25	75	133	70
2014	**6,544**	486	224	49	10	13	91	17	78	**323**	77
Total	47,248	12,095	2,792	1,556	511	767	2,482	469	2,217	1,905	837

Note: Highest number in **bold**; second highest in **_bold italics_**; lowest in _italics_. Source: CBS. The table is divided into two different sections into separate pages.

Year	GRE	BUL	ROM	HUN	CZS	POL	LIT	LAT	EST	TOT	Grand total
1990	6	*844*	1,201	299	22	151	3,100	3,517	520	11,485	199,516
1991	8	914	*520*	231	33	*121*	1,144	1,410	*293*	6,614	*176,100*
1992	9	306	472	145	*41*	67	490	695	151	4,582	77,057
1993	*18*	211	393	212	29	91	650	954	163	5,504	76,805
1994	11	222	510	214	30	50	430	630	108	5,221	79,844
1995	8	195	344	*272*	17	64	353	541	60	4,928	76,361
1996	10	198	364	226	38	54	335	705	99	5,165	70,605
1997	23	413	322	156	30	35	332	599	75	5,012	65,990
1998	14	277	204	98	36	29	194	446	40	3,879	56,693
1999	7	148	284	115	16	33	198	326	55	3,373	76,766
2000	4	221	268	165	*44*	20	300	390	100	3,386	60,192
2001	9	140	201	103	24	31	308	300	60	2,824	43,580
2002	6	80	102	59	17	22	176	166	27	3,261	33,567
2003	3	62	69	*37*	21	18	68	78	22	2,836	23,268
2004	3	81	88	112	11	23	41	69	7	3,219	20,898
2005	6	45	89	94	16	43	38	74	11	3,741	21,180
2006	3	22	50	63	16	36	23	41	6	3,717	19,269
2007	5	33	37	49	19	22	21	26	4	3,549	18,131
2008	*1*	21	29	54	9	24	17	31	5	*2,628*	*13,681*
2009	9	17	35	90	10	18	12	68	12	3,022	14,564
2010	6	30	54	87	10	*15*	15	76	30	3,366	16,631
2011	8	33	41	128	7	17	21	67	8	2,996	16,892
2012	10	17	51	110	14	16	19	57	10	3,143	16,557
2013	7	*15*	41	148	*5*	25	32	36	*3*	4,319	16,882
2014	15	24	*22*	122	11	29	*10*	*28*	6	*8,179*	24,064
Total	209	4,569	5,791	3,389	526	1,054	8,327	11,330	1,875	109,949	1,315,093

Note: Highest number in **bold**; second highest in ***bold italics***; lowest in *italics*. Source: CBS. The table is divided into two different sections into separate pages.

Julius H. Schoeps
Renewal or Regression?
Jewish Self-Assertion and Re-Orientation in Twenty-first Century Central Europe

When Adolf Hitler led Nazi-Germany into World War II, the extermination of Europe's Jews was a distinct aim of his ideological impetus. In fact, the Jewish populace in Europe, persecuted by the Nazis with an almost infernal hate, stood at the top of the list of the Nazis' "mortal enemies". Liberation from Judaism was considered to be the Germans' path to salvation from all evil. As can be gleaned from his concoction *Mein Kampf*, Adolf Hitler, who stylized himself as a savior, an instrument of God, was completely convinced that the extermination of the Jews would not only bring salvation and deliverance to Germany, but to the whole world.[1]

There were only a few European countries where Jews were spared from Hitler's racial mania and persecution during the war; for example, in unoccupied Great Britain or in neutral Sweden, where, however, only a very small number of Jews could save themselves due to highly restrictive admission policies. In other countries such as Poland or the Baltic States, almost no one was able to escape this murderous system. A few were able to survive in hiding. Some escaped persecution by fleeing into the woods and joining partisan units.

That in the end, two-thirds of the European Jewish population – i.e., six to nine million Jews – fell victim to the genocidal mania of Hitler and his followers just because they were Jews is something we all, not only historians, are still trying to come to terms with today. We try to comprehend the motives and reasoning taken by Hitler and his followers, yet sense that this is beyond the limits of our understanding.

How could it be, we continue to ask ourselves, that a continent so proud of its Christian roots did not feel able to put a stop to the million-fold industrial slaughter of Jewish men, women, and children? What is the reason, we ask ourselves, for the complete lack of solidarity shown from the non-Jewish population towards the persecuted Jews almost all over Europe? Why did so many men and

---

1 See "Erlösungswahn und Vernichtungswille. Der Nationalsozialismus als politische Religion", in Julius H. Schoeps, *Das Gewaltsyndrom: Verformungen und Brüche im deutsch-jüdischen Verhältnis* (= *Deutsch jüdische Geschichte durch drei Jahrhunderte*, Vol. 8) (Hildesheim, 2012), 333–354.

women refuse to stand up for their Jewish fellow citizens? This silence, the failure to act at the time, continues to hurt today.

We still only have incomplete and unsatisfactory answers to these questions. Up to now, the only thing like a consensus that has been reached is that the mass murder and its aftermath have fundamentally transformed the European continent. The German-Israeli historian Dan Diner has even called what happened between 1933 and 1945 a "break in civilization,"[2] thereby perfectly encapsulating the circumstances.

The situation became particularly difficult for the Jews in Eastern European countries, where the Jewish communities had effectively ceased to exist after the Nazi occupation and the Shoah. The remaining members of these communities were merely scraping by after 1945. This is particularly true for the Baltic states, where almost nothing remains that could evoke the formerly flourishing Jewish life there. At most there are some weathered inscriptions on buildings and some scattered tombstones in the cemeteries. *A seyfer is a jeder Schteyn* – 'Each stone a book' is how the Yiddish-language poet Moshe Kulbak (1896–1940) metaphorically described the situation on the eve of the catastrophe.[3]

In the countries of Eastern Europe, the systematic murder of the Jews was either ignored, trivialized or instrumentalized for politicians' transparent aims. Since World War II, religious and cultural Jewish life had come to a complete standstill almost everywhere. Even worse, in some countries, areas, and towns, a new murderous wave of antisemitism broke out only weeks or months right after the war.

The pogrom in Kielce, Poland that took place shortly after the end of the war was particularly shocking in this context. Kielce is a small town to the south of Warsaw where 25,000 Jews had been living when the Wehrmacht invaded. Around two hundred Jews returned after the end of the war and were anything but welcome. The reasons and motivations for this were obvious.

The Polish populace of Kielce feared that they would have to return the property they had stolen from the deported Jews, and began a pogrom on July 4, 1946 in which forty-one Jews were murdered and more than fifty badly injured by firearms, axes, and rocks.[4] These events have troubled Jewish-Polish

---

[2] See Dan Diner, *Zivilisationsbruch: Denken nach Auschwitz* (Frankfurt/Main, 1988), 9–13.
[3] Julius H. Schoeps, "A seyfer is a jeder schteyn", in Eberhard Schürmann, Horst Zeller, Fritz Schmidt (eds.), *... und die Karawane zieht weiter ihres Weges: Freundesgabe für Jürgen Reulecke, den Vorsitzenden des Mindener Kreises, zum 75* (Geburtstag, Ebersdorf, 2015), 300–310.
[4] See Bozena Szaynok, "The Pogrom of Jews in Kielce, July 4, 1946", *Yad Vashem Studies* 22 (1992): 199–235; see also Werner Röhr, "Massaker an Überlebenden. Zum antijüdischen Pogrom

relations up to today.⁵ The entire extent of what took place demands further clarification.

Open hostility toward Jews in Eastern Europe did slightly decline in the following decades. However, through the underlying nationalistic atmosphere along with repressive measures enacted by the state, socialist regimes saw to it that Jewish life de facto grounded to a halt. Fewer and fewer synagogues, religious schools, Jewish schools, and kindergartens were able to remain in operation.

Eastern Europe, which at the beginning of the twentieth century still had the world's largest and most vibrant Jewish population, lost almost all of its Jewish centers through the Shoah and in the years thereafter. The world of Jewish life in the shtetl is irrevocably destroyed. The only things left are faded, mostly faceless memories.

Only after the end of the Cold War, at the beginning of the 1990s, were Jewish communities able to form in some cities in Eastern Europe like Budapest, St. Petersburg, and Kiev. However, these communities remain dependent on permanent support from international aid organizations. More time is necessary for self-sufficient Jewish life capable of regenerating itself to emerge, if it will be possible at all.

Returning to the period immediately after the war, it remains remarkable that a large percentage of the uprooted and homeless Jewish Shoah survivors were drawn to Germany. These so-called *Displaced Persons* gathered in temporary camps set up by the Allies. Their numbers were estimated to be in the several hundred thousands.⁶ They were usually people who were waiting to be able to emigrate to America, Australia or possibly Palestine.

This was not the case for all *Displaced Persons*; some remained in Allied occupied West Germany. In cities like Cologne, Frankfurt, Hamburg, Munich, and, of course, Berlin, Jewish communities were re-founded – and Jews who had returned from exile even settled in the Soviet occupied area, the future German

---

in der polnischen Stadt Kielce am 4. Juli 1946", *Bulletin für Faschismus- und Weltkriegsforschung* 29 (2007): 1–32.

5 The events in Kielce should be seen as aftereffects of the Shoah. For more on the discussion held on the Shoah and the participation of parts of the Polish civilian population, see Stephanie Kowitz-Harms, *Die Shoa im Spiegel öffentlicher Konflikte in Polen. Zwischen Opfermythos und Schuldfrage (1985–2001)* (Berlin, 2014).

6 See Angelika Königseder, Juliane Wetzel, *Lebensmut im Wartesaal. Die jüdischen DPs (Displaced Persons) im Nachkriegsdeutschland* (Frankfurt/Main, 1994); see also the articles by Angelika Königseder, Joachim Schroeder, and Angelika Eder in Julius H. Schoeps (ed.), *Leben im Land der Täter. Juden im Nachkriegsdeutschland (1945–1952)* (Berlin, 2001), 33 ff.

Democratic Republic or East Germany. From the 1950s to the end of the 1980s, the number of registered Jewish community members in West Germany ranged between just 20,000 and 30,000, with the number steadily decreasing. In East Germany the number of Jews registered in the community in the 1950s was under 5,000[7] – and continued to fall. With the fall of the Berlin Wall in 1989, there were just four hundred and fifty Jews still in East Germany.

From this perspective, up until the end of the 1980s, Germany was the classic example of a disappearing diaspora community. In 1996, the British historian Bernard Wasserstein referred to this diaspora community as a *Vanishing Diaspora*[8] in his widely acclaimed book. Already then, Wasserstein pointed out that the number of European Jews – at least in the traditional sense – had been consistently decreasing since 1945 and would most likely continue to do so.[9]

According to Wasserstein, there are incredibly complex reasons and factors for the perceptible decrease in the Jewish population in Europe beginning with the unfavorable demographic pyramid and the extremely low birth rate of the Jewish community. Additional factors include the process of increasing assimilation and secularization, and of course immigration to Israel, the U.S. or other countries where Jews believe they can live more safely than in their country of origin.

At least in Western Europe there were and continue to be trends and countertrends. France, for example, which lost thousands of Jews by occupation, Nazi terror, and deportations to the death camps, saw a relatively large immigration of Jews from former North African colonies in the 1950s and 1960s.[10]

In cities like Paris, Marseille or Strasbourg, something like a new rudimentary Jewish life developed in the late 1950s and early 1960s with synagogues, community centers, schools, kindergartens, and grocery stores. However, to qual-

---

[7] On the demographic development see Lothar Mertens, *Davidstern unter Hammer und Zirkel: Die jüdischen Gemeinden in der SBZ/DDR und ihre Behandlung durch Partei und Staat* (= Haskala. Wissenschaftliche Abhandlungen, Vol. 18), Hildesheim et al. 1997: 27 ff.

[8] The original was published in 1996 by Hamis Hamilton Ltd. in London, while the German translation appeared in 2000 entitled *Europa ohne Juden: Das europäische Judentum seit 1945*, by Kiepenheuer and Witsch in Cologne. See here the discussion by the author, "Hat Hitler doch gesiegt?", which appeared in *Die Zeit*, January 20, 2000.

[9] Wasserstein, *Europa ohne Juden*, 327 ff.

[10] In 1948 the Jewish population in the Maghreb countries (Morocco, Algeria, Tunisia, Libya) measured between 475,000 and 550,000. Rounding this number off to a half a million, then in 1948 between 265,000 and 285,000 Jews were living in Morocco, 140,000 in Algeria, 105,000 in Tunisia, and 38,000 in Libya. Most went to the newly founded Jewish state in the 1950s and early 1960s, while those who had French citizenship, mainly the Algerian Jews, settled in France. Today there are only an estimated 3,500 Jews living in the Maghreb.

ify, it should be said that while this life raised hopes, right now it is increasingly endangered by attacks. In this way Jewish life in France is transitory.

Today, in France the main threat to Jews is no longer coming solely from right-wing radicals, but rather as well from radicalized Muslim milieus which have emerged especially in suburbs ("banlieus") at the margin of metropolitan cities. The brutal Islamist terrorist attacks of recent years have been directed at general public targets (e. g., newspaper offices, restaurants, music clubs), but frequently also at Jewish sites. There is, in fact, a huge fear in the Jewish communities across Europe that the "French developments" might spill over to their countries of residence as well.

As long as liberal and democratic trends are not able to prevail in Muslim circles, there will be no true acceptance of Islam in Europe. And certainly not as long as there are those preaching hate in the mosques, demanding that the devout go to war against the infidels. There is also the danger that perspectives coming out of the Muslim community will influence the thinking of the majority population and change it.

One of the consequences of the hate sermons that many do not wish to hear, yet which remains an undisputed truth, is that right now young, radicalized Muslims as well as non-Muslims were rushing to leave their hometowns and to join up with the Islamic State troops in Syria and Iraq, with Al-Quaida, and with other radical organizations in the Middle East. Now, after the defeat of the Islamic State in the Middle East, there is huge anxiety – among all population groups – that radicalized, returning Islamic warriors, who often possess French, Belgian or even German citizenship, might try to escalate tensions on the "Old Continent."

However, despite all of the dangers and threats, the Jewish communities in European countries like France, Great Britain, Germany, Sweden, and Italy are living their individual and communal Jewish lives. There is, though, a rising trend today that Jews are increasingly avoiding the provincial and small-town regions, instead favoring to live and remain in big cities – like Paris, London or Berlin.

Apart from the fact that Jews are "city people" *per se* and have traditionally preferred city life to country life, one reason for this may be that they believe that they can lead safer lives in the anonymity of the big city as opposed to the provinces. Cities hold the promise of greater protection not to be recognized as a Jew, at least that is what one hopes. There is no current research on this topic, yet it can be presumed that those who chose to live in a city consider such things when making their decision.

That sometimes things turn out completely differently than expected is a well-known fact. In Germany up until the end of the 1980s, people assumed that Jewish life would no longer continue to exist there in the foreseeable future. However, with the surprising influx of Jewish immigrants from the countries of the former Soviet Union in the 1990s, things turned out differently. While deciding where to go, many of these people decided in favor of Germany, which was willing to accept Jewish immigrants.

There is a wide variety of reasons for Jews to leave their home countries. These range from the antisemitic resentments endured to the desire to lead a self-determined Jewish life. Other reasons include economic considerations, as well as the perfectly understandable desire to provide their children with an adequate education, enabling them a better future. The immigrants' arrival in Germany was welcomed by politicians and the media, at least in the beginning. However, right now it is impossible to say if this influx will become a success story in the long-term. We can only speculate as there is no reliable data at the moment to back up such an assertion.

It is estimated that over 225,000 Jews from the former Soviet Union (Russia and various republics) immigrated to Germany between 1989 and 2012.[11] Since 2007 the annual influx per year has constantly decreased. However, a little noted side-effect of this immigration is that only about forty percent of the newcomers have become members of a Jewish community. The majority have more or less become absorbed by the broader German society.

The integration of the new arrivals into local Jewish communities did not proceed without its problems. The reason for this was that most communities were only willing to accept those immigrants who could prove their Jewish heritage.

The majority of communities continue to refuse to recognize an exclusively patrilineal lineage.[12] The immigrants were required to prove that they were the sons or daughters of Jewish mothers, which led to some consternation amongst those concerned, as only very few of them were willing or able to do so.

The common practice of accepting immigrants from the former Soviet Union into a Jewish community in Germany can be described as thus: the Liberal communities are less restrictive when it comes to the requirements for accepting immigrants. On the other hand, the Unified communities, which are mostly Orthodox, or at least define themselves as Orthodox, erected high obstacles to accepting new members.

---

[11] Sergio DellaPergola, "World Jewish Population, 2013," in Arnold Dashefsky and Ira M. Sheskin (eds.), *The American Jewish Year Book* 113 (Dordrecht: Springer, 2013), 279–358.
[12] See Barbara Steiner, *Die Inszenierung des Jüdischen: Konversion von Deutschen zum Judentum* (Göttingen, 2015), 139 ff.

Those in the communities who were in the position to decide on who would be accepted as a new member based their decisions on the Halakha, the regulations of Jewish religious law, which means that those wanting to become members were subject to rigorous examination. In some cases, an immigrant's application would be denied on the grounds that the applicant cannot be considered Jewish under the Halakha. This policy, justifiably seen by many as arbitrary, stands in stark contrast to the self-perception of many immigrants who see themselves as Jewish, even when one parent was not. They point to the ruling in the countries of the former Soviet Union where being Jewish did not denote a religious affiliation, but rather a nationality in the sense of ethnic origin and appeared on personal documents as such.

In addition, there were also demographic problems[13] caused by the general decline in the Jewish population in European countries. The influx of Jews from the former Soviet Union did nothing to change this. For example, the immigrants who arrived in Germany as contingent refugees according to the 1990 provision were generally older people who found it difficult to become integrated into the Jewish communities and German society.[14] In contrast to the German population with an average age of 42.1, this group is considerably older. The average age of immigrants from the former Soviet Union is 57.2. The consequences of this, that the group as a whole is aged, coupled with a low birth rate, are clear and do not imply any genuinely positive perspectives.

In Germany, just like in other European countries, there are also certain structural problems which make it difficult for a true continuity in Jewish communities. According to a hypothetical model from Sergio Della Pergola, a well-known Jerusalem demographer at the Hebrew University, only those local Jewish communities with a membership of at least 4,000 have a real long-term "chance of survival".[15] If Della Pergola and his prognosis are proven right, then this means that in twenty to thirty years in Germany, here standing in for other European countries, there will only be seven or eight communities remaining out of the more than one hundred existing today. These are serious considerations that

---

13 See here Sergio DellaPergola, "Die demographische Entwicklung der europäischen Juden vom 12. bis zum 20. Jahrhundert", in Elke-Vera Kotowski, Julius H. Schoeps, Hiltrud Wallenborn (Hg.), *Handbuch zur Geschichte der Juden in Europa*, Vol. 2: *Religion, Kultur, Alltag* (Darmstadt, 2001), 15–28.
14 See Bernhard Vogt, Willi Jasper, and Julius H. Schoeps, *Ein neues Judentum in Deutschland? Fremd- und Eigenbilder der russisch-jüdischen Zuwanderer* (Berlin/Potsdam, 1999).
15 Sergio DellaPergola, "Jews in Europe: Demographic Trends, Contexts and Outlooks", in Julius H. Schoeps and Olaf Glöckner (eds.), *A Road to Nowhere? Jewish Experiences in Unifying Europe* (Leiden/Boston, 2011), 33f.

must be acknowledged. However, it is also possible that things could turn out differently. As we all know, history sometimes takes a sudden, sharp turn.

Much will depend on if and how the next young generation of European Jewry treats their own heritage and tradition. At the moment, it does not look like much will change. One of the effects of the aforementioned general assimilation process, or more to the point the pressure of secularization, is that Jewish young people, just like young people in other faiths, are increasingly turning away from traditional community life and are beginning to go their own way.

Despite all of the negative prognoses, over recent years the beginnings of a quite distinct and varied Jewry have developed in Germany. One example of this is that Germany is one of the few countries in Europe which has not only secondary schools, colleges, and research institutes, but even three rabbinical seminaries: the Liberal *Abraham Geiger Kolleg*, the conservative *Zacharias Frankel College* in Potsdam, and the *Rabbinerseminar zu Berlin* in the tradition of neo-Orthodox Esriel Hildesheimer.

The question arises whether what is emerging is a specifically Jewish culture. In Germany, just like in many other European countries, there are annual, well-attended Jewish theater and film festivals, literature weeks, concert series, museum openings, and much more. This would seem to prove that there is a nascent revival of something like Jewish life in Europe. This should not, however, obscure the fact that we are here dealing with a pseudo-prospering. These are generally the activities of non-Jews who are purported to be Jewish, but very rarely actually are. At the same time, we can confirm that something has been set in motion, at least in certain European countries. Organizations like the *European Union for Progressive Judaism*, *United Jewish Appeal*, or – on the Orthodox side – *Chabad Lubavitch* and the *Ronald Lauder Foundation*, are committed to supporting the rejuvenation of their local Jewish communities. They actively support the establishment and development of Jewish kindergartens and schools, as well as the founding of youth centers and other community activities.

An additional phenomenon of the most recent years is the continuous influx of young Israelis especially to Berlin. Within a few years the number of Israeli women and men permanently living in the former "Reichshauptstadt" has increased to more than 10,000, while some media even estimate a number between 20,000 and 30,000. The reasons to settle here seem quite varied: searching for better career chances, start-up ventures, avoiding military experiences or just searching for a more tolerant climate for lesbians and gays. Berlin is very hip for all these groups, but also the ideal gate to Europe in general. Right now, it is not clear to what extent "Diaspora Israelis" will also be interested to join local Jewish communities in Germany and elsewhere.

Seen from this angle, it may be that we are experiencing something like a "Jewish" revival in Europe. We can only speculate. The term "Jewish Renaissance" was coined for this context a while ago by the French-Jewish intellectual Diana Pinto. For a long time she has also been a champion of the idea that European Jewry forms a "third pillar" in contemporary Judaism alongside American Jewry and Israel.[16] Pinto has been like a particularly positive, optimistic polar opposite of Bernard Wasserstein, who fears that the "Diaspora will disappear" in the near future. However, there is the strong objection that over the last few decades the sense of cohesion in the Jewish community was more or less defined by the avowal to remember the Shoah and solidarity with Israel. For many today, Jewish or non-Jewish, Judaism seems like a faded memory, a fiction, a figment of the imagination that people are trying to fill with life.

People still believe in Judaism as a (law-based) religion yet suspect that this is a faith that is not based on Jewish tradition, but rather mainly on the bitter experience of the Shoah. This is also true of the fight being fought against the new-old antisemitism.[17] This fight has a similar function as the faith derived from the Shoah: the fight against antisemitism and never forgetting is supposed to contribute to a sense of identity. Opinions vary as to whether this is actually the case.

An authentic Jewish culture, however, no longer exists in Europe. There are hardly any Jews today who speak Yiddish, which was the *lingua franca* of around ten million people in the late 1930s, or Ladino, the Jewish-Spanish language, not to mention those who are fluent in ancient Hebrew. Prayers are still said in Hebrew in the synagogues, but most of those attending the services recite the prayers more or less mechanically, without understanding their meaning. In this regard, according to Bernhard Wasserstein, we could compare the Jews to average Catholics, who are in a similar situation. They attend a Latin mass, but are unable to make anything of the prayers in Latin or follow the point of the rituals. Judaism and Christianity are facing similar problems, if you like. Both have lost the sense of meaning of the traditional rituals.

In the last few decades, Jewish communities, regardless of whether they are located in London, Paris, Stockholm, Kiev, Rome or Berlin, have unfortunately not understood how to develop a distinct collective identity oriented towards traditional Jewish cultural heritage while also exhibiting a European-oriented per-

---

[16] Diana Pinto, "A new Jewish Identity for post-1989 Europe", in *JPR policy paper* 1/1996; Bertelsmann Stiftung (pub.), *Deutsch-Jüdischer Dialog 1992–2002* (Gütersloh, 2003), 26–31.

[17] For more on the terminology see i.a. Klaus Faber, Julius H. Schoeps, Sascha Stawski (eds.), *Neu-alter Judenhass: Antisemitismus, arabisch-israelischer Konflikt und europäische Politik*, 2nd Edition (Berlin, 2007).

spective for the future. Instead, like it or not, non-Jews run a deeply commercialized counterfeit culture which has developed in the non-Jewish world. This includes performances of Anatevka, klezmer music concerts and the telling so-called "Jewish" jokes, that actually are not. We have to ask ourselves if European Jewry as it is today is even in a position to renew itself independently. There is room for doubt. The necessary reference points are gone. For example, German-Jewish culture as we know it from the time of the Weimar Republic in Germany or during the Austrian monarchy in Vienna, Lemberg or other places no longer exists. It has been consigned to history once and for all.

Moreover, and as already mentioned, the increasing number of terrorist attacks recently – like those in Toulouse, Brussels, Paris, and Copenhagen – intensified the perceived threat to life and limb. A study commissioned by the European Union, *Discrimination and Hate Crime against Jews in 9 selected EU Member States* (2012/13), revealed that around twenty percent of the Jewish respondents avoid going to Jewish venues purely out of fear of assault or attack.[18]

According to the study, another twenty percent have actively begun to try not to stand out as being Jewish in public over the last few years. They are quite simply afraid of being harassed or even physically attacked by a random passer-by when wearing a yarmulke or a Star of David. These are not justifiably tailored fantasies, but the painful experiences of real people, which need to be taken more seriously by current antisemitism research than they have been up to now. Currently, a follow-up study supported by the European Union, conducting a survey among Jews and Jewish communities, now in fourteen EU member states, is in the works. Based on the results, it is aimed to formulate suggestions for a more efficient fight against Jew hatred across Europe.

But for the moment, as it seems, more and more Jews are leaving France,[19] Belgium, and even Sweden.[20] The situation has even worsened during very aggressive and partly violent anti-Israeli demonstrations across Europe in the summer of 2014. Sweden, widely considered to be a liberal and tolerant country, has recently had massive antisemitic attacks, which were, however, largely ignor-

---

[18] European Union Agency for Fundamental Rights (FRA), Discrimination and Hate Crime against Jews in EU Member States: Experiences and Perceptions of Antisemitism.
[19] According to statistics from the Israeli Immigration Ministry, around 1,900 Jews emigrated from France to Israel in 2012; in 2013 the number was 3,288; already in 2014 apparently more than 7,000 Jews decided to emigrate. According to the 2013 study by the European Union Agency for Fundamental Rights, eighty-eight percent of French Jews had observed an increase in hostility toward Jews and forty-six percent had thought about emigration because of this.
[20] See Jörg Lau, "Bewusste Juden müssen sich darüber klar werden, dass sie hier keine Zukunft haben", in *Die Zeit*, January 29, 2011.

ed by the international public. The level of danger is not the same everywhere, and more prevalent in some countries than in others. However, on no account should this situation be downplayed or underestimated.

The number of Jewish immigrants is, in total numbers, still relatively low, but it shows that some European countries, not all to be sure, have a massive problem and need to work harder to protect all – not just their Jewish – minorities. Jews in France, Belgium, and Sweden are feeling highly anxious. The aforementioned European Union study documents the extent to which fear and apprehension are increasingly dictating the mood within the Jewish community.

In the case of Germany, it is an undisputed fact that a consistent part of the population holds strong prejudices against Jews. Already in the mid-1980s, the sociologist Alphons Silbermann spoke about the fact that around twenty percent of the West German population had strong antisemitic prejudices. Silbermann went even further and then determined that another thirty percent held latent antisemitic views. This means that in certain situations, such as in a movie considered controversial or in a book that has caused a stir, latent prejudices can erupt as manifest antisemitic actions. This was new and triggered a debate not only amongst researchers, but also in the general public.

Silbermann's revelations, based on empirical studies, have hardly changed up to today. The most that can be said is that new motives and reasons for antisemitic bias have emerged, for example, anti-Zionist and anti-Israeli attitudes. To be sure, these existed thirty years ago, but at that time they only played a marginal role in the general discussion. These attitudes can also include an antisemitic undertone, as we have had to acknowledge.

If Europe will be able to, in the end, provide its Jewish minority – now totaling one and a half to two million people out of around 500 million E.U. citizens – with a feeling of security and belonging despite this development felt by many to be dangerous, then we will have made considerable progress. However, this does not actually appear to be the case in any European country at the moment. At the same time, a dangerous situation does exist, even if it should not be overrated. The threat level is higher in certain European countries and lower in others.

If the circumstances for Jews were to stabilize in the near future and the threat level does not exacerbate, what could or should the real existing Jewry contribute to the future of Europe specifically on its own? The first thing I think of is getting involved in public matters, committing to civil society, and most of all, exercising criticism wherever freedom and human rights are under attack. It is no coincidence that it is the Jewish intellectuals, like Bernard-Henri Lévy, who call out European patriotism and stand in the forefront in striking back against the enemies of democracy. Lévy was in Kiev in 2014 when the Maidan movement disposed the Ukrainian dictator Yanukovich and Russian forces invaded the Crimea and strongly supported separatist forces in the Eastern Ukraine – till today. Bernard-Henri Lévy is also one of the most-heard, prominent intellectual voices calling for the European countries to stand together and to fight for the future of the EU.[21]

Today a number of Jewish intellectuals, essayists, and academics once again consider Europe to be a place for Jews to call their home. This is the reason why Jews in Paris, London, Warsaw or Jerusalem engage with the public and spark debate with like-minded people. Their influence, although at times exaggerated, is usually exerted below the surface; however, it is noticeable with a little effort.

The Judeo-Christian heritage is an integral part of European culture. No reasonably informed person would deny this. Jews belong to Europe just as Europe belongs to the Jews. While this fact has been widely forgotten, even contested by some, it is a legacy closely intertwined with European development and the European idea. We should always keep this in mind.

In past centuries, Jewish influence has left a clear imprint not only on the arts, sciences, and the economy, but also on Europe's intellectual life – not only in Germany, it should be noted, but in all European countries. A Europe which is not conscious of this heritage cannot correctly understand its own roots. If Europe refuses to acknowledge this heritage, it will remain a soulless construct. As we have all come to realize at our own costs, Europe has long been on a desperate search for its own identity and must not become fixated only on economic questions. The politically accountable, and not only them, but all of us, should refrain from talking about Europe solely as a unified economic and monetary zone. It is equally, if not more, important for us to treat the continent's cultural roots and traditions with greater respect in the future.

For the time being, Europe's post-War Jewry has survived, and at several places in Western and Central Europe it was even able to stabilize. For some Jewish migrants from other continents (and even for some from Israel), the European Union and re-unified Germany have become a favored place of destination. At the same time, certain trends of disintegration, radical Islamist terror, and new forms of Judeophobia have intensified Jewish fears of the future on the "Old Continent." The coming years will reveal whether Europe's Jews have found their homes – or will be pushed into restless circumstances once again.

---

21 Interview with Bernard-Henry Levy in *The Guardian*, July 14, 2016, https://www.theguardian.com/film/2016/jul/14/bernard-henri-levy-europe-without-the-british-spirit-cannot-be-europe.

Bibliography

DellaPergola, Sergio. "World Jewish Population". In *The American Jewish Year Book* 113, edited by Arnold Dashefsky and Ira M. Sheskin, 279–358. Dordrecht: Springer, 2013.

DellaPergola, Sergio. "Jews in Europe: Demographic Trends, Contexts and Outlooks." In *Road to Nowhere? Jewish Experiences in Unifying Europe*, edited by Julius H. Schoeps and Olaf Glöckner. Leiden/Boston, 2011.

DellaPergola, Sergio. "Die demographische Entwicklung der europäischen Juden vom 12. bis zum 20. Jahrhundert." In *Handbuch zur Geschichte der Juden in Europa, Vol. 2: Religion, Kultur, Alltag*, edited by Elke-Vera Kotowski, Julius H. Schoeps, and Hiltrud Wallenborn, 15–28. Darmstadt, 2001.

Diner, Dan. *Zivilisationsbruch: Denken nach Auschwitz*. Frankfurt/Main, 1988.

European Union Agency for Fundamental Rights (FRA), Discrimination and Hate Crime against Jews in EU Member States: Experiences and Perceptions of Antisemitism. Vienna 2013. https://european-forum-on-antisemitism.org/report/discrimination-and-hate-crime-against-jews-eu-member-states-experiences-and-perceptions. Accessed October 9, 2018.

Faber, Klaus, Julius H. Schoeps, and Sascha Stawski (eds.). *Neu-alter Judenhass: Antisemitismus, arabisch-israelischer Konflikt und europäische Politik*. 2nd Edition. Berlin, 2007.

Königseder, Angelika, and Juliane Wetzel. *Lebensmut im Wartesaal. Die jüdischen DPs (Displaced Persons) im Nachkriegsdeutschland*. Frankfurt/Main, 1994.

Kowitz-Harms, Stephanie. *Die Shoa im Spiegel öffentlicher Konflikte in Polen. Zwischen Opfermythos und Schuldfrage (1985–2001)*. Berlin, 2014.

Mertens, Lothar. *Davidstern unter Hammer und Zirkel: Die jüdischen Gemeinden in der SBZ/DDR und ihre Behandlung durch Partei und Staat*. (= Haskala. Wissenschaftliche Abhandlungen, Vol. 18), Hildesheim et al., 1997.

Pinto, Diana. "A new Jewish Identity for post-1989 Europe." In *JPR policy paper 1/1996*. London, 1996.

Röhr, Werner. "Massaker an Überlebenden. Zum antijüdischen Pogrom in der polnischen Stadt Kielce am 4. Juli 1946." *Bulletin für Faschismus- und Weltkriegsforschung* 29 (2007): 1–32.

Schoeps, Julius H. "A seyfer is a jeder schteyn". In *...und die Karawane zieht weiter ihres Weges. Freundesgabe für Jürgen Reulecke, den Vorsitzenden des Mindener Kreises, zum 75*, edited by Eberhard Schürmann, Horst Zeller, and Fritz Schmidt. Geburtstag, Ebersdorf, 2015.

Schoeps, Julius H. *Das Gewaltsyndrom: Verformungen und Brüche im deutsch-jüdischen Verhältnis* (= Deutsch jüdische Geschichte durch drei Jahrhunderte, Vol. 8). Hildesheim, 2012.

Schoeps, Julius H. (ed.). *Leben im Land der Täter: Juden im Nachkriegsdeutschland (1945–1952)*. Berlin, 2001.

Steiner, Barbara. *Die Inszenierung des Jüdischen. Konversion von Deutschen zum Judentum*. Göttingen, 2015.

Szaynok, Bozena. "The Pogrom of Jews in Kielce, July 4, 1946". *Yad Vashem Studies* 22 (1992): 199–235.

Vogt, Bernhard, Willi Jasper, and Julius H. Schoeps (eds.). *Ein neues Judentum in Deutschland? Fremd- und Eigenbilder der russisch-jüdischen Zuwanderer.* Berlin/Potsdam, 1999.

Wasserstein, Bernard. *Europa ohne Juden: Das europäische Judentum seit 1945.* Köln, 2000.

Section II: **Breaks, Changes, and Continuities in Austria and Hungary**

Vladimir (Ze'ev) Khanin
"Russians," "Sephardi", and "Israelis": The Changing Structure of Austrian Jewry

Austrian Jewry could be defined as a sort of "new immigrant" community that shows a typologically indicative example of multi-dimensional trends among the world and European Jewry. This community reflects parallel trends of internal integration and diversity of various religious, sub-ethnic, and cultural-communal groups. Three groups are most prominent among them: the "indigenous" Austrian Jews, or Alt-Wiener (the Holocaust survivors and their descendants); the "Hungarians" (Jewish immigrants from Eastern Europe); and the "Bukhara Jews" or "Russian Sephardi," who compose the most visible group of Jewish immigrants from the USSR and FSU of the 1970s, 1980s and 1990s. The newest migration waves have brought to Austria also immigrants from Israel and Iran. However, Jewish immigrants to Austria with a Soviet or post-Soviet background still play a specific role. This group of immigrants consists of three to four thousand persons who stayed in Austria, which was a transit point for Jewish emigration from the USSR to the West in the 1970s, or returned there from Israel, as well as for the FSU Jews who joined them in the 1990s. Besides the fact that this people represent the "Russian-speaking" Jewish community of Austria, and in this capacity also the Austrian segment of the transnational Russian Jewish Diaspora, the overwhelming majority of this group is composed of representatives of "oriental" Jewish communities of the (former) Soviet Union – mostly Bukharian, as well as Georgian and, to lesser extent, Caucasian ("Mountain") Jews. A significant number, or even majority, of the Austrian Jewish immigrants with roots in the former USSR spent a certain period of their lives in Israel, and thus are Israeli passport holders. All these groups exemplify various models of ethnic, religious and civic identity, social mobility as well as vision of the host country and the State of Israel, migration dynamics, and they quite often have their own institutional infrastructure.

Introduction

Two new transnational Jewish Diasporas, Israeli and Russian-Jewish, formed in the recent few decades and had already become a notable factor of contemporary Jewish life and an important element of the ethnic cultural mosaic of the various countries of the world. However, besides their "metropolis," meaning Israel and Russia/the FSU respectively, both groups still enjoy a peripheral status of vis-à-vis "indigenous" Jewish communities; the whole common element of these two transnational groups – Russian Israelis out of Israel – almost nowhere became either a bridge or a leadership element of them.

The Jewish community of contemporary Austria, albeit relatively small but very typologically important for the understanding of Jewish and general transnational-diaspora trends, is among very few and may even be only an exclusion from this rule.

The contemporary Jewish community of this country, which shared the destiny of European Jewry of modern times, may be defined as a classical example of the "neo-immigrant" Jewish entities. Not more than 5,000 of about 200,000 Jews that lived in Austria (including ninety percent in its capital of Vienna) on the eve of the World War II survived by 1945. However, indications of rebirth of this Jewish community were already obvious by the end of that decade. In the course of time, those who survived the Holocaust in Austria or returned there from emigration in Israel, USA and Great Britain, and their descendants, were joined by Jewish immigrants from Eastern Europe (mainly Hungary, Czechoslovakia, and Poland, and later on, the USSR and post-Soviet states), and in the recent decades, also from Israel and Iran. At the moment, the Jewish population of Austria is estimated between 10,000 to 12,000 persons, however just about 7,500 of them are officially registered in the local Jewish community of the country, including ninety-five percent in Vienna, and the rest in small Jewish communities (a bit more than a hundred persons each) of Linz, Salzburg, Innsbruck, Graz, Baden, and four other cities.

Despite obvious internal integration trends, Austrian Jewry still represents the diversity of sub-communal and ethnic culture groups. Among them were representatives of "indigenous" Austrian Jews and of two later waves of Jewish immigration to this country: Ashkenazi (European) and Sephardi (Oriental) Jews, as well as members of a few religious Jewish denominations, including Chabad Lubawitsch, Modern Orthodox and Reform, together with traditionalists and secular Jews, and many other groups.

Hence, the categorization and sociocultural typology of Austrian Jewry remains an uneasy task. The Israeli sociologist of Austrian origin Susanna (Suzy) Cohen-Weisz believes that this community could be divided into three "generations" of Jews ("returnees" and "new immigrants") and their descendants[1]:

---

[1] Susanne Cohen-Weisz, *Like the Phoenix Rising from the Ashes: Jewish Identity and Communal Reconstructions in Austria and Germany.* Ph.D. Dissertation, Submitted to the Senate of the Hebrew University of Jerusalem, December 2010, 97.

- "Viennese" – "indigenous" Austrian Jews, or *Alt-Wiener*, which by 1953 composed about fifty percent of the local Jewish population;
- "Hungarians" (Jewish immigrants from East-Central Europe); and
- "Bukharian," or "Russian Sephardic" Jews, who in terms of demography and culture were prominent among the USSR and the CIS immigrants of the 1970s – 1990s.

This division, of course, may not be seen as totally sufficient due to the fact that each of these three groups in turn are internally diverse, and immigrants from the USSR and post-Soviet states are even more so than others. This group consists of three to four thousand former Soviet Jews that stayed in Austria which was a transit point for Jewish emigration from the USSR to the West in the 1970s, or returned there from Israel, as well as of those FSU Jews that joined them in the 1990s.[2] In turn, this group belongs to at least three categories of local Jewry. Firstly, they represent the "Russian-speaking" Jewish community of Austria (and in this status also the Austrian part of the transnational Russian Jewish Diaspora). Secondly, the overwhelming majority of this group is composed of representatives of "oriental" Jewish communities of the (former) Soviet Union – mostly Bukharian, as well as Georgian and, to a lesser extent, Caucasian ("Mountain") Jews.

Even more important, in the context of this research, a significant number or even the majority of Austrian Jewish immigrants with roots in the former USSR spent a certain period of their lives in Israel, and thus are Israeli passport holders.[3] In this way, Austria presents an interesting example of the Israeli Diaspora community – "Israeli Sephardi Russians." This group, together with a few hundred "Israeli Ashkenazi Russians" and some two thousand Israeli passport holders that were born either in Israel or in the Diaspora beyond the FSU, now comprise one-third to forty percent of the Austrian Jewish population, and to a large extant define its sociocultural and political appearance.

Paradoxically, this extremely interesting sociological and historical phenomenon was hardly analyzed in academic literature. In fact, with the important exclusion of the above-mentioned dissertation of Suzan Cohen-Weis and Evelyn Adunka's history of official Jewish institutions of Austria by the early 1990s (which thus

---

[2] Alexander Friedmann, "Psycho-Socio-Cultural Rehabilitation in an Ethnic Subgroup: A 30-Years Follow-Up." *World Cultural Psychiatry Research Review* April/July (2007): 88 – 89.
[3] Susanne Cohen-Weisz, *From Bare Survival to European Jewish Vision: Jewish Life and Identity in Vienna* (Jerusalem: Center for Austrian Studies, Hebrew University of Jerusalem, 2009), 32 – 33.

did not review the dramatic changes of the recent two decades),[4] systematic comparative analyses of social, cultural, political, and identity trends in the Jewish community of Vienna and Austria after 1945 were yet to be published. As for the "Russian" and Hebrew-speaking Israelis in Austria, both groups remain almost *terra incognita*. To the best of our knowledge, the only comprehensive research of Israeli immigration to Austria was published four and a half decades ago,[5] while the academic studies of the phenomenon of Russian-Israeli immigration to this country, compared with other Russian and non-Russian-speaking Jewish and Israeli immigrant groups, have not been done at all.

This sociological study of Jewish immigrant communities of Vienna (which in turn was a part of a project of studies of new transnational Jewish Diasporas)[6] aimed to fill this gap. Two hundred respondents were questioned in the Austrian capital via face to face interviews based on standard structured questionnaires. In the absence of thoroughly accurate demographic data on all local Jewish population categories in which we are interested in this study, meaning (a) "Russian" Israelis; (b) Russian Jews that never lived in Israel; and (c) Israelis of "non-Russian" origin, construction of the classical representative quota sample was hardly possible. Due to this, respondents were selected with the help of the snowball method and ranged according to the broad random sample, considering their numbers and internal structure (i.e. Ashkenazi-Sephardi, "indigenous-immigrant," "affiliated-unaffiliated," and other subgroups).

In addition, with the aim of getting comparative data, a group of Vienna Jews without roots in the USSR/FSU or Israel were also interviewed in the course of the study. As a result, the structure of our sample was pretty close to the structure of registered membership of the Jewish community of Vienna and to the estimated structure of Austrian Jewry (including unaffiliated persons) in general. Our aim, however, was to understand principal differences between various Austrian Jewish ethnic cultural groups rather than numerical generalizations. To specify these parameters, besides the quantitative poll, our study also included eight in-depth interviews aimed at collecting expert analyses and typical life-stories, based on the "oral history" methodology (this method had already been effectively used for the study of history, and, to a lesser extent, the sociology of

---

4 Evelyn Adunka, *Die Vierte Gemeinde – Die Wiener Juden in der Zeit von 1945 bis Heute* (Berlin: Philo, 2000).
5 Friederike Wilder–Okladek, *The Return Movement of Jews to Austria after the Second World War – With Special Consideration of the Return from Israel* (The Hague: Martinus Nijhoff, 1969).
6 This project was initiated by the Ministry of Aliya and Immigrant Absorption of Israel and fulfilled in cooperation with the Lookstein Center for Jewish Education in the Diaspora, Bar-Ilan University and the "Lubimyi Gorod" Foundation, Moscow.

Jews in Austria and Vienna).[7] This way we were able to identify seven basic categories of Austrian Jews which we compared in terms of their social, cultural, and civic identification, as well as their organizational infrastructure, migration plans, and other parameters.

Table 1: Community subgroups.

Categories of respondents	N	%
1 Born in Israel	29	14.5
2 Born in the USSR/Russia and CIS, lived in Israel	30	15.0
3 Born in any other country, lived in Israel	14	7.0
4 Born in Israel to USSR-born parents	17	8.5
5 Born in the USSR, never lived in Israel	70	35.0
6 Austrian non-Russian Jews, did not live in Israel	29	14.5
7 Austrian-born to USSR-born parents, did not live in Israel	11	5.5
Total	200	100.0

Of course, one can also observe internal divisions in the categories of "Austrian non-Russian Jews that did not live in Israel" and "born in any country beyond the USSR/FSU and lived in Israel" between those who were born in Austria and those who were born abroad. However, these internal divisions proved to be much less important than the differences between the seven basic groups mentioned above.

---

7 See, for example, the project "Jüdische Lebensräume" (Jewish Habitats) of the Institute for Jewish History in Austria (Das Institut für jüdische Geschichte Österreichs) in St. Pölten, aimed at collecting memories of Jews during the period between the two World Wars (http://www.injoest.ac.at/projekte/abgeschlossen/lebensraeume_erinnerungen_lebensgeschichten_oes terreichischer_juden/), as well as recording the life stories of Austrian Jewish entrepreneurs after World War II, in M. John and A. Lichtblau, "Jüdische Unternehmer in Österreich nach 1945: Oral History und ihre Forschungsperspektiven für die postfaschistische jüdische Geschichte" [Jewish Entrepreneurs in Austria to 1945: research of oral history and its prospects for the post-fascist Jewish History], in *Studien zur Geschichte der Juden in Österreich* [*Studies in the History of the Jews in Austria*] (Wien, 1994), 166–191.

Ethnic, Religious and Civic Identity

Although at first glance the Jewish community of Austria and Vienna is rather homogeneous in ethnic and religious terms, a more careful look inside may lead us to the conclusion of substantial diversity within this entity.

Three-quarters of our sample were representatives of ethnically homogeneous (meaning, Jewish in all their generations) Jewish families, which in my mind corresponds to the real structure of the whole community.

As the table above shows, according to the respondents' statements, both parents of the members of four out of seven categories of Vienna Jews we highlighted were ethnic Jews in ninety-three to one hundred percent of the cases, and yet eighty percent in another category. Only in two categories – immigrants from the former USSR and "Austrian Jews," both of them having had no experience of living in Israel – was the share of those whose both parents were ethnic Jews somewhat lower (correspondingly, over fifty-seven percent and approximately sixty-eight percent). But even in these categories, the vast majority of other members were born to Jewish mothers. That brings the Jews' share, according to the criteria of Halakha (traditional Jewish law) which is actually used by local Jewish organizations, in these two groups to over ninety percent and eighty-five percent respectively. As to the male-line descendants of Jews, the only significant group of theirs (slightly more than fourteen percent) is present in the category of respondents consisting of Austrian Jews that had no "Russian" roots and never lived in Israel on a permanent basis. It is no accident that, according to the estimates of the interviewed observers, a good number of representatives of this particular cluster of Vienna's Jewish population are members of the city's only reformist synagogue whose demographic and political influence in the community is not yet that great.

If so, the very Judaic religious criteria must be the basic structure-forming factor of Jewish identity in this country. However, despite the fact that our research had recorded a strong Jewish identity in all categories of Austrian Jews without exception, the share of religious and non-religious respondents in our sample turned out to be roughly equal (over forty-one percent versus forty-three percent) and almost sixteen percent found it difficult to answer the question on religiosity. The smallest number of religious persons is demonstrated by those who originated from the former USSR and did not live in Israel, which is understandable in view of the Soviet Jewish national identity having established

Table 2: Respondents' Jewish family background, according to respondents' country of origin.

	Respondents' country of origin							Total
	Born in Israel	Born in the USSR/Russia and CIS, lived in Israel	Born in any other country, lived in Israel	Born in Israel to USSR-born parents	Born in the USSR, never lived in Israel	Austrian non-Russian Jews, did not live in Israel	Austrian-born to USSR-born parents, did not live in Israel	
Both parents of Jewish origin	100.0%	80.0%	92.9%	100.0%	57.4%	67.9%	100.0%	76.8%
Only father of Jewish origin	.0%	.0%	.0%	.0%	4.4%	14.3%	.0%	3.6%
Only mother of Jewish origin	.0%	13.3%	7.1%	.0%	33.8%	17.9%	.0%	17.0%
Only one of grandparents	.0%	6.7%	.0%	.0%	2.9%	.0%	.0%	2.1%
Only spouse	.0%	.0%	.0%	.0%	1.5%	.0%	.0%	.5%
Total N	27	30	14	16	68	28	11	194
%	100.0%	100.0%	100.0%	100.0%	100.0%	100.0%	100.0%	100.0%

itself on a secular basis, against the nearly complete suppression of outer manifestations of the Jewish self-identification.[8]

We will definitely not be far from the truth by supposing that a large part of non-religious persons in this category falls into the Ashkenazi share. While living in the USSR among peoples with a strong ethnic and religious identity, "Russian Sephardi" communities were, together with them, a focus of the authorities' greater tolerance towards local religions. Due to this, Bukharian, Mountain, and Georgian Jews managed to retain their Jewish religious and cultural heritage to a greater extent than was customary in other parts of the USSR. As a thirty-five-year old Viennese Georgian Jew, Eka, pointed out while conversing with us, the settlement in Georgia where she had lived with her parents was commonly called "a small Georgian Yerushalayim" and was always notable for its Jewish tradition.

> The tradition was observed through and through – the Shabbat, the kashrut and the holidays. Locals resisted the authorities that wanted to close down the synagogue back in the Soviet time, and locals won. They didn't let them destroy either the synagogue or the Jewish community. [That is why] the Jews from [eastern communities] of the former Soviet Union who live in Vienna are mostly of traditional origins. In the USSR, you couldn't [lead a Jewish life] otherwise. Now here they can become religious orthodox, although in their mass, urban and Georgian Jews are still religious [traditionalists], but not orthodox.

In any case, nearly ninety percent of our respondents (twice as many as the share of "strictly religious" ones) stated that, regardless of their religiousness level, Judaism is a religion they consider "theirs." In fact, we detected no visible differences in this point between all respondent categories, except for the ones originating from the former USSR that never lived in Israel (they also had the smallest share of religious persons), almost sixteen percent of whom named either Christianity, or Judaism and Christianity together, or any other religion as "their" religion, while over seven percent turned out to be dedicated atheists. It seems that in this instance, as in others, religion (while not the only part) acts as a very important part of the ethnic and cultural background for virtually all "new immigrant" categories that form a majority of the Jewish community in Austria.

By applying a known statement made by Zvi Gitelman to the Austrian situation, that religion plays a role of the "window of ethnicity" for many Jews in

---

[8] For details see Vladimir (Ze'ev) Khanin, "Russian-Jewish Ethnicity: Israel and Russia Compared", in E. Ben-Rafael, Y. Gorny, and Y. Ro'i. (eds.), *Contemporary Jewries: Convergence and Divergence*, 216–234.

"Russians," "Sephardi", and "Israelis": The Changing Structure of Austrian Jewry — 81

Table 3: Religious identity, according to respondents' country of origin.

			Respondents' country of origin					Total
	Born in Israel	Born in the USSR/ Russia and CIS, lived in Israel	Born in any other country, lived in Israel	Born in Israel to USSR-born parents	Born in the USSR, never lived in Israel	Austrian non-Russian Jews, did not live in Israel	Austrian-born to USSR-born parents, did not live in Israel	
Religious identity*								
Feel like a religious person	37.9%	41.4%	57.1%	52.9%	24.3%	67.9%	54.5%	41.4%
Do not feel like a religious person	51.7%	44.8%	35.7%	29.4%	57.1%	17.9%	18.2%	42.9%
No exact answer	10.3%	13.8%	7.1%	17.6%	18.6%	14.3%	27.3%	15.7%
Total, N	29	29	14	17	70	28	11	198
Cultural religious identification**								
Judaism	93.1%	96.6%	100.0%	100.0%	76.8%	93.1%	100.0%	89.3%
Christianity	.0%	.0%	.0%	.0%	7.2%	.0%	.0%	2.5%
both Judaism and Christianity	.0%	.0%	.0%	.0%	7.2%	3.4%	.0%	3.0%
Other	.0%	.0%	.0%	.0%	1.4%	3.4%	.0%	1.0%
None	6.9%	3.4%	.0%	.0%	7.2%	.0%	.0%	4.1%
Total, N	29	29	14	17	69	29	10	197
Total, %	100.0%	100.0%	100.0%	100.0%	100.0%	100.0%	100.0%	100.0%

*Missing: N = 2 (1%) **Missing: N = 3 (1.5)

western countries, same as in Eastern Europe,[9] we would like to point out that it is true both in relation to the "new generation" of immigrant Jews and, probably, regarding "indigenous" Austrian Jews as well. This fact also erases a clash between the official status of "Jewry" in Austria primarily as a religious entity and a perception of Jewry as an ethnic and confessional or singularly ethnic category that is predominant in all groups, except for the "indigenous" Vienna Jews, especially among (ex-)USSR immigrants and the Israelis.

Eventually, as can be seen from the collected data, the "proper" Jewish or universal Jewish identity is pronounced in all seven categories of Austrian Jews, and this is the only identity platform that unites all sub-ethnic and sub-communal groups of Jews residing in Vienna and other Austrian cities into a single community.

However, if we look at the affiliation with other ethnic, sub-ethnic and civil identification types, representatives of the seven categories of Austrian Jews that we have accentuated demonstrate major differences, despite the marked ethnic homogeneity of the Jewish community.

Austrian and Jewish Austrian Identity

This identity is developed to the greatest extent among those persons who have been born outside the former USSR and Israel and lived in the Jewish state (over forty percent of this category were born in Austria) and Austrian-born children of immigrants from the (former) USSR that had no experience of living in Israel. Interestingly, among the "proper" Austrian Jews that have not previously lived in Israel either, the share of those who consider themselves primarily "Austrians" and/or "Austrian Jews" is noticeably lower – correspondingly twelve percent and twenty-three percent respectively. There are probably three reasons to explain it.

Firstly, based on the respondents in our sample, less than forty percent (39.3%, to be exact) of "Jewish Austrians" that had not lived in Israel were actually born in Austria. It is significant that approximately this number of people – 42.9% – in this group named their resident county "their country in the fullest extent." Another important point is that one-third of the respondents in this category have been living in Austria for less than five years. Incidentally, nearly a

---

[9] Zvi Gitelman, "Jewish Religion, Jewish Ethnicity – The Evolution of Jewish Identities", in Z. Gitelman (ed.), *Religion or Ethnicity? The Evolution of Jewish Identities* (New Brunswick, NJ: Rutgers-University Press, 2009), 1–5.

"Russians," "Sephardi", and "Israelis": The Changing Structure of Austrian Jewry

Table 4. To what degree you feel: (4 – Very much; 5 – Mostly or exclusively).

		Respondents' country of origin							Total
		Born in Israel	Born in the USSR/ Russia and CIS, lived in Israel	Born in any other country, lived in Israel	Born in Israel to USSR-born parents	Born in the USSR, never lived in Israel	Austrian non-Russian Jews, did not live in Israel	Austrian-born to USSR-born parents, did not live in Israel	
Jewish	5	85.7%	76.9%	72.7%	100.0%	68.3%	76.9%	77.8%	77.1%
	4	10.7%	15.4%	18.2%	.0%	23.3%	15.4%	22.2%	16.6%
Austrian	5	3.8%	.0%	27.3%	.0%	.0%	12.0%	33.3%	6.3%
	4	7.7%	4.8%	.0%	8.3%	1.9%	16.0%	22.2%	7.0%
Austrian Jew	5	3.8%	4.5%	36.4%	8.3%	1.9%	23.1%	60.0%	12.4%
	4	11.5%	9.1%	9.1%	.0%	1.9%	15.4%	20.0%	8.1%
Israeli	5	80.8%	28.6%	16.7%	61.5%	.0%	.0%	.0%	23.3%
	4	11.5%	9.5%	8.3%	15.4%	.0%	8.0%	11.1%	6.9%
Russian	5	–	.0%	.0%	.0%	11.3%	.0%	.0%	4.5%
	4	–	33.3%	.0%	.0%	28.3%	.0%	.0%	14.1%
Russian Jew	5	–	22.7%	10.0%	8.3%	41.5%	.0%	.0%	18.5%
	4	–	27.3%	.0%	.0%	30.2%	.0%	11.1%	14.6%
Sub-Ethnic Identity	5	3.8%	34.8%	36.4%	33.3%	29.8%	12%	44.4%	25.2%
	4	0%	4.3%	9.1%	8.3%	17.5%	8%	11.1%	9.8%
N		29	30	14	17	70	29	11	200

quarter of the respondents in this category state that they permanently live, whether psychologically or actually, not in Austria, but in another foreign country (other than Russia and Israel), or live "in two or three countries together." Therefore, in a situation where their residence in Austria is perceived more as a civil and religious given rather than an ethnic and national belonging, the universal Jewish identity for this group is dominant without doubt and includes all other identity types as a peripheral element.

On the contrary, for those emigrants from Israel who were born in the Diaspora, identification with a country and diasporic Jewish identity certainly seems to have a much greater significance. Their range includes 42.9% of Austrian natives (about the same or even a slightly greater number than that among "local" Jews that have not previously lived in Israel) and 28.6% of people that have lived twenty or more years in Austria, while the absolute majority of this entire group (64.3%) spent less than five years in Israel. Taking into consideration that, in its turn, an absolute majority (seventy-eight years) of respondents that lived in Israel for less than five years in this category consists of natives and long-term residents of Austria (over twenty years in Austria), it is no wonder that it is in this particular category that the share of those who consider the Republic of Austria "their country to the greatest extent" (over 69.2%) appeared higher than in any other categories of the respondents. It also shows the greatest share of those who are not engaged and are not going to be engaged in any pro-Israel activities (42.9%). It would be easy to assume that the identification complex of these people reflects their status of migrants that have made their choice (even if not a final one for many of them) not between one *galut* and another, but between the Jewish state and a diaspora.

As for the children of immigrants from the former USSR, the relatively high degree of their identification with Austria by all accounts is a result of a combination of two factors: their birth in the country and the influence of their parents' experience (mostly immigrants from the 1970s who "escaped from the "Evil Empire" and dropped out from the Israeli track then). Let us add that the same group demonstrates the highest level of "rootedness" in Austria. Thus, 72.7% of respondents in this category (2.5 times more than in the overall sample and 1.5 times more than among Austrian Jews whose roots are not connected with CIS and who have not previously lived in Israel) declare that their entire family lives in Austria. These circumstances, combined with the fact of Jewish identity that dominates in this group, same as in others, gives while compared to other highlighted categories the highest percentage of those who have declared their Austrian communal and sub-ethnic Jewish identity (sixty percent feel "exclusively or mainly an Austrian Jew" and twenty percent more feel as such "to a great extent.")

Most probably, the majority of them might support a definition given by one of our respondents that was born in Vienna in a family of immigrants from the former USSR: "As I was born in Vienna, I think that my home is here. In the cultures of both Austrian and Jewish society I feel at home. I have no emotional link to Russia, the Soviet Union, or Ukraine (from which his parents immigrated first to Israel and then to Austria)." The share of holders of the same identity among those who were born outside the USSR and CIS and lived in Israel for some time was not so significant, but more evident as compared to other categories of Vienna Jews. No wonder that to the members of this category, the ethnic belonging as a "given" turned out to be much more important than the "activist" aspect of Jewish identification (children of immigrants from the former USSR visit a synagogue and actively participate in the community's life two to three times correspondingly more seldom than Austrian non-Russian Jews who did not live in Israel).

Israeli Identity

This identity, as expected, is dominant in the community of immigrant Jews that were born in and (in most cases) grew up in Israel. There were over eighty percent and eleven and a half percent of the holders of "exclusively" and "to a great degree" Israeli identity there. Of greater interest is that almost two-thirds of representatives of the second generation of Jewish repatriates from CIS and Baltic states that ended up in Austria declared their "exclusively Israeli" identity (another fifteen percent of this group feel Israeli "to a great extent"). And this is assuming that the degree of Austrian "rootedness" for "Russian" natives of Israel is significantly higher than that for "non-Russian" *Sabras*. Thus, two-thirds of the members of the first group have spent over five years in Austria (one-third has spent over twenty years there), as opposed to indigenous "non-Russian" Israelis, two-thirds of which have lived in Austria for less than five years.

Accordingly, members of the first group, while being almost as proficient in Hebrew as indigenous "non-Russian" Israelis, speak German better than the latter (one-third of them are also proficient in the Russian language). The immediate family of nearly half (47.1%) of "Russian" *Sabras* also live in Austria – that is, twice as much of a share as that of "non-Russian" natives of Israel (24.1%). So, it is no accident that 82.4% of representatives of the second generation of "Russian" repatriates residing in Austria name Austria, as opposed to some other country, the country of their permanent residency, that is 1.5 times more often than indigenous "non-Russian" Israelis (sixty-nine percent). Nevertheless, "Russian" natives of Israel, same as their parents, are second after indigenous "non-Russian" Israelis (sixty-nine percent) in terms of those who considered Israel their country to the greatest extent (forty-seven and fifty percent correspondingly).

Speaking of Israeli identity in former and current holders of Israeli citizenship among Jews that were born in a diaspora, the ultimate factor for them possessing or lacking such identity is commonly believed to be the period of their stay in Israel. This particular statement is popular in the expert community in regard to repatriates that have both remained in Israel for good and committed the second emigration – naturally, with all amendments as to certain models of their integration into Israeli society, especially within the past twenty to twenty-five years.

True, the relatively big share of those who declared that they felt Israeli to a "great" extent (thirty-eight percent) among our respondents out of former Soviet Jews born in a diaspora, yet with experience living in a Jewish state, is, at a first glance, comparable to their community containing a similar share of those who had lived in Israel for more than ten years prior to their emigration to Austria. A very weak Israeli identity, or its complete absence, in three-quarters of "non-Russian" natives of the diaspora that experienced living in Israel (forty-three percent of whom were Austrian natives, it should be reminded) whom we questioned is also in line with the fact that over sixty-four percent of this group's members have been living in Israel for less than five years. Therefore, in theory, it fits into the same paradigm.

In its scope, it is a harder task to explain an existence of over forty percent of those who possess an Israeli identity "to some extent" and over ten percent of those who possess it "to a great extent and mostly" among children of immigrants from the former USSR that were born in Austria and never lived in Israel at all. It is probably incidental not so much to them having Israeli national and/or civil self-awareness, as to the understanding of Jewry as a primarily national and cultural or ethnic phenomenon that is inherent in Soviet and post-Soviet Jewry (including its second generation, as our and other studies have demonstrated[10]). Naturally, Israel plays an important symbolic role in this.

One of the experts we had interviewed – the son of Lvov Jews who was born in Vienna and whose parents emigrated to Israel in the 1970s and then, nine months later, back to Austria – explained the meaning of this phenomenon as related below:

---

[10] Vladimr (Ze'ev) Khanin, Alek Epstein, and Dina Pisarevskaya, *Post-Soviet Jewish Youth: Identity, Community and Relations with Israel* (Moscow and Ramat-Gan: Institute for Oriental Studies, Russian Academy of Science, and the Lookstein Center for Jewish Education in the Diaspora, Bar-Ilan University, 2013), originally in Russian.

My parents returned to Vienna after such a short period, because they were very dissatisfied with Israel, particularly its culture. They actually wanted to come back to the Soviet Union, but as [the Soviets authorities] did not permit that, they remained in Vienna [with which they are now] very happy... What is Israel to me, a Vienna Jew? It is our people and our country. Israel must be very important for every Jew, because it is a country of every Jew. [Although] I never thought of going there to work or study. I do not visit Israel frequently. And I have very few relatives in Israel. In Austria, or Germany, or America either...

Russian-Jewish and Russian Identity

Such identity is mostly pronounced (and no wonder) in the community of the first generation of emigrants from the (former) USSR: primarily, among Russian-speaking Jews and their family members that have arrived to Austria directly from the USSR and CIS countries (41.5%) and, to a lesser extent, among those who lived in Israel for some time after having abandoned the Soviet and post-Soviet territory (22.7%). In aggregate with those who identify as Russian Jews "to a great extent" (about one third of each category), it means that over seventy percent and fifty percent of these groups' members correspondingly possess a quite stable communal and ethnic identity.

In other words, the process of forming new post-Soviet (sub-) ethnic Jewish communities and, at the same time, of a "transnational" Russian-Jewish diaspora that was related to the collapse of the USSR and the former Soviet Jewry entering into a complex system of relations with the changing "local communities" (i. e. host societies), both at the initial residence location and in the emigration destination countries, definitely casts a projection on its European part – tellingly, in pretty much the same proportions as in CIS and Israel correspondingly. Unfortunately, our sample does not enable us to follow to which extent this identity that is also typical for a rather large part of representatives of the so-called "one and a half" generation of post-Soviet Jews residing in CIS and Israel is common for Austria as well. However, it is obvious that, judging from the data of our research, this subject is of no interest yet to the second immigrant generation in this country.

As for their parents, their weak Austrian (Jewish-Austrian and civil) identification against the background of them retaining a dominant Russian-Jewish identity is due to the fact that, as to the date of the interviews, sixty percent of the members of the group were residing in Vienna for less than five years, and a mere fifteen-odd percent lived there for more than twenty years. It should be noted that among those natives of the former USSR that had an experience of living in Israel, there turned out to be almost twice as many Austrian "veteran residents" than among those natives of the former USSR that never lived in Israel before.

An additional difference between the above two groups of Jews residing in Austria and born in the USSR and FSC (the so-called "direct emigrants" and "secondary emigrants" from Israel) that are actually quite similar in many other parameters is the degree to which a factual Russian component is present in their identity. In the course of the questioning, we did not specify which version of the "Russian" identification was meant, that is, the one with the Russian ethnos, with Russia (in a broad sense) as a country, or with the Russian culture. However, most respondents deducted from the context that we mostly speak about the first of these three notions, although the other two were not excluded either.

As one can see, none of the Austria-residing Russian Israelis of the first (or, even more so, the second) generation called himself/herself "exclusively or mostly Russian," clearly meaning (as follows from the elucidating questions and several in-depth interviews that were conducted at our request) the ethnic and national aspect of this notion. Nevertheless, one-third of the respondents in this category mentioned that they possess a Russian identity "to a great extent." It seems that many respondents, if not their majority, meant both the country of origin and the cultural background – i.e., more or less those parameters that immigrants from the former USSR are dictated by the "host environment" in Israel as well.

Approximately the same was the share of those who possess "Russian identity to a great extent" among natives of the former USSR that have not previously lived in Israel. However, a group was found in them, even though a small one (11.3%), that insisted on their Russian identity as their first priority, while meaning, by all accounts, the state and national identification. It is no accident that "direct emigrants" were the only category in our sample where there were fewer holders of "predominating" Jewish self-awareness (68.3%) than in others and where there was a significant share (41.2%) of those who called Russia or another CIS country "theirs to the greatest extent" (in other categories, this share varied from zero to less than six percent).

Sub-ethnic, Communal, and Cultural Groups

While the Jewish community of Austria is very diverse in terms of the traditional ethnic and cultural groups represented there, our research has demonstrated that self-identification by this factor is quite modestly expressed in the local Jewish environment. It manifests itself in the lowest degree in the community of indigenous Israeli, which is understandable in relation to the fact that in their mass, our respondents within this category are a product of the Israeli "melting pot" that have been actively erasing communal differences until recently, not to mention that the Israeli "national" diasporic identity itself, as the study by Steven Gold showed, often serves as a sufficient alternative to the "communal and religious" identity of local Jews[11].

But the sub-ethnic identity is manifested just as slightly in the community of Austrian Jews that had no roots in the USSR and have not resided permanently in Israel. A mere twelve percent of this last group stated that they evaluate themselves exclusively or mainly through this prism and another eight percent pointed out that their belonging to Bukhara, Mountain, Georgian or Ashkenazi Jews is of significance to them. This is three times less frequently than, for instance, among Austrian-born children of immigrants from the former USSR. The share of those who pointed out that the fact of their belonging to traditional Jewish ethnic cultural communities was exceptionally important for their identity turned out to be the highest in that group.

Eka, for instance, the above-mentioned Georgian Jewish female who had immigrated to Vienna back in her youth and had been living there for nineteen years by our conversation date, while speaking of the identity of her teenage children born in Austria, emphasized: "As for religion, they are just Jews.[12] Today this is vaguer than some fifteen years ago, when there was a clear differentiation by groups (Georgian, Bukhara, and Mountain Jews). But my children consider themselves Georgian Jews – but not [ethnic] Georgians. The children have a formal historical connection with Georgia (which they only visited once), but not an emotional one."

Correspondingly, 44.4% and 11.1% of respondents in this group feel predominantly or to a great extent as Ashkenazi, Mountain, Georgian or Bukhara Jews. Equally, those who were born in the USSR and CIS and either immigrated to Austria directly or lived in Israel before moving to Austria had a stable sub-ethnic identity manifested 1.5 times more often than those who were born in other countries, including Austria and Israel. All this is encouraged by the retaining of communal and cultural traditions, both at the public and family level, despite the fact that, as Eka pointed out, "90% of them are not observed as strictly as they were ten years ago."

---

11 Steven J. Gold, *The Israeli Diaspora* (London: Routledge, 2002), 183–185.
12 It was clear from the conversation that our respondent meant what sociologists normally define as the "ethnic confessional identity."

Table 5: Sub-ethnic Identity among Vienna Jews of the FSU and non-FSU Origin.

Feel as a Bukhara, Mountain, Georgian or Ashkenazi Jew		Soviet/Non-Soviet natives		Total
		Born in any country except the USSR	Born in the USSR/Russia and CIS	
Not at all		66.3%	41.3%	54.0%
A little, a bit		2.4%	5.0%	3.7%
To some extent		6.0%	8.8%	7.4%
To a great extent		6.0%	13.8%	9.8%
Mostly		19.3%	31.3%	25.2%
Total	N	83	80	163
	%	100.0%	100.0%	100.0%

Dominant Identity and a Sense of Motherland

Making a general conclusion, we can note the following: all seven (or, more precisely, nine) of the categories of Austrian Jews we have distinguished differ from each other by a whole range of parameters and therefore, as was noted above, are difficult to combine into more general groups.

These differences have a direct relation not only to the place of birth, but, in a number of instances, as was noted before, to the period of residence in Austria (and for those of diaspora natives who had an experience of living in Israel – the period of residence in this country as well). However, in other instances we found no direct and obvious correlation between Jews that dominate in particular categories and were not born in Austria, identity types and the time they spent in this country or Israel.

The difference between these groups is much more noticeable in their sense of motherland, meaning the attitude towards one country or another as "theirs" to the greatest extent. Thus, the biggest share of respondents that consider Israel such a country is present not only among two groups of Israeli-born people, "Russian" and other origins of their parents (forty-seven percent and sixty-nine percent correspondingly), but, as was noted above, among those immigrants born in the former USSR that previously lived in Israel (fifty percent). An opinion of a young immigrant from South Caucasus who spent fifteen years in Israel and then relocated to Austria can be considered typical in this sense:

> A European country is very convenient for living in. Austria is a more or less multicultural people (sic!); one can do anything here, find everything, and obtain all sorts of information

and services, without any problem. But to say that I feel at home here – no, not at all... And as for Israel, I grew up there, so essentially, it is my home. Last time I visited the country of the first exodus 20 years ago, it is my motherland [in the sense that] I was born there, but speaking of big feelings – it is unlikely. There are some feelings, stemming from my childhood, something that's rather mythical than real. Therefore [the only] country of all that I can name as my home is Israel.

On the other hand, those who consider Russia or other CIS countries their motherland dominate among the natives of the former USSR that previously did not live in Israel and that are similar to this last category by other parameters. In their turn, the natives of the (former) USSR that came to Austria via Israel differed by this parameter from the natives of other countries that also have lived in Israel for some time (over forty percent of which were born in Austria, as was noted before). These latter ones considered Austria, not Israel, their country to the greatest extent, same as forty percent of Austrian-born children of Russian-speaking Jews that did not previously live in Israel, and forty-two percent of Jews that also never lived in Israel and whose origins were not related to the former USSR.

As a side note, in the last two categories, there was a smaller, yet visible share of respondents that checked other options: in the first case, Israel, in the second one, another country (other than Austria, Israel and Russia). Overall, it is comparable to the first category's share of those whose USSR-born parents had spent relatively much time in Israel prior to their immigration to Austria or those tied with Israel by especially strong emotional and family connections. In the second category, the share of those who chose "another country" option was comparable to the share of those born outside Austria (i.e., Hungary, the Czech Republic, Poland), which in this sense makes them similar to those who moved directly from the former USSR to Austria.

All this enables us to use modern materials to adjust the notions of the identity types of Austrian Jews after World War II[13] that currently exist in the academic literature. In particular, returning to the more general categorization of "immigrant" and "indigenous" Jewish groups of Austria and Vienna after all that has been said, we find it possible to highlight the following four "macro-categories":
- **The "Israelis."** All respondents that were born in Israel of both "Russian" and other origin, which share "Israeliness" as their dominant identity. Internal differences in this group are related to the "traditional sub-ethnic iden-

---

[13] See for instance Christoph Reinprecht, "Jewish Identity in Postwar Austria: Experiences and Dilemmas", in A. Kovács and E. Andor (eds.), *Jewish Studies at the Central European University: Public Lectures 1996–1999* (Budapest: CEU Press, 2000), 203–215.

Table 6: Sense of motherland, according to the respondents' country of origin.

The most important country for me is:	Born in Israel	Born in the USSR/ Russia and CIS, lived in Israel	Born in any other country, lived in Israel	Born in Israel to USSR-born parents	Born in the USSR, never lived in Israel	Austrian non-Russian Jews, did not live in Israel	Austrian-born to USSR-born parents, did not live in Israel	Total
Israel	69.0%	50.0%	23.1%	47.1%	1.5%	14.3%	30.0%	27.5%
Russia	.0%	3.6%	.0%	5.9%	41.2%	.0%	.0%	15.5%
Austria	10.3%	35.7%	69.2%	35.3%	25.0%	42.9%	40.0%	31.6%
other country	10.3%	3.6%	.0%	.0%	22.1%	32.1%	10.0%	15.0%
more than one country	10.3%	7.1%	7.7%	11.8%	10.3%	10.7%	20.0%	10.4%
Total	100.0%	100.0%	100.0%	100.0%	100.0%	100.0%	100.0%	100.0%

tity" parameter that is present as a non-dominant element in the first subgroup and is virtually absent in the second one. In our sample, this entire group was twenty-three percent.
- **The "Russian (Jews)."** This includes almost all immigrants that were born in the USSR/CIS, who either directly immigrated to Austria or another western country, or first lived in Israel and then moved on to Austria. Their dominant identity is "Russian and Russian Jewish," while their internal differences are determined by the first sub-group having a more expressed traditional sub-ethnic component than the second one. In our sample, this group consisted of fifty percent respondents.
- **The "Austrians."** Respondents that were born in any country outside the USSR/CIS and Israel, lived in Israel (with a large share of Austrian-born), and Austrian-born children of Russian Jews that possess an expressed Austrian and Austrian Jewish identity, without any significant differences between sub-groups in this category. The share of respondents in this category was twelve and a half percent in our sample, though it is difficult to say how many respondents of this kind there actually are.
- **The "(Universal) Jews."** Fourteen and a half percent of our respondents, born in any country outside the USSR/CIS and Israel and never lived in Israel. The dominating component of their identity is only (or universally) Jewish. Interestingly, it is this particular categorization that has a rather clear correlation with what the language respondents preferred to use while conversing with our interviewer or filling out the questionnaire in the course of the poll – that is, in a situation where a person is facing an urgent choice, that person follows the first impulse that, regardless of his/her subjective evaluation of language proficiency, as a rule, matches his/her spontaneous cultural and lingual choice. According to the opinion that dominates in professional literature, this is usually a result of the cultural and educational environment where the respondents' personality was formed.[14]

Social and Institutional Infrastructure of New Immigrant Communities of Vienna

Despite the relatively small headcount of the Jewish population of Austria and Vienna, this overall community and the immigrant component that dominates

---

14 Marina Niznik, "How to be an Alien? Cross–Cultural Transition of Russian-Speaking Youth in Israeli High Schools," *Israel Studies Forum* 23, no. 1 (2008).

therein (including the groups we are most interested in – Russian Jewish and Israeli immigrants and their descendants) are characterized by a rather complex formal and informal internal structure. This structure includes three main levels that sometimes exist in parallel to each other, sometimes cross, and sometimes, so to say, are "built into" each other.

The first level is formed by formal organizations and institutions of the organized Jewish community of Austria, particularly the Federal council of Jewish communities, where the leading role is played by the heads of the Viennese community and its cultural, educational, and charitable organizations and media projects, thus following the tradition that was established back in the late nineteenth century. Other participants include a number of "umbrella" organizations that appeared after the Higher Court of Austria nullified the law that had been in force since 1890 in September 1981, according to which only one Jewish community in each city had an official status. Thus, starting in May 1982, an independent religious Jewish orthodox community, and starting in 1990, a reformist organization "Or Khadash," are both functioning in Vienna. In addition to them, there is a Union of Sephardi (i.e., Georgian, Mountain and Bukharian) Jews, a Union of Jewish students, and a Zionist Federation, as well as various religious, veteran, sport, cultural, educational, research, political, and advocacy institutions, along with representative establishments of international Jewish organizations.

The second level consists of sub-ethnic communities and their formalized groups that are structured around sectoral religious, cultural, and communal centers and functional associations related to them. A special role in this category is played by twelve (according to other sources, fourteen) Viennese communal synagogues, including five Ashkenazi orthodox synagogues (mostly belonging to various Hasidic streams), worship houses of Georgian, Caucasian, and Iranian Jews, and the Jewish Reform community. There is also a Sephardi center that was founded in 1992, that serves as the umbrella for the operation of two "Bukharian" synagogues – a "traditional" one and a neo-orthodox one (the second synagogue is actually a project of "Or Avner" center sponsored by the World Chabad Movement, whose major donor is Levi Levaev, an Israeli millionaire of Bukharian Jewish origin).

The third level is formed by informal groups of social, professional, business, cultural, family, clan, and friends' communication and interests that aggregate both on the basis of the first and second-level structures and around informal "quasi-communal institutions" of various sorts, such as commercial enterprises, "home interest circles," clubs, specific "ethnic" restaurants, web-forums and other platforms for group socialization. The grounds for these groups' functioning are intercrossing personal contacts and horizontal ties that form links of wider social networks, as a rule, within those seven to nine social and cultural categories we have highlighted and their individual clusters. The consolidation mechanism of informal associations within these communities includes a number of options and levels of their "separation" from the overall external Austrian environment, from other categories of local Jews and from other groups within their category, including mental and organizational differences between Ashkenazi and "Sephardi," and between those who have lived in Israel and those who have no extensive experience of residing in this country.

The evidence of our respondents and our own observations demonstrate that representatives of different groups do not come across each other much, and it primarily applies to the immigrants of the latest surge, especially Bukharian and Georgian Jews. The same is largely confirmed by the data of our questionnaire. Thus, the majority in the environment of both groups (those that have and have not previously lived in Israel) of natives of the former USSR, where "Russian Sephardi" dominated – that, as was noted before, possess a mostly Russian Jewish and sub-ethnic Jewish identity – belongs to those who declared that the sphere of their primary communication did not exceed the limits of these same groups. Typically, the same answer was given by one-third of their children born in Israel.

Olga, whose grandfather was Jewish, and who immigrated to Austria directly from Ukraine, told us that despite the fact that she was representing the third generation of descendants of mixed marriages, her social circle in Vienna is limited by "highly educated Russian Israelis." She elucidated later that she meant not only Israeli, but all "Russian (i.e. Ashkenazi) Jews from the former Soviet Union." According to her observations, people from her social circle "communicate mostly in their environment; if they work, then it is mostly in Russian companies, and they are not particularly integrated into the local Jewish and non-Jewish community." In this, they differ from Bukhara Jews, for instance, as, according to Olga, they "are not well-educated, engaged in jobs that require no high professional skills, are different in terms of their appearance and are better consolidated within their community."

These impressions are quite in line with estimations of our experts who claim that a special role in the inner structure of communities of "Russian Ashkenazi" is played by social and professional parameters. Such communities are small yet influential groups of highly qualified Russian-speaking scientists, engineers, and experts that are united by quite intense personal contacts and reside in Vienna. Members of this group received a good education and worked in big cities of the former USSR (Yekaterinburg, Kharkov, Minsk, etc.) and then, in mass, relocated to Vienna using a labor contract and received Austrian citizenship in due course.

There is another category of Russian-speaking immigrant Jews that also socialize more actively within their own community. They are more numerous than the previous group, consisting of large- and medium-scale Jew entrepreneurs from Russia, Ukraine, and other former USSR republics that obtained an Austrian residence permit subject to business immigration. In recent years this community was strengthened due to the trend that emerged in the threshold of this and previous decades – making Vienna the second (after London) "backup aerodrome" for heavy CIS businessmen that eagerly buy expensive real estate in the vicinity of *Stephansplatz*, the central square, and its neighborhoods of the city's historical center. Observers also add to this category such Russian-speaking Jews as top managers and specialists employed in commercial ventures founded by these businessmen and, subsequently, tapped into their communication and interest networks.

As a result, a specific subculture of an Austrian-Russian-Jewish business community emerged with a lifestyle that is inherent in the majority of its members. Meanwhile, their families, as a rule, permanently live in Vienna and these people's constant business trips between CIS and Europe turn them into a type of "reciprocal migrants." Their infrequent leisure is almost entirely filled with select clubs, expensive stores, and other elements of the conspicuous consumption sphere of the Austrian capital city. No wonder that the system of social bonds that is determined, in addition to the above circumstances, by their status and business interests, seldom also steps over the limits of the same community.

Ethnic-social and ethnic-professional communities that are equally closed are formed by Georgian (with Mountain Jews often linking up to them) and Bukhara Jews whose majority is engaged in small- and medium-scale entrepreneurship, including commerce in shops and markets, footwear and clothes repair, key cutting, rendering real estate services, etc. It is only lately that a group of people of intellectual occupations emerged among them. At the same time, their children often get a higher education in Vienna and completely fit into the outer environment. As Albert Kaganovich, an Israeli researcher of modern Bukhara Jewry, points out, the break-up of old ties, on one hand, promotes acculturation and update of these Jews' minds, and on the other hand, forces them to create new communities with a purpose of uniting [themselves] by an ethnic and cultural principle in order to preserve the usual environment in unusual circumstances.[15]

---

[15] Alexander Kaganovich, "Bukhara Jews: Community-building in the Three Continents." Paper presented at "Identity and Memory: Achievements and Goals in the Study of Oriental Jewish Communities on the Verge of the 21st Century." Ben-Zvi Institute for Oriental Jewish Studies, Jerusalem, May 22–25, 2000 (originally in Hebrew).

According to our studies, the basic unit of these Jews' self-organization in Vienna, same as in other places, is constructed out of solid "extended family groups" whose significant inter-relations' element is an observation of communal and religious traditions and customs, while the socialization platform is made of family and cultural events (exhibitions, concerts and similar), communal synagogues as well as their business facilities (as a typical example, our expert pointed at a Central Asian carpet store located in the center of Vienna. Almost daily, this store hosts owners of neighboring shops and artisan enterprises, their relatives, and customers that gather for tea and meanwhile discuss business, public, and political issues that are of interest to them).

An additional factor of the internal organizational division in this community is an existence or an absence of a residence experience in Israel that supposedly has a serious impact on the transformation of norm-and-value behavioral standards of these communities. "There is an opinion that, once a person has lived in Israel, he or she changes, and it really is so, especially if we talk about people that were born or grew up there," Eka holds. In her opinion, such people "were born or lived for a long time in a free country in a sense that they depend less on the others' opinion, they are more open and look at this world with different eyes. The downside [of this situation] is the fact that those who have come here directly back in the 1970s continue observing the communal traditions more closely than those who lived in Israel for some time."

Informal Networks

The cross-points between various informal groups that perceive each other as "their people" within the framework of a wider Russian-Jewish (or even wider, a new immigrant) community exist as well. Examples include the interaction and cooperation between representatives of various groups in settling legal and administrative issues and in culture-related and business projects, and a balanced system of mutual business recommendations. All these fit well into the above-mentioned phenomenon of a much greater stability of the sub-ethnic identity in the community of the former USSR immigrants and their descendants compared to those who were born in other countries, primarily the "non-Russian" Israelis and Austrians.

The informal infrastructure of Israeli communities is no less complex. Leaving aside the above-mentioned "indigenous" Austrian Jews that moved to Israel in 1930s and returned in the 1950s, the Israeli community includes three main groups. These are Israeli citizens – *Sabras* and natives of the diaspora's various countries that immigrated to Austria within the past twenty to thirty years, mostly medium-scale businessmen dealing in wholesale commerce, investments, banking operations and, partly, high-tech and specialist areas – a sort of "labor migrants" that often live in two countries simultaneously.

Then, there are a few hundred Israelis – *Sabras*, mostly students and security guards of the community facilities that declare that they intend to go back to Israel once their studies are over or their employment contracts expire, which is not necessarily true at all. Finally, there is a group of Israeli-born children of the repatriates from the former USSR and a small number of representatives of the first and one and a half generations of "Russian" Jews (as a rule, of Ashkenazi origins, with a few exceptions) that managed to acculturate radically in the Israeli environment prior to their emigration from Israel to Austria.

In a very general sense, the life of the majority of Israelis in Vienna goes on in three public planes. Economic and professional operations are performed within the scope of the local community, while superficial communication with its members and groups that are a part of this community provides a necessary amount of personal contacts. Jewish religious and ethnic, or ethnic confessional identification of *jordim* (Hebrew for émigrés from Israel), is provided through sporadic visits to a synagogue (as a rule, it happens on major holidays and sometimes on Saturdays) and by a formal membership of some of them in official Jewish communities. The remaining social, cultural, and psychological separation from "indigenous" Austrian Jews at the behavioral level is implemented in the limited contacts with the "local" Jewish population (and, subsequently, in a low level of involvement in the activities of official Jewish organizations) and in a tendency to communicate more intensely with members of other new immigrant communities, primarily the Russian-speaking ones. In the cases of Israeli-born children of repatriates from the former USSR (primarily of the Russian Sephardi origin), there is also some participation in the "sub-communal" activities. Finally, a substantial part of the private and group cultural space functions within the framework of specific organized communities of Israelis, that in most cases are not separated by strict limitations from the outer Jewish and non-Jewish environment, but, same as everywhere worldwide, possess particular "separating" characteristics.

As a result, informal communities and networks that are founded by the Israelis, as a rule, do not exceed the margins of the said sub-groups and use various platforms for their consolidation, such as their own commercial ventures and offices, kosher restaurants of "Mediterranean cuisine" and educational institutions and, much more seldom, synagogues and other formal institutions of the urban Viennese and "sectoral" Jewish communities. An example of such a platform is the previously mentioned Ron Lauder Business School in Vienna. The student body has highly heterogeneous origins and includes a large group of Jews, three quarters of which, according to the data at hand, are Russian-speaking Jews from the FSU countries, Israel and Europe, while the rest of them are non-Russian-speaking Israelis and Jews from a range of countries from Western and Eastern Europe and North and Latin America.

According to our subjects, in the networks and groups of informal communications and mutual support that emerge in this environment, "everyone finds some interests, they communicate quite well, but virtually do not mix with each other." As a result, there exist a number of "informal associations" of Russian-speaking students, a small group of immigrants from Bulgaria, Poland, and other countries of Eastern Europe ("that communicates only with its own kind and sometimes mixes with the Russian team"), etc., each of them "leading its own life," while Israelis usually keep a distance from all others.

Main Conclusions

In summary, the data we have obtained and its analysis enable us, in our opinion, to come to number of conclusions that are important for understanding today's state of Austrian Jewry and the entire Jewish diaspora.

First, an obvious and somewhat paradoxical imbalance is present between a relatively high ethnic homogeneity of the Jewish community(-ies) of Vienna, and a heterogeneity of groups within it at a level that is more typical for immigrant and transnational communities rather than for a well-established community of a diaspora.

Each of the identified seven cultural identification categories, in addition to the fact that they all feel as Jews to a great extent, possesses, as a rule, one or two dominating identification types. Thus, persons of "non-Russian" origin that were born in Israel have a dominant Israeli identity. The second generation of "Russian" Israelis have a major Israeli and, to a lesser extent, traditional sub-ethnic identity. Immigrants from the (former) USSR that previously lived in Israel have a Russian-Jewish and, to a lesser extent, traditional sub-ethnic identity. The USSR-born "Russian" Jews that never lived in Israel have a Russian-Jewish identity. Immigrants and re-immigrants that previously lived in Israel and have no roots in the former USSR have Jewish-Austrian, Austrian, and traditional sub-ethnic identities. Austrian-born children of Russian Jews (same as in group 5) have a Jewish-Austrian, Austrian, and traditional sub-ethnic identities. And only Austrian Jews that have no connections with the USSR and did not permanently live in Israel have an expressed Jewish identity alone and no any other ones.

Second, it explicitly follows from the data obtained that three of the seven groups of Austrian Jews that we defined in our study are a part of new, transnational Jewish diasporas – Israeli, Russian-Israeli, and Russian-Jewish – and, while differentiating substantially from one to another, possess a number of very important similar features. It separates them dramatically from the Austrian Jewry environment, primarily by such parameters as an attitude towards Israel and an identification with its status.

Third, our data supports the conclusion as to the deep changes in the subjective feeling of "rootedness" in favor of the residence country that have occurred within the past two and a half decades. At the same time, the majority of representatives of all respondent categories but one lack a desire to consider Austria "their country to the greatest extent" and a complex of a conscious or subconscious absence of a belief that Austria is the end of their migration journey that is typical for many members of the Jewish community of Vienna, and especially for both generations of Jew immigrants of the latest surge.

And fourth, our data enables us, even partly, to answer the question as to the existence of a single Russian-speaking Jewish or Russian-Israeli community in Vienna. In our opinion, such a community that includes a multitude of formal and informal associations, communication networks, and interest groups really exists, but rather as a virtual reality now that cannot be necessarily called "Russian-speaking" in every sense of the word. In this community, this includes immigrants from the former USSR and their descendants, regardless of whether they did or did not live in Israel, the status languages are German and (in some instances even to a greater extent) Hebrew, while the Russian language acts as kind of a collective symbolic marker and, if necessary, as a means of intra-communal intergroup communication.

Bibliography

Adunka, Evelyn. *Die Vierte Gemeinde – Die Wiener Juden in der Zeit von 1945 bis Heute*. Berlin: Philo, 2000.

Cohen-Weisz, Susanne. *From Bare Survival to European Jewish Vision: Jewish Life and Identity in Vienna*, 32–33. Jerusalem: Center for Austrian Studies, Hebrew University of Jerusalem.

Cohen-Weisz, Suzanne. *Like the Phoenix Rising from the Ashes: Jewish Identity and Communal Reconstructions in Austria and Germany*. Ph.D. Dissertation, Submitted to the Senate of the Hebrew University of Jerusalem, December 2010.

Friedmann, Alexander. "Psycho-Socio-Cultural Rehabilitation in an Ethnic Subgroup: A 30-Years Follow-Up". *World Cultural Psychiatry Research Review* (2007): 88–89.

Gitelman, Zvi. "Jewish Religion, Jewish Ethnicity – The Evolution of Jewish Identities". In *Religion or Ethnicity? The Evolution of Jewish Identities*, edited by Z. Gitelman, 1–5. New Brunswick, NJ: Rutgers -University Press, 2009.

Gold, Steven J. *The Israeli Diaspora*, 183–185. London: Routledge, 2002.

John, M., and A. Lichtblau. "Jüdische Unternehmer in Österreich nach 1945: Oral History und ihre Forschungsperspektiven für die postfaschistische jüdische Geschichte" ["Jewish Entrepreneurs in Austria to 1945: research of oral history and its prospects for the post-fascist Jewish History"]. In *Studien zur Geschichte der Juden in Österreich* [*Studies in the History of the Jews in Austria*], 166–191. Wien, 1994.

Khanin, Vladimir (Ze'ev). "Russian-Jewish Ethnicity: Israel and Russia Compared." In *Contemporary Jewries: Convergence and Divergence*, edited by E. Ben-Rafael, Y. Gorny, and Y. Ro'I, 216–234. Leiden and Boston: Brill Academic Publishers, 2003.

Khanin, Vladimir (Ze'ev), Alek Epstein, and Dina Pisarevskaya. *Post-Soviet Jewish Youth: Identity, Community and Relations with Israel*. Moscow and Ramat-Gan: Institute for Oriental Studies, Russian Academy of Science, and the Lookstein Center for Jewish Education in the Diaspora, Bar-Ilan University, 2013 (originally in Russian).

Kaganovich, Alexander. "Bukhara Jews: Community-building in the Three Continents." Paper presented at "Identity and Memory: Achievements and Goals in the Study of Oriental Jewish Communities on the Verge of the 21st Century." Ben-Zvi Institute for Oriental Jewish Studies, Jerusalem, May 22–25, 2000 (originally in Hebrew).

Niznik, Marina. "How to be an Alien? Cross-Cultural Transition of Russian-Speaking Youth in Israeli High Schools." *Israel Studies Forum* 23, no. 1 (May 2008).

Reinprecht, Christoph. "Jewish Identity in Postwar Austria: Experiences and Dilemmas." *Jewish Studies at the Central European University: Public Lectures 1996–1999*, edited by A. Kovács and E. Andor, 203–215. Budapest: CEU Press, 2000.

Wilder-Okladek, Friederike. *The Return Movement of Jews to Austria after the Second World War – With Special Consideration of the Return from Israel*. The Hague: Martinus Nijhoff, 1969.

Ildikó Barna and András Kovács
Jewish Religious-Cultural Traditions and Identity Patterns in Post-Communist Hungary

In 1999 and 2017, questionnaire-based sample surveys were carried out in Hungary among Hungarian Jews.[1] Jews were defined as those who have at least one Jewish grandparent or those who converted to Judaism according to halacha. We interviewed 2,017 people in 1999 and 1,879 people in 2017. In the surveys, many identical questions were used, making longitudinal comparisons possible. They covered various topics, such as social-demographic background, social mobility, identity, relations between Jews and non-Jews, personal network, attitudes toward Israel, perception of antisemitism, assimilation, social and political attitudes, media consumption, and participation in Jewish religious and secular organizations. Both surveys included questions on ten religious-cultural traditions: Sabbath observance, fasting on Yom Kippur, Passover Seders, keeping kosher, cooking cholent, mezuzah, bar/bat mitzvot, Jewish burial, circumcision (brit milah), and observing Hanukkah. We asked whether the respondents kept these traditions in their childhood and/or in their present family. In this paper, we aim to present how these traditions are passed down from generation to generation. The questioning of two – parental and present – generations, in 1999 and 2017, makes it possible to grasp a very long period ranging from the 1910s until today. The analysis also sheds light on the Jewish revival of Central and Eastern Europe after the Transition, which is often called the "Jewish renaissance" (Kovács 1994, Kovács et al. 2011, Papp 2004, Vincze 2012).

One of the main questions of the 1999 survey was whether survey data supported the widespread observation regarding a reemergence of a vivid Jewish life

---

[1] One of the greatest challenges is generalization. Since there are no reliable statistical data about the Jewish population of Hungary, drawing a proper representative sample was impossible both in 1999 and 2017. A further difficulty of sampling was to access the "hiding Jews," those who think it is better to conceal their Jewish origin, as well as those who are Jews by origin, but it does not mean anything to them. Thus, neither of the surveys are statistically fully representative of Hungarian Jewry. The more affiliated Jews and those who have stronger Jewish identity are overrepresented in both samples compared to their proportion in the total Hungarian Jewish population. Nevertheless, the wide variety of sources for sampling makes it possible to describe every important subgroup of Hungarian Jewry, even though their relative sizes cannot be exactly estimated.

https://doi.org/10.1515/9783110582369-006

in post-Communist Hungary. Another question of our research was about the nature of the "Jewish renaissance": did our data indicate a renewal of Jewish religious tradition in the strict sense, or in other forms of Jewish identity construction? The results of the survey justified the expectations: we could identify a relatively large group of mostly young people who have reverted to Jewish tradition. However, a complete revival of religious tradition affecting all aspects of life has only been characteristic for a smaller subgroup. For the majority, the elements of revived tradition seemed destined to serve as a token of ethnic group consciousness – a tool for creating an ethnic identity (Kovács 2003, 234–239). In the following analysis, we intend to investigate whether or not the tendencies observed eighteen years ago are durable. Do Jewish religious traditions occur more frequently in the youngest generation or in their parents' generation, and are they indicators of a religious revival or an ethnic one? To assess this last question, we classified the yearly participation in Hanukkah celebrations and Passover Seders as markers of relatively "thin" and ethnic rather than religious identification, while on the other hand, Sabbath observance, keeping kosher, brit milot and bar/bat mitzvot were classified as markers of a relatively "thick" religious identification, which demand a larger and longer lasting "investment" in Jewish identity.

As presented in Figure 1, we observed in 1999 that all traditions were kept more frequently in the childhood families than in the present ones. In the case of circumcision, celebrating bar/bat mitzvot, cooking cholent, and Jewish burials, the differences exceeded 20 percentage points, and there was no single tradition where the difference was less than ten percentage points. Between 1999 and 2017, less than one generation passed. Therefore, it is not surprising that those ratios measured for the present families in 1999 are close to the numbers for the childhood families in 2017. In 2017, respondents reported cooking cholent (48%) and having Jewish burials in the family (46%) as the most frequently observed traditions in their childhood. Their prevalence remained almost the same in the present (43% and 48% respectively). In the childhood families, 20–25 percent of the respondents held Hanukkah celebrations, Passovers Seders, brit milot, bar/bat mitzvot, fasted on Yom Kippur, and/or had mezuzahs on their doorposts. Apart from brit milot and bar/bat mitzvot, the prevalence of all the others is much higher in their present families. Approximately 50 percent of the respondents keep Hanukkah and Seder, and one-third of them fast on Yom Kippur and use mezuzahs. There is also a great increase in Sabbath observance: while 10 percent of the childhood families kept the Sabbath, now 21 percent observe it. There is also a slight increase in eating kosher, from 7 percent in the past to 12 percent in the present. Thus, the data shows that the increase in the prevalence of ethnic markers is much larger than in the case of religious ones.

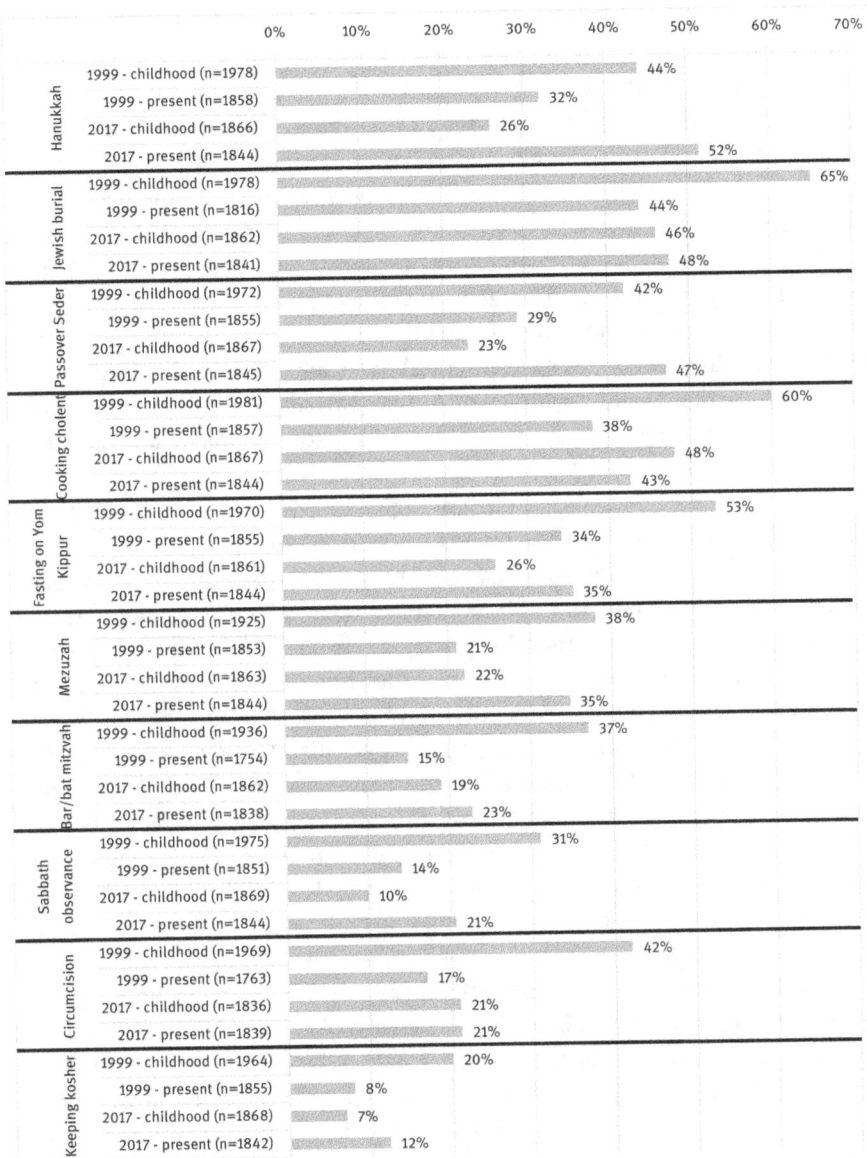

Figure 1: Religious and cultural practices in the childhood and present families in 1999 and 2017 (base: full sample, n = 1836–1981).

There is a U-trend if we look at the changes from the childhood families in 1999 up to the present. There are only two exceptions: Jewish burial and circumcision, whose prevalence remained the same after a sharp decrease. In the case of almost all traditions and festivals, the ratio of those observing them have increased since the 1990s, but their present prevalence lags behind the 1999 data for the childhood families, when many of those who were children before and during World War II were still alive. Hanukkah and Seder are the only exceptions in that they are slightly more frequently observed now than in the past.

As previously shown, the prevalence of cultural traditions and festivals kept in childhood and present families in many cases changed considerably over time. When scrutinizing different age groups, these changes are also visible (Figure 2 and 3). The respondents of the 2017 survey were born between 1922 and 1999, representing a very diverse period full of societal changes in the life of the country and in that of Hungarian Jewry. In the analysis, we used four age groups: (1) 18–24-year-olds; (2) 25–44-year-olds; (3) 45–64-year-olds; and (4) those aged 65 and older.

The end of the 1980s and the beginning of the 1990s brought about fundamental changes not only in the political system of Hungary, but also, consequently, in the life of Hungarian Jewry. After a century of voluntary and forced assimilation, persecution, and anti-religious and "anti-Zionist" Communist policy, second- and third-generation Holocaust survivors, being a maximum of thirty years of age at the time of the collapse of the Communist system, wanted to get rid of their parents' stigmatized identity (Kovács 2003, 237–238). In order to construct positive identity patterns, they started to re-discover their Jewish heritage and include Jewish customs and tradition into their everyday lives. For the members of the youngest age group in our survey, those who were born after the Transition, living a Jewish life became possible, and the presence of the religious-cultural tradition in their childhood families depended solely on their parents' choice (mostly members of the age group 45–65 in 2017), and in their present family, on their own.

Looking at the data from the 2017 survey in a breakdown by age, these changes become apparent (Figure 2). In the case of Hanukkah, Passover Seder, Mezuzah, Sabbath observance, and fasting on Yom Kippur, the overall difference in their prevalence in childhood and the present families is large (i.e. around ten percentage points or more). The tendencies are very similar when the past and the present are compared by age groups. These customs and traditions were more prevalent in the childhood families of those aged sixty-five and over than in the case of the younger ones, because most of them were a child before, during, or right after the war. This difference is especially large in keeping Sabbath and fasting on Yom Kippur. In the age groups of 25–44 and 45–64 years, these Jewish traditions were much less prevalent, since most of them were children during the communist and socialist regimes of the fifties, sixties and seventies. Traditions became prevalent again in the parental families of the youngest ones. However, a fundamental difference lies behind the similar ratios of the 18–24 year-olds and those sixty-five and older. On the one hand, for the youngest age groups, the traditions with high childhood prevalence serve as ethno-cultural rather than religious markers. On the other hand, in many cases of the oldest generation, these traditions were yet present before fading away after the Holocaust, while in the lives of the youngest ones they were already present as a consequence of the Jewish revival.

In the case of the youngest age group, although the prevalence of religious-cultural traditions in the current families is somewhat higher than that of the parental ones, the difference is smaller on average than in other age groups. Members of the oldest generation, those aged sixty-five and older, display ethno-cultural markers (such as Hanukkah and Seder) much more often now than in their childhood, while the prevalence of the especially religious ones is almost the same. However, based on the survey of 1999 and the socio-political processes described above, the rediscovery of Jewish identity most likely lies behind this apparent stagnation in the case of those aged 65 and older. The greatest increase can be seen regarding the two middle age groups, the 25–44 and 45–64-year-olds: they keep these traditions – even the religious ones – much more now than in their childhood.

The structure of Figure 3 is very similar and shows those religious-cultural traditions where the difference between childhood and present prevalences were smaller than in the case of those in Figure 2. Kashrut and bar/bat mitzvot observance is a bit more frequent now than in the childhood families, except in the case of the oldest respondents. The prevalence of Jewish burial, circumcision, and cooking cholent are almost stagnant, again with a slight decrease in the case of those sixty-five years and older.

As in the analysis of the 1999 survey (Kovács 2003, 224–226), we created a bi-generational model based on the number of traditions kept in the parental and in the current family. Seventeen percent in 1999 and twenty-one percent in 2017 of the total sample fell into the fully assimilated group where none of the traditions were present in the parental family nor in the current family (Figure 4). On the other extreme, in both years, traditions were observed by both generations in the case of fourteen to fifteen percent of families. We created three more groups: one of the groups abandoned traditions or moved away from them; while traditions were present in their childhood families, they follow none or much less of them in their current family. The size of this group decreased drastically: while almost half of the respondents fell into this group in

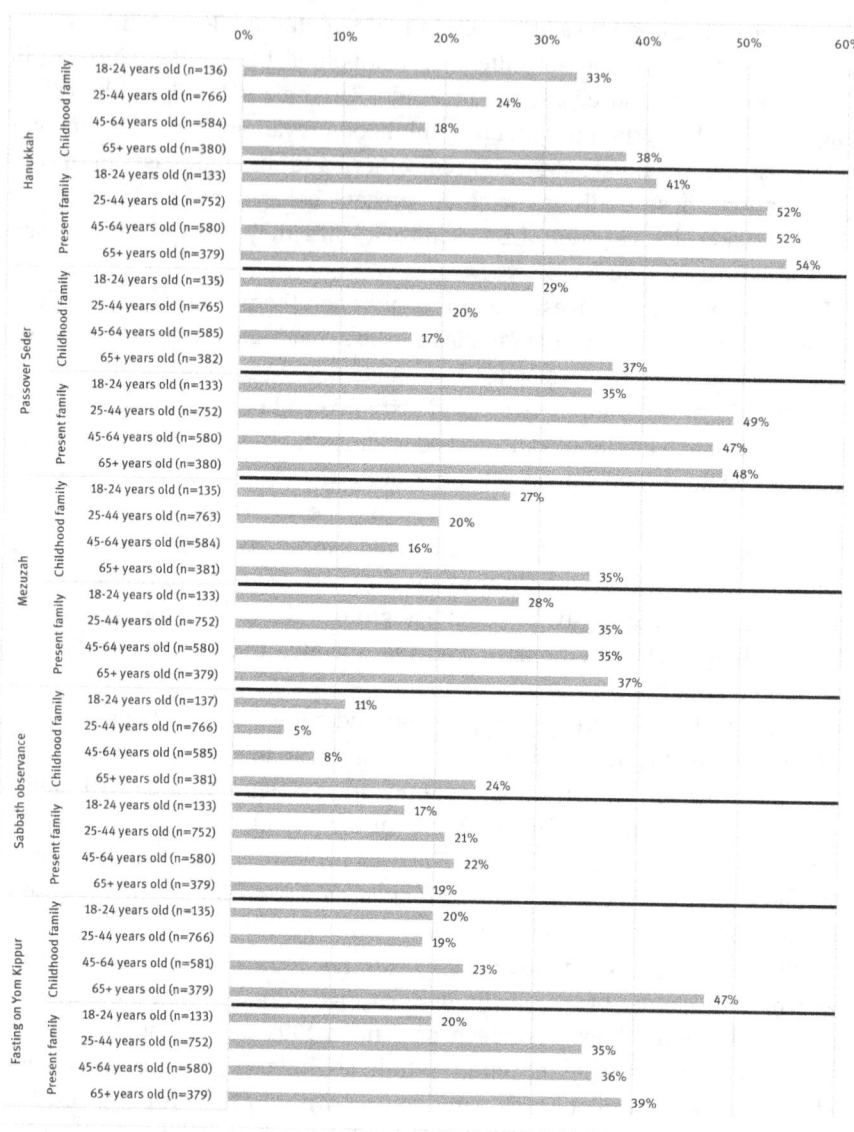

Figure 2: Religious and cultural practices in the childhood and present families in 2017 in the full sample and by age groups (Part 1) (base: full sample, n = 1836–1869).

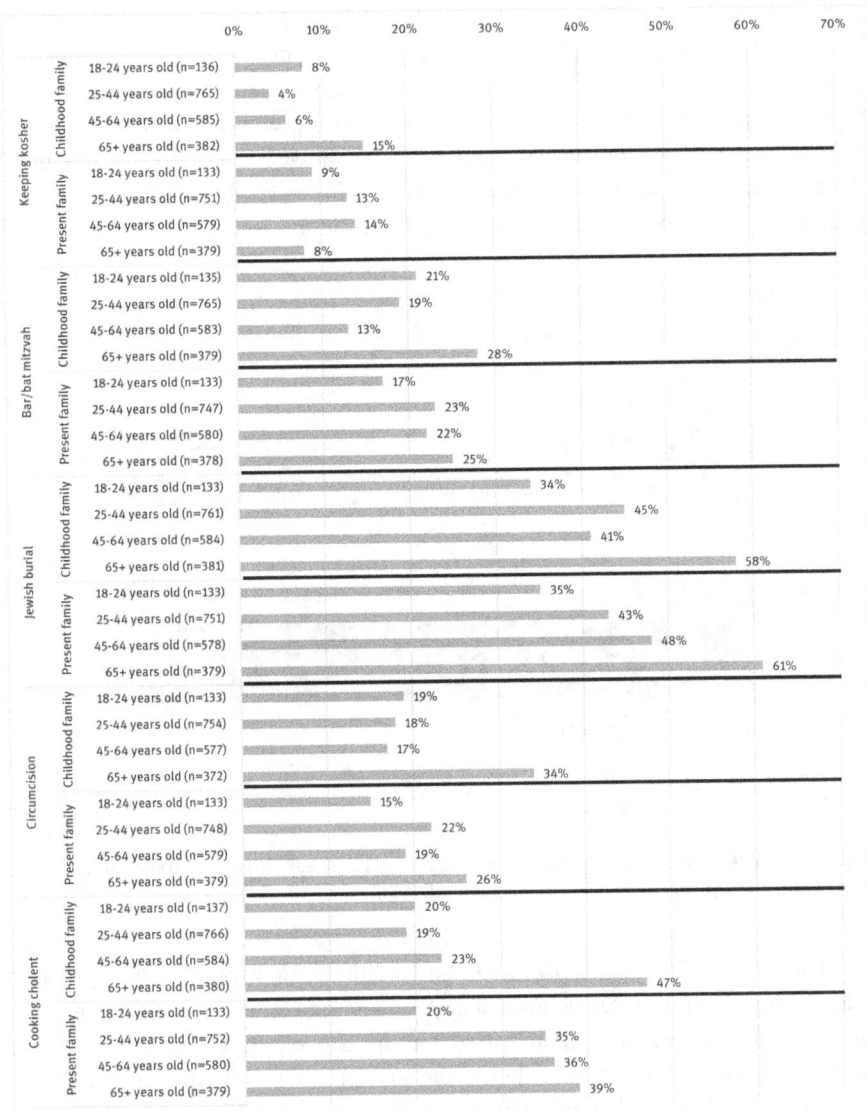

Figure 3: Religious and cultural practices in the childhood and present families in 2017 in the full sample and by age groups (Part 2) (base: full sample, n=1836–1869).

1999, only sixteen percent were members of it in 2017.[2] In the next group, traditions were symbolically present in both generations, mainly by celebrating some of the main Jewish festivities once in a year. In 1999, their ratio was twelve percent, in 2017 twenty percent. There is another group whose size increased considerably over time: the revivalist group, where Jewish tradition is stronger in the current family than had been in the parental one. While in 1999, twelve percent of the respondents fell into this group, in 2017 this increased to twenty-eight percent.

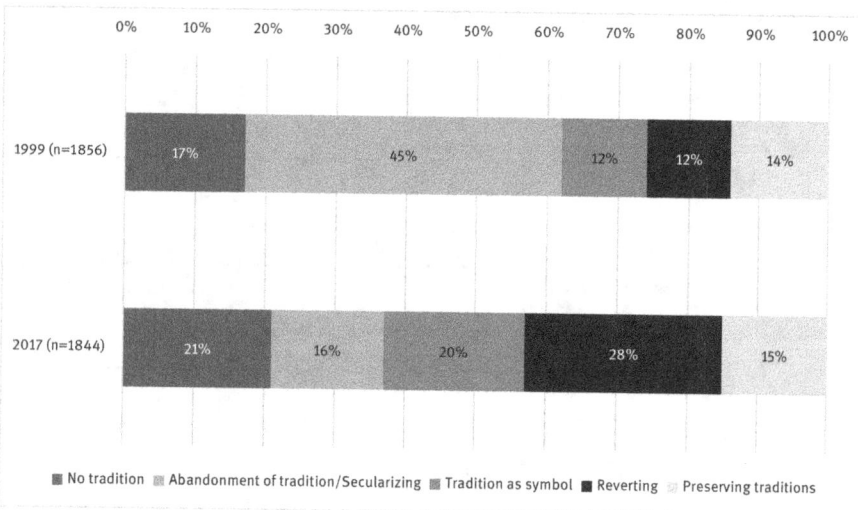

Figure 4: The relationships of parents and respondents to tradition: bi-generational model in 1999 and 2017 (base: full sample, n = 1856 and 1844).

Those who did not keep any of the traditions, or kept only the symbolic ones, are overrepresented among the youngest age group. These are exactly the groups that are underrepresented among the oldest respondents. As we have already seen, in many of their childhood families, Jewish traditions were present. Data shows that one-quarter of them abandoned or detached from Jewish customs, while almost the same number preserved them. The Jewish revival is most prevalent among those aged 25–44 and 45–64. Approximately one-third of them keep more traditions now than in their parental families.

---

2 In 1999, we could distinguish those who abandoned traditions and those who moved away from them. However, in 2017 the low proportion of these respondents made this differentiation impossible.

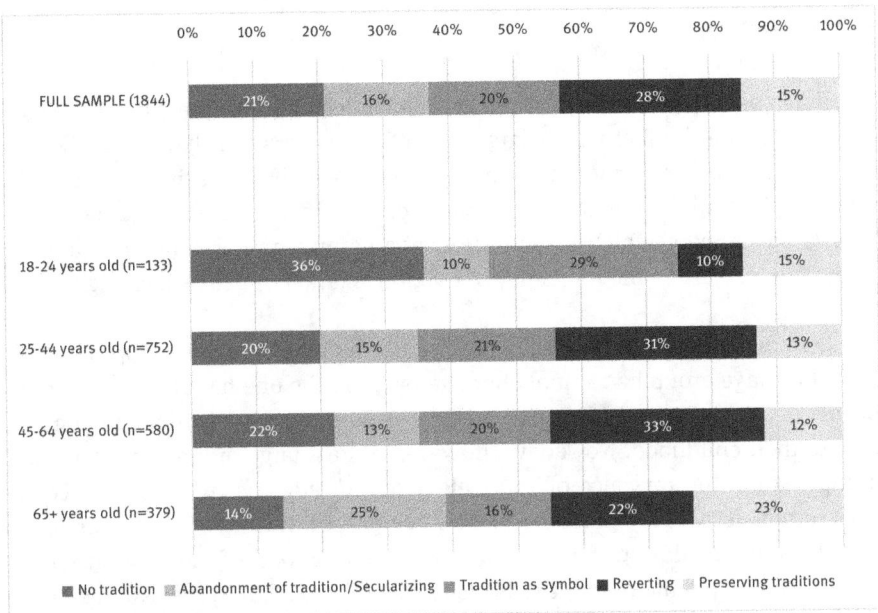

Figure 5: The relationship of parents and respondents to tradition: bi-generational model in the full sample and by age groups (base: full sample, n = 1844).

In this paper, we analyzed only one possible explanatory factor: the homogeneity or heterogeneity of the family background of our subjects. We constructed a religious-ethnic homogeneity index already used in the analyses of the 1999 data (Kovács 2003). The "homogeneous group" consisted of those whose grandparents (all four of them) were considered (by the respondent) to be Jewish regarding descent and/or religion. Survey participants with one non-Jewish grandparent were put into the "partially homogeneous group." Finally, the group of "mixed descent" was formed comprising respondents with at most two Jewish grandparents. Differences in these types are greater in the parental family than in the current one. The average number of traditions kept in the childhood family by those with homogeneous backgrounds was 3.8, with partially homogeneous ones 1.7, and with heterogeneous ones 1.2 (it has to be noted that homogeneous backgrounds and the presence of tradition in the childhood family is much more frequent in the oldest generation than in the younger ones). In the current families, these averages are 4.1, 3.1, and 2.5, respectively. These data indicate that those with heterogeneous family backgrounds observe many more traditions now than in their childhood family, and, moreover, even more than the partially homogeneous group kept in their childhood.

Conclusion

The data show that the revival tendencies recorded eighteen years ago have continued. Despite the fact that a large part of the generation that either lived or were socialized before the Shoah have passed away by now, the prevalence of almost all traditions has increased, the most significant ones being Hanukkah and Passover Seder, which are indicators of an ethnic, rather than a religious revival. The increase in the frequency of religious markers, such as keeping kosher, was low.

Looking at these tendencies by age groups elaborates these changes further. The oldest age group has a dual characteristic: on the one hand, there are many of them, especially the older ones, who moved away from traditions in comparison to their childhood. Nowadays, however, quite a large proportion of this age group reports the presence of some elements of religious-cultural practices in their family. It is highly probable that a large part of the latter group abandoned Jewish traditions during the Communist era, and as part of the "Jewish renaissance" reverted to them after the Transition. In this age group, revival meant especially an ethnic one since the prevalence of religious markers decreased.

The Jewish revival most affected the age groups of 25–44 and 45–64-year-olds. All religious-cultural practices we asked about were more frequently lived in their present family than in their childhood. The increase is the highest in ethnic markers; however, returning to systematic religious practices is also frequent. In these age groups, prevalence of religious traditions, such as keeping Shabbat, kashrut, fast on Yom Kippur, brit milot, and bar/bat mitzvot increased.

The data of those between eighteen and twenty-four show that the Jewish revival has continued, however at a slower pace. For one-third of this age group, neither in childhood nor in the present family were religious-cultural practices observed. In the case of those who have returned to some Jewish traditions, the renewal is an ethnic turn rather than a religious one: only the prevalence of ethnic markers increased among them, while religious ones in many cases decreased. On the other hand, the proportion of those who practice only a few selected symbolic elements of tradition is relatively high.

When we scrutinized the effect of one of the possible explanatory factors of the indicated differences, we found that homogeneity and heterogeneity of the family background correlates significantly with the results: it is not surprising that growing up in ethnically/religiously homogeneous families correlates positively with the presence of religious-cultural practices in both the childhood and current families. However, the results indicate that intermarriage does not necessarily lead to the disappearance of tradition, since those who have had ethnically/religiously heterogeneous backgrounds practice twice as many religious-cultural traditions in their current family than their Jewish parent did in their childhood.

Bibliography

Kovács, András, Barna Ildikó, Sergio DellaPergola, and Barry Kosmin. *Identity à la Carte: Research on Jewish Identities, Participation and Affiliation in five Eastern European Countries*. Oxford: JDC International Centre for Community Development, 2011.

Kovács, András. "Jewish Groups and Identity Strategies in Post-Communist Hungary." In *New Jewish Identities: Contemporary Europe and Beyond*, edited by Zvi Gitelman, Barry Kosmin, and András Kovács, 211–243. Budapest: Central European University Press, 2003.

Kovács, András. "Changes in Jewish Identity in Modern Hungary." In *Jewish Identities in the New Europe*, edited by Jonathan Webber, 150–160. London/ Washington: Littmann Library of Jewish Civilization, 1994.

Papp, Richárd. *Van-e zsidó reneszász? Kulturális antropológiai válaszlehetőségek egy budapesti zsidó közösség életének tükrében.* [*Is There a Jewish Renaissance? Cultural Anthropological Answer Alternatives in View of the Life of a Jewish Community in Budapest*]. Budapest: Múlt és Jövő Kiadó, 2004.

Vincze, Kata Zsófia. "Re-learning Tradition: Jewish Baal Tesuva Identity in Post Socialist Hungary." In *Identities, Ideologies and Representations in Post-transition Hungary*, edited by Mária Heller and Borbála Kriza, 221-261. Budapest: ELTE Eötvös Kiadó, 2012.

Zsófia Kata Vincze
The "Missing" and "Missed" Jews in Hungary

When I'm asked, as I often am, how many Jews live in Hungary, in order to avoid a complex explanation, I respond with a joke: "it depends upon where we are applying for funds." The question this paper will address is how to count the missing, potential or imagined "greater Jewish community" in Hungary, more literally an "imagined community" than in Andersen's characterization (Andersen 1983). By today there is a common assumption that, in Hungary, about 85 to 90 percent of those who might possibly be considered Hungarians with some sort of Jewish heritage are not committed or affiliated with any kind of Jewish organization. Most outreach programs seek relentlessly, hoping to find these "missing" Jews, in order to engage them in Jewish activities, and "to give them back" their Jewish identity, or to guide them to explore and find their Jewish identity themselves, often offering a large spectrum of Jewish identity definitions.

My paper addresses an emic question that is one of the main concerns of the Jewish community: the missing, and deeply missed Jews. The focus group of this paper are those who, from the Jewish community's perspective, have the potential to be counted as part of the Jewish population, but, for various reasons, have no affiliation with any formal or informal Jewish community at all. My assumption is that one of the reasons these "potential Jews" may be missing can be found in Bauman's contention that "if the modern 'problem of identity' was how to construct an identity and keep it solid and stable, the postmodern 'problem of identity' is primarily how to avoid fixation and keep the options open." (Bauman 1996, 18).

In 2016, the president of the recently restructured "Hungarian Federation of Jewish Faith-based Communities" (MAZSIHISZS), András Heisler, said after a re-election acceptance speech given in November 2014[1]:

> In Hungary the Orthodox, the Neologue and the 'Status Quo' have historical traditions. I would like to preserve their values in the long run. This is possible only if we are able to reach out to significantly more people from what is considered to be a Jewish population

---

[1] This statement was repeated in János Gadó's interview with Andras Heisler published in the monthly Jewish magazine *Szombat* (March 12, 2016), http://www.szombat.org/politika/a-zsido-felekezetek-kozott-a-mazsihisz-a-legdemokratikusabban-iranyitott-szervezet.

of 100,000. We want to reach out to a much higher percentage than we are reaching out to today.

Like Schrödringer's cat, if one is able to directly observe and enumerate someone as a member of the Hungarian Jewish community, the state of that person's identity becomes fixed, whereas, as long as it is unobserved, as long as the Jew is considered missing, then the state of that person's identity is ambiguous and, in that the options are open, the person, albeit unobserved (or perhaps because he/she is unobserved), may be counted as part of the Hungarian Jewish community. Given this ambiguity, the mere existence of Hungarian Jewry as one of the considerably larger European Jewish communities with a strong Jewish Federation is a contested issue, let alone whether building a new Jewish life after the fall of Communism could be understood as a "revival" movement.

The question of my paper is twofold. First, why do most Jews, activists, journalists and even occasionally researchers hold on to the perception that Hungary has a potential Jewish population of 80,000 to 150,000 or even 260,000 people? Second, why do foreign activists, journalists and even scholars have the impression that a "Jewish revival" or "Jewish renaissance" is taking place? Can we explain the seemingly paradoxical and ambiguous character of a post-socialist Jewish life, which appears to be vibrant in some aspects, but struggles with commitment?

The perspective of this paper is also twofold: a cultural/anthropological and ethnological one, based on ethnographic principles and hypothetical assumptions, given the fact that the focus group is basically an invisible one. I draw conclusions and questions based on one hundred and fifty in-depth interviews, qualitative informal conversations, and anthropological participant observations conducted over the last thirteen years in different Budapest communities. Due to the attempt to look for the "missing" Jews, the methodology is also based on discourse analysis, component analysis, contextual semantic discourse analysis of about four hundred articles, blog posts, innumerable comments, and informal, "accidental" conversations. I also draw from surveys carried out between 2009 and 2015 on students who signed up for two university courses taught at ELTE University in Budapest, Hungary. The first course was entitled "Tradition and Modernity in Judaism" and had one hundred and twenty enrolled students who completed my survey. The second course was entitled "Modern Israel" and had one hundred and sixteen students in the first year, and two hundred and fourteen students in the second year, who filled out my questionnaires. Because of the variation in perspectives with regard to determining "who is Jewish," as well as the diversity of people that completed the forms, and because I am not a sociologist, I do not consider any of the surveys I have conducted as representative of any particular demographic. However, hopefully the results of my research will shed a light on existing tendencies and perceptions.

My hypothesis is that, in Hungary, when members of the Jewish community acculturate or almost fully assimilate into the mainstream social majority, while they may seem to disappear from the active Jewish community scene itself, they often continue to play a very important role in the imagination of the active, affiliated Jews in the center of the community. The "missing Jews" are deeply "missed" especially by those activists, professionals, shlichim, and practitioners whose goal is to reach out to the margins of the community (affiliated, but much less so) and especially to the unaffiliated, "missing" potential Jewish population. This "85–90 percent" of the Hungarian Jewish population is painfully "missed," like the phantom pain known in medicine, when a person feels pain in a limb, even after the limb has been amputated.

The academic focus on the "missing Jews" cannot be underestimated given the fact that we could state that the engagement of the "missed" Jews is one of the main points of focus of almost every organization. The reason for the immense investment in Jewish outreach activity in Hungary is, I assume, precisely the constant hope that these "missing Jews" might be finally reached. Most Jewish religious or secular programs are designed specifically with the goal of attracting more people from the margins of the known Jewish community, but especially attracting new people from the "missing" non-group. A lot of new organizations have been established, and new projects and initiatives hold on to the hope that their results will be different this time. There is also a big emphasis on constantly modernizing and improving the manner in which people are pursued, invited, and called in, as well as on portraying Jewishness as something positive, diverse, flexible, constructive, inclusive, and definitely attractive.

However, the size of the effort and investment in what has become a huge variety of community building organizations, supported and run mainly "from above" and from outside, seems to be disproportionate to the desired results from the perspective of the activists. For example, the sponsored communities that were supposed to have become independent and self-maintaining are still unable to sustain themselves without artificial life support; most organizations have not created the commitment level and not attracted the numbers that had been hoped for. While there are a few thriving religious communities in Budapest, most struggle with the "minyan problem," as they call it. At the same time the focus of most secular organizations appears quite modern, in order to attract new members, or to bring the members that are loosely affiliated closer to the body of the community. As several other scholars, I have been writing about the "Jewish renaissance" listing and analyzing the quality and quantity of those organizations that are or were centers of a vibrant community life, some for longer (JCC – Jewish Community Center, Szarvas summer camp, Sirály-Aurora-Marom), some for shorter periods of time (Pesti Shul, UJS – University Jewish Students).

The Gap Between the Committed and the Other Hungarian Jews

For getting a clearer picture, it is quite helpful to adopt DellaPergola's demographic differentiation between a "core" Jewish population and an "enlarged" Jewish population. The concept of core Jewish population has been used since the 1990s, and includes all those who, when asked in a survey, identify themselves as Jews. DellaPergola adds those as well "who are identified as Jews by a respondent in the same household, and do not have another monotheistic religion." He extends the core Jewish population to all those who identify as Jews by religion, as well as others who are not interested in religion but see themselves as Jews by ethnicity or by other cultural criteria, regardless of their commitment level, affiliation or halakhic standards. DellaPergola distinguishes the core Jewish population from an enlarged Jewish population that includes the sum of: (a) the core Jewish population; (b) persons reporting they are partly Jewish; (c) all others of Jewish parentage, who—by core Jewish population criteria—are not currently Jewish (non-Jews with Jewish background); and (d) all respective non-Jewish household members (spouses, children, etc.). Non-Jews with a Jewish background, as far as it can be ascertained, include: (a) persons who have adopted another religion, or otherwise opted out, although they may claim to be also Jewish by ethnicity or in some other way; and (b) other persons with Jewish parentage who disclaim being Jewish. DellaPergola creates a different marginal circle for those who meet the minimal requirement of Israel's Law of Return. Israel's distinctive legal framework for the acceptance and absorption of new immigrants awards Jewish immediate citizenship and other civil rights to new immigrants (Della Pergola 2014, 13–15).

DellaPergola's categories support my orientation to this ghost hunting research for the "missed Jews", but I also incorporate affiliation, commitment, and visibility as identity markers. Based on András Kovács' study in 2001, and his 2012 overview authored jointly with Aletta Forrás-Biró, ten years after the "revival," not more than ten percent of the Jewish population is affiliated with a Jewish organization. Less than a decade ago, scholars estimated the total number of Jews in Hungary to be approximately 70,000–100,000 based on the commonly accepted number of post-Holocaust Jewish survivors, and accounting for the known or estimated numbers of births, deaths, and emigration since 1945. To break this down, in 1946 there were 143,000 people in Hungary who identified their religion as "Israelite," and 260,000 people that were defined as Jewish. In 2000, Tamás Stark, taking the Holocaust survivors' ages, anticipating birth rates and taking the number that might have entered into mixed marriages into account, estimated the number of halachically Jewish people as being between 64,000–117,000 (Stark 2002, 101–27).

The survey data from the Kovács and Forrás-Bíró approach indicates that the number of Jewish Hungarians rises to 150,000 adults (over eighteen years of age) and increases still further when persons under the age of eighteen are added. If we include everyone that had at least one Jewish grandparent, the number rises to higher than 260,000 (including people under eighteen). In defining a Jew as a person with "one Jewish parent" however, they also extended this definition so that that the "one Jewish parent" also needed to have only one Jewish parent, either mother or father (Kovács and Forrás-Bíró 2011, 7–8).

Sergio DellaPergola estimates the number of Jewish Hungarians to be only 48,200. He claims that Stark underestimated the emigration as the enlarged Jewish population in Hungary is assessed by DellaPergola at about 95,000 in 2014" (DellaPergola 2014, 63). In 2004 and 2006, 30,000 individuals, and their close relatives, applied for and received compensation for being persecuted as Jews after 1939 (Kovács and Forrás-Bíró 2011, 9). In the 2011 national census, 10,965 people identified themselves as "Israelite" (using the census terminology). Ten years earlier this number was 12,871. This means that, on the national census, only about 0.1 percent of the total population identified themselves as following the Jewish religion.

In 2013, an analysis of the number of people making a mandatory one percent tax contribution to any recognized religious organization indicates that the number of taxpayers who chose Jewish religious organizations was 7,082 to the Neologue "Federation of Jewish Faith-Based Communities" (MAZSHISZ), 1,450 to Chabad (EMIH), and two hundred and thirty-two to the Autonomous Orthodox community. While less than 9,000 people chose to designate that one percent of their taxes go to Jewish religious institutions, 23,257 people designated one percent of their taxes to go to the Krishna Consciousness church.

If we compare the estimated number of Jews from András Kovács's recent analysis (100,000–160,000) to the number of those who demonstrate any active commitment to the Jewish community, as measured by a) those who direct that one percent of their income tax be contributed to Jewish religious institutions, b) the number of children enrolled in Jewish elementary and high schools, and c) the number of participants in Jewish religious or cultural events (e.g. a Holocaust commemoration, or a Yom Kippur service), it appears that only about 7,000–8,500 people, or five and a half to ten percent of the estimated 100,000 to 160,000 Jews, are "actively Jewish." Also, from participant observation, for the past decade, we see the same twenty to sixty people participating in cultural events, and the same one hundred and fifty to two hundred people participating in religious or Holocaust-related events. My estimation from participant observation is that about 7,000–8,000 people, in total, attend synagogue services on Yom Kippur, go to the March of the Living, or visit a program during Jewish cultural festivals.[2] In 2017 on Yom Kippur, students and scholars working on an unpublished project under the supervision of András Kovács, historian in Jewish studies, counted in total 6,500 people who attended services in the Capital.

This gap between the estimated 100,000 to 160,000, who might possibly be considered Jewish based on parentage, and the 13,000 who self-identify as religiously Jewish in official censuses, and the 7,000 to 8,000 who show some commitment to the Jewish community, makes obvious that a large proportion of the potential Jewish population (Kovács and Forrás-Bíró 2011, 9) in Hungary is indeed "missing". Yet religious, cultural, secular, and academic reports refer to a magically, much larger number of Jews which, in a certain hypothetical framework, supports the "third largest community in Europe" image, which seems to be a misleading perception. While scholars carefully distinguish between the actual and potential, the real and virtual numbers, the Jewish outreach rhetoric creates and constantly reinforces in the Jewish consciousness the hope for the existence of a virtual greater Jewish population.

As it seems, not only Jewish activists overestimate the current number of Hungarian Jews. The independent social research institute, TARKI, based on a representative survey conducted in 2012, estimates that Hungarians estimate the number of the Jewish Hungarians to be between 9.2 and twelve percent of the Hungarian population.[3] Based on this perception, the Jewish community in Hungary would number approximately one million people – eight to ten times more than the commonly estimated figure.[4] In 2017, after a governmental

---

2 This article was submitted as part of a conference presentation in Prague in 2016, reviewed in Spring 2018, just before András Kovács and Ildikó Barna's significant and comprehensive publication about Jews in Hungary was published in Summer 2018.
3 http://www.rubicon.hu/magyar/oldalak/a_magyarorszagi_zsidok_a_szamok_tukreben/.
4 Jewish Hungarians are generally opposed to ethnic nationalism and recognize the jeopardy of their own tenuous situation as members of the Hungarian nation if they define themselves as a separate nation. This is why the Jewish community rejected a 2005 proposal made to the government by a group calling itself "Társaság a Magyarországi Zsidó Kisebbségért" (Society for Hungarian Jews as a Minority) that the Jewish community be defined as a "national minority." The

"hate campaign" against the philanthropist George Soros, and a proposed bill called "stop Soros," out of twenty second-year BA students, fifteen thought there are about three million Jews in Hungary and about three to four billion in the word. "They are definitely not a minority," said one of them. Obviously, this data does not represent any kind of mindset, but it could illustrate the immediate response of a two-year-long campaign against the "Jewish billionaire puppet master" who wants to allegedly undermine the nation by financing the Muslim invasion.

Why do these potential members of the Jewish community go missing? And how can we consider them as a non-group people, as a group that is connected by similar liberal views but, however, do not consider themselves to being to any group, let alone any Jewish group? Ironically based on a new national concept, they become the "others", often labeled as "Jews".

Becoming Non-Jewish

Given emigration, constant threats, and perceived antisemitism, as well as the aftermath of the assimilation contract and the general secularization that was very high even before the Communist period, plus the increased number of mixed marriages after 1945, and the severe silence about the Holocaust, Jewish family stories and Zionism (Hanák 1984, 6; Erős 1993, 23, 186–187; Száraz 1984, 295–355), we can assume that there are many additional reasons today prompting people to hide their Jewishness.

The Holocaust meant a "brutal violation of the assimilation contract" (Kará-dy 1993, 33–66; Vörös 2003: 229–255) that caused a confused situation. There is an extensive body of literature about the identity crisis of Jews, who were formally considered Hungarians, faced after the war, describing how many chose to deny or conceal their ancestry during the Communist period. Most probably, many continue to conceal their Jewish ancestry until today. "The strategy of silence" led to decades in which one or two generations grew up without knowing anything about their family history and Jewish heritage. These strategies of silence often meant leading a double life in which members of the Jewish community were forced to keep their private and public lives very separate, or even to hide their Jewish heritage from their children.

Chabad Lubawitsch leadership described the proposal as beneficial for the diverse Jewish community yet noted that Judaism is more than just a nationality, ethnicity or religion.

Ferenc Erős, András Kovács, and Katalin Lévai analyzed one hundred and seventeen interviews recorded in the 1980s and examined the manner in which the second generation dealt with the taboos regarding their heritage. They noted that some discovered their family history only as adults, for example from their neighbors, or from classmates who kept track on other people's Jewish heritage. Many of these memories recall "finding out" about their origins in an antisemitic incident. More than half viewed their Jewishness as a mark of shame (Száraz 1984). Fifty percent of the respondents regarded their Jewish heritage as so stigmatizing that they wanted to avoid acknowledging it to outsiders and taught concealment strategies to their children (Erős, Kovács, and Lévai 1985, 129–144; Erős 1992, 91; Karády 2001, 121; Kovács and Vajda 1992, 116–118; Kovács 1992, 280–291; Kovács 1987, 55–66).

However, from today's results mentioned above, we can assume that, despite these interviews, many people never found out that they have Jewish roots; these are the people in the above mentioned sixth non-group group. Or we can assume that many inherited the family's "silence" strategy and continue to practice it. After the fall of Communism, what seemed to allow a euphoric "Jewish revival" or an ethno-cultural "coming out," embracing more Jewish traditions, was a general phenomenon of the late 1990s, not only in former communist countries like Hungary, but also in Poland and Germany.[5]

---

[5] See: Zvi Gitelman, "Becoming Jewish in Russia and Ukraine," in *New Jewish Identities: Contemporary Europe and Beyond*, ed. Zvi Gitelman and András Kovács (Budapest: CEU Press, 2003), 105–37; about Eastern and Western European observance see Gitelman, Kosmin, and Kovács (eds.), *New Jewish Identities: Contemporary Europe and Beyond* (New York: CEU Press, 2003); Diana Pinto (ed.), "A Third Pillar? Toward a European Jewish Identity," in *Public Lectures 1996–1999*; András Kovács, *Jewish studies at the Central European University* (Budapest: CEU Press, 2000); Diana Pinto, "Towards a European Jewish identity," in *Golem. Europäisch-jüdisches Magazin*, February 25, 2010. In North America and England the so-called "return movement" started in the 1960s and scholars linked it to the hippie movements, to different counter cultural movements, the new age atmosphere (Danzger, 79–95), and feminism (Kaufmann 1993; Davidman 1991). The "return" movement is also often related to the Six Day War and explained by Jews as becoming "proud" of, or "inquiring" about their own religion (Danzger 1989: 78–79). While the return movement, even though religious, often included an ethno-cultural identity construct for most born again Jews, in Israel, where a basic tenet was the separate ethnocultural/national nature of the Jewish people, this "return" focused more on spiritual "conversion" or shift to a religious lifestyle (see Janet Aviad's 1983 book *Return to Judaism: Religious Renewal in Israel*). Herbert Danzger in his book *Returning to Tradition* reviews the history of the return movement in Israel and in the United States, reflecting especially on the big hippie waves and the newly found Baal Teshuva Yeshivas (Danzger 1989). Danzger called the phenomenon "returning to the tradition"; Lynn Davidman defined it as "religious re-socialization" (Davidman

But then, several scholars (Gitelman 2012, Pinto 2006, 2013, 2015) and activists described disillusionment just ten to fifteen years after the fall of Communism. János Kőbányai, the editor of a prominent Jewish magazine, *Múlt és Jövő*, stated that Hungarian Jewish cultural-religious life struggled like a "beard growing on a dead person." The Jewish schools witnessed a declining number of pupils already in 2000, and the Jewish Cultural Association "diminished to be a weekend old age club". Kőbányai complained that most "subscribers of books and periodicals cancelled their subscription[s]," and would not show support for Jewish intellectual prosperity anymore. As the editor of a Jewish magazine, he added, "We must admit that our hopes, intellectual and financial investments have failed. Neither the number and standard of authors nor the sponsorship of our readers proved to satisfy our expectations in 1989" (Kőbányai 2001, 31, 41). In 2011, András Kovács wrote that "the commitment-free festival Judaism (…) and ruin pub renaissance in the Jewish quarter creates an individualistic Jewish community and does not nurture the classic Jewish community feelings." (Kovács 2011, 33). The large spectrum of communities that offer, according to scholars, only "some sort of Jewishness" – "rather than Judaism" (Liebman's terminology, Liebman 2003, 322), or as Webber puts it, are "to a great extent self-made cultural bricoleurs, constructing their Jewishness […] as they go." (Webber 2003, 323). Liebman believes that the transmission of Jewish heritage in the future is weakening further by an already thinning Jewish culture: "Over the long run the absence of boundaries not only renders the maintenance or construction of a community impossible, but also undermines efforts to inject meaning or substance into one's Jewishness" (Liebman 2003, 346–347). The thick and thin Jewishness seems to be a double-edged sword; it might attract more people, but it also raises the question "in what sense are these groups Jewish at all?" I will now refer to the post-2010 "revival", after prime minister Viktor Orbán was elected (and re-elected three times, recently in 2018) that changed these tendencies.

State Nationalism and the Jewish Exclusivity

2010 was a break in Hungary's political, cultural, and economic scene. While this year was marked as the real "regime change," new terms were coined in order to explain the new system: Viktor Orban proclaimed the new "illiberal democracy" and "unorthodox economy." Political and civil critics consider this new admin-

---

1991). Since the 1980s many scholars have written extensively about the "return" movement (see Shaffir 1983; Kaufmann 1993; Caplan 2001, 369–398; Benor 2012).

istration to be an ethnic nationalist, authoritarian regime. Focusing only on culture and the politics of culture, it is important to mention that between 2010 and 2015 there were several public institutions, national strategy research institutes, new history and cultural centers, new museums, and new public universities established, as well as hundreds of new monuments erected (many glorifying antisemitic "heroes" from the Horthy era). Overall, public administration significantly centralized education[6] and textbooks' content was changed.[7] The declared goal of several quickly established new "academic" institutions is to "take back" and "re-write" history so that it tells "the truth," and teaches "back" the "real Hungarian" culture that was allegedly hijacked and suppressed by Communist and left-liberal academia for sixty-five years. Not surprisingly, the name of one of the recently created historical research institutes is just "Veritas Institute."

The strong nationalist rhetoric used by the Hungarian government implicitly includes antisemitism, while some extreme political parties have practically legitimized open antisemitism in political and public discourse. Worldwide media headlines about the purported increase of antisemitism in Hungary motivated donors to support additional Jewish organizations and educational proj-

---

[6] See, on Hungary's regime change, centralisation, and the controversies around the new constitution a deep insight and a meticulous analysis by Gergely Bárándy, a legal expert and member of the parliament for twelve years and member of the Lawmaking Committee Törvényalkotó Bizottság. Gergely Bárándy, *Centralizált Magyarország – Megtépázott Jogvédelem. A hatalommegosztás változásairól (2010 – 2014)* [*Centralized Hungary – The Destroyed Protective Rights; Changing the division of power between 2010–2014*] (Budapest: Scholar Publishing-house, 2014); Gergely Bárándy's *Centralizált Magyarország – Megtépázott Jogvédelem. A hatalommegosztás változásairól (2014 – 2018)* [*Centralized Hungary – The Destroyed Protective Rights; Changing the division of power between 2014 – 2018*] (Budapest: Scholar Publishing-house, 2018).

[7] See regarding this the 2015 Freedom House "Nations in Transit" report which considers Hungary as a semi-consolidated democracy. Diagramed in a graph of democracy levels in Central Europe, Hungary has declined most in recent years. It is still slightly better than Romania, but Romania does not display the steep downward trend displayed by Hungary. The 2015 report edited by Balázs Áron Kovács, who, through close analysis and monitoring of different variables, rated Hungary's lack of national democratic governance as 3.75 in 2015 (on a scale from 1 to 7, where 1 is the best and 7 is the worst), versus the rating of 2 that Hungary received in 2006 (just to compare, the Czech Republic with a similar population scored 2.75, and Slovenia scored 2.0). In the 2016 report (edited by Dániel Hegedűs), Hungary's score for democratic governance decreased 0.25 from the previous year to 4.0. This indicates that in the ten years since 2006, the quality of democratic governance in Hungary has deteriorated significantly. The decline in media independence, judicial framework, independence, and corruption seems to be significantly steeper than in previous years. The report uses the World Bank's World Development Indicators 2015, https://freedomhouse.org/report/nations-transit/2015/hungary, accessed April 10, 2016.

ects as well as several initiatives that attempt to combat antisemitism, as for example the "Action and Protection Association."[8] Since 2010, there are more Jewish organizations in existence than ever before. As an example of one of such new programs, *Minyanim* seeks to build a new cohort of young Jewish leaders in Central and Eastern Europe. It is jointly funded and run by the Jewish Agency, the Israeli Cultural Institute in Budapest, and the UJA Federation of New York. It specializes in professional community leadership training, and tries to be inclusive, while adhering closely to Jewish tradition. From secular to orthodox, a broad spectrum of institutions, such as the Israeli Cultural Center, Moishe House, Dor Hadash, Lativ Kollel, Tikva Hungary and many, many more, offer a variety of Jewish programs every single day. My assumption is that the influx of donations and the multiplication of Jewish associations, organizations, and hubs could be due to the fact that Jews feel excluded and treated as a separate "subgroup" based on the new concept of the nation expressed in the constitution and translated into every day Hungarian political practice.

Rather than portraying the "revival" as a response to increased insecurity, it is often considered as a positive indicator of an allegedly strong Jewish community within a supportive environment. For example, an article in a February 2015 edition of *Haaretz* described a "friendly government" and a new "Jewish renaissance." It must be mentioned, most secular and especially religious communities are fully dependent upon funding from the Hungarian state, from the USA or from Israel. However, visiting researchers, journalists, international outreach organizations, and prospective donors usually meet only the same few so-called "professional Jews" who regularly repeat the same enticingly hopeful line about the "100,000 Jewish soul" potential of Hungary. Many of these visitors continued to report optimistically about a vibrant "Jewish life" and "Jewish renaissance" even though it may have only been based on a "Potemkin village" facade. From the Israeli newspaper Haaretz to the New York Times,[9] reporters as well as academics were fascinated by the Sirály club's (now Aurora club and

---

8 Action and Protection (http://tev.hu/regional-v4/) monitors antisemitism in Hungary as a watchdog organization reporting monthly about antisemitism while they also deal with individual complaints and work together with the local police and law enforcement organizations. The professional leadership, oversight, and background to research and monitoring activities of the Brussel Institute is given by András Kovács, sociologist, Ildikó Barna, sociologist, and Andrew Srulevitch, expert on Eastern Europe for the Anti-Defamation League (ADL). However, the organization is closely linked to Chabad Lubavich Hasidic group that is considered to be the most loyal Jewish group to the Hungarian government.

9 See https://www.haaretz.com/jewish/fighting-for-jewish-identity-in-budapest-1.5305339, accessed April 10, 2016; http://www.nytimes.com/2009/12/09/arts/09iht-GRUBER.html?_r=1&.

the Mosaic group)[10] innovative creativity, run by Marom (a youth organization within the global Masorti movement), and the "revived" gentrified "new" Jewish quarter in Budapest (Gruber 2000).

However, in spite of the Jewish institutional boom, and in spite of the more inclusive change, more diverse identity concepts, the number of Hungarians identifying as Jews, or their level of commitment to Jewish communal life did not change significantly. The "missing Jews" were targeted again (for their protection to maintain or re-discover their "own" identity, Jewish identity, to help them find their way to a Jewish community), but most of them remained missing (perhaps for their protection of their own privacy or deliberate choice of not declaring themselves Jewish, among other reasons).

On one hand, while described in many reports and highly condemned by diplomatic and political officials from outside Hungary, the manifestations of the new extreme ethnic nationalism (e.g. the changes to the constitution, the national curriculum, the Holocaust museum, the remembrance of the victims of communism, the monuments that deflect Hungarian responsibility for the Holocaust, glorifying and whitewashing the Nazi era) generated heated academic and public debates. The disagreements were unresolved and the conflict between the government and the Jewish community became very open. The press described this conflict as "the Jewish boycott." Indeed, in the last twenty-five years, this was the first time that the Jewish Federation (MAZSIHISZ) had not participated in government-sponsored events for ten months and officially stopped any communication with the government.

On the other hand, while the "revival" of new Jewish organizations and programs did not really produce more committed communities, the changes to Hungary's democracy and the discussions related to responsibilities during the interwar period and the Holocaust generated heated debates, discussions, and dialogues that made many people speak out and articulate opinions about Jews in Hungary. These discussions resulted in thousands of academic articles, discourses, blog posts, editorial opinions, open letters, comments, statements, tweets, protests and petitions. I quoted above a few of those manifestations where in interviews, conversations or on the sidelines of different debates, people confessed or mentioned briefly their Jewish heritage and expressed an opinion about the Hungarian "Jewish question."

I distributed a questionnaire to my own students who were enrolled in either my "Tradition and Modernity in Jewish Culture" or my "Modern Israel" course.

---

10 These are hubs or clubs where there are daily cultural, religious, political, and academic activities. They also offer space for NGOs intimidated or restricted by the government.

Perhaps, not surprisingly, in response to the open-ended question *Why did you choose this course?*, about twenty-five percent of the students mentioned that they have some sort of Jewish background. "My aunt is Jewish, but we are not, and I definitely wouldn't like to be Jewish today. Nor truly deeply Hungarian," someone answered.

Interestingly enough, during about three weeks of volunteering to help refugees crossing Hungary in August and September 2015, I had the impression that I met more people in a few hours who might fit into this "missing Jewish" non-group than in most outreach events. Obviously, I am aware that this is a contextual perception. The way people "came out" during these activities was simple. People working together in such a critical, emotionally difficult, and chaotic situation, whether during preparing food, or folding and distributing clothes side by side, eventually start chatting, starting with easy question like, "Where are you from?" and "What do you do for work?" After mentioning that my research area is within Jewish studies, they asked me "why?" and soon several (selecting from these people, I made only seventeen informal interviews contextualized in the refugee crisis in addition to my previous more formal interviews) people mentioned that they have "some Jewish roots." I focused only on those that expressed, in some way, their deliberate intention not to be part of the offline Jewish life.

The Possible Branches of Missing Jews

Analyzing a "virtual population," the "missing Jews," and attempting to classify them into groups may be merely an attempt to validate a hypothesis, but hopefully it will lead to a better understanding of the nonaffiliated people who, if they exist, make up eighty-five to ninety percent of the Hungarian Jewish community.

Regarding Jewish identity, my primary approach is the confessional one; however I do also take into consideration the emic perspectives from the essentialist, halachic one, to family heritage, to the constructivist identity concept, the sense of belonging to the Jewish society (in-group preference), active commitment, or even Sartre's definition of being Jewish by the eye of the "Other," or in an antisemitic approach of assigning someone a Jewish identity label that he or she rejects.

I will not focus on those people who participate in Jewish events (often even orthodox ceremonies) but who have absolutely no Jewish background and fabricate Jewish origins for themselves. Nor on those who also do not have Jewish roots, but while they construct for themselves a Jewish identity do not fabricate a Jewish heritage; they do not feel the need to convert either. They consider themselves Jewish enough to feel that they belong to certain Jewish groups. Since they are part of and belong to the visible Jewish scene, I do not take consider them "missing Jews".

While in this paper I do not focus on the "tangible," even minimally committed Jewish Hungarians, I touch upon the transitional identity (on the way in, or on the way out from Judaism), but mainly I will attempt to describe the "missing Jews" based on above-mentioned factors, hypothetically lining them up in six branches as non-group groups that might not show more than certain dominant tendencies. The main principle is not whether they are Jewish or not, but the fact that they refuse to affiliate with any Jewish organizations.

These "non-groups" potentially could be:

1. Those who know about their Jewish family roots, and have participated in Jewish programs, educational, cultural or religious events but deliberately decided that they do not want to "deal with Judaism" anymore (they did not feel rejected, it was their choice to step out from any Jewish group, activity or support anything that has Jewish content). I also count in this first non-group those who are halachically not considered Jews and therefore were confronted with this "disadvantage" in a negative context. In fact, they had a bad experience of not being accepted in certain Jewish groups which define Jews only halachically; therefore these people disappointedly never want to give Judaism another chance.

2. Those who also know about their Jewish family roots but have not participated yet in any Jewish events, who have no basic knowledge about Judaism, and a) are not interested in the "Jewish question" debates in the press or public discourse; however, they cannot avoid observing some visible Jewish life and "Jewish political discourse" in Hungary. Some might have adopted certain antisemitic stereotypes. The b) subgroup of this second non-group are those who do care about the "Jewish question" debates and react in one way or another to antisemitism. Here I could mention those people who never participate in any Jewish activity, but support the so-called MTK (Magyar Testgyakorlók köre, Hungarian Physical Educations Club established in 1888); their football team had several Jewish members, including some who were killed in the Shoah. The team is considered "Jewish" even today. These group could be called the "reactive Jews".

3. Those who know they have Jewish family background, but do not consider themselves Jewish.

4. Those who may say "I have Jewish roots" in absolute privacy or in a safely anonymous environment, but are carefully and deliberately in the closet. Most have never participated in any Jewish community events. They do not have a deep knowledge of Judaism; however, they follow the socio-political debates on "Jewish matters" or about the other "others." They react in particular to antisemitism, and they vote accordingly. Some might also be MTK supporters as well. These could be, for example, the political blogger, closet Jews.
5. Those who perhaps have or do not have any Jewish background, and reject being considered Jewish, but in the "eyes of outsiders" are considered and labeled Jewish because of their alleged family background or family connections, their views, political choices, occupation, behavior, "Jewish sounding name," or even based on their "looks" or other characteristics or stereotypes.
6. The sixth non-group is the most hypothetical, but at the same time its existence must be real. They are the completely "lost Hungarian Jews" who were probably fully assimilated and absorbed into a non-Jewish society (in Hungary, Australia, Canada etc.). Now, this third or fourth generation after the Holocaust indeed does not know, and no one else knows anything, about their possible Jewish heritage. Therefore, they have probably "completed" the assimilation process for the time being.

These six groups could constitute the rather unknown "greater imaginary Jewish community," each on different levels. While people in the first four non-groups have some awareness, suspicion or blurry assumption about their existing Jewish family roots, the common element of all these six groups is that they deliberately do not want to participate in or be part of any explicitly Jewish group.

The first four non-groups, regardless of their family heritage, consciously reject any affiliation with a denominational, non-denominational, cultural or any specifically Jewish group. However, they do relate to Judaism and Israel in one way or another (from ignorance to "outsiders'" individual support). As opposed to the fifth or sixth non-group, if we relate to Judaism, we can only speculate and reinforce stereotypes by "recognizing" in "them" inherited characteristics, choices, and behaviors that are considered "Jewish".

At this point I will illustrate the conscious rejection with quotes from a few interviews conducted from 2013–2016 that could fit in the different groups mentioned above.

48-year-old man who is a medical doctor from Budapest:

I have been married to a Calvinist woman for 25 years. We have three children who we raised in a liberal spirit. We value every culture, but we do not want them to build their self-worth based on national pride rather than on their own achievements. In the last few years we went to the Jewish summer festival and a few times for Yom Kippur, to the Dohany St. Synagogue (...) My wife encouraged me to go to Jewish events, seminars, lectures. At most of these seminars the topic was pretty much the same: we should be proud to be Jewish. We should be Jewish, much more Jewish. We should not let our children abandon their ties to the Jewish community, assimilate and 'God forbid' marry a non-Jew. We should not let them assimilate? You mean mix with society? What? Will we become infected by non-Jews? At almost every lecture and seminar I could hear openly that the Jewish culture is distinct. We are different. Our contribution to the Hungarian culture is enormous. Hungarian culture would be nothing without the Jews. (...) and Israel is unique, Israel has the most startups, the most talented, the most Nobel prizes, basically Israel is a superpower... I felt more and more uncomfortable hearing this, especially in front of my wife. I felt ashamed in fact. We condemn Jobbik when its members talk about Hungarian nationalism; but we do the same. I don't go to Jewish events anymore.

A 22-year-old student:

The obsession with Jewish identity makes me sick... we judge the right-wing nationalist Hungarians, but we do the same. I went to Szarvas, I went to Lauder, and now, when I want to apply for Masa program, I can't get it, because I don't have any papers. My family wanted me to go to a liberal school. I grew up in Lauder, I speak Hebrew, I know the culture. But I am not Jewish by blood. I decided to go to England and hopefully I will forget all my Jewish childhood.

A 43-year-old journalist:

I am saying this (...) because I do have Jewish roots, both from my mother's mother's side and my father's mother's side. I don't care. I am not going to the March of the Living, or any other Holocaust commemoration. The Holocaust is not my problem. It is the Hungarian society's problem. I would rather go to a Roma demonstration.

38-year-old attorney married with an observant Jewish man:

My grandfather was a Holocaust survivor. I grew up all my life hearing about his Holocaust experience, yet my kids are not accepted in the Jewish kindergarten (Chabad), because my mother is not Jewish. What is this blood purity approach again?

50-year-old medical doctor:

I am Jewish. I never cared, I don't care what my kids choose to be. I think we never discussed religion with them. I don't know. (...) I know which of my colleagues look at me with an evil eye. But it makes me sick to park my car next to a colleague that has 'Greater Hungary' map sticker on his car.

42-year-old university professor:

I don't know what Jewish means. When I witness antisemitism, then I'm a Jew, when I sense homophobia, then I'm gay, when I see racism against Roma, then I speak up as Roma. I have nothing to do in a synagogue, because I am not just Jewish. I don't want my kids to become [sic] only Jews either as I don't want to close myself in boxes. I have nothing to do with where and what family I happened to be born into.

33-year-old stay-at-home mother:

Who has time to be busy with these kinds of questions? I think Jews – and I must say that both my parents come from Jewish families – are fixated on antisemitism, and to keep the Jewishness on the table. They are a very loud minority. Protesting against every monument, every annoying sentence (...). Why should I care about the Jewish programs or any events? In our little free time, we go to places that are entertaining and we are all welcomed. We just want to live and let others to live. I don't ask my friends what religions they have.

23-year-old student:

My grand grandparents were Jews, but my family is absolutely not Jewish. I have a Jewish aunt in Israel, but my parents had a Catholic wedding. My grandparents had a Catholic funeral. We all were raised Christians, or rather non-religious. I personally don't care about any of them; all are crazy.

21-year-old student:

I think I am eight percent Jewish, but to be honest, my family never considers this stupid eight percent. I know more about New Zealand than about Jewish stuff, except all the Holocaust things.

45-year-old man, volunteer helping refugees:

I never been in 'The' (!) Synagogue (the Great synagogue), I know I'm partly Israelite but I don't care.

Why do people choose to not affiliate with any Jewish groups? The conversation with the last quoted respondent showed that he has no idea that there is Jewish life beyond the one big synagogue, the Dohany Street Great Synagogue (that everyone in Budapest knows about). He cannot even say the word "Jew" or "Jewish" (*zsidó*), since he feels that it is a negative term (a legacy of Communism); instead he uses archaic terminology that has been used in the past to mark someone's religion. Most of these people saw Chabad in the media, and they connect them to "the Synagogue," and have no idea that most Orthodox and Chabad refuse to pray in the Dohány Synagogue since it has been built in the spirit of assimilation (with an organ, towers, the bima in front and pulpits) and follows the non-Orthodox, Neologue practice of Judaism. The research on "missing Jews" has found that these opinions are not uncommon at all. What could be the reason behind this and what social dynamics have led to this point?

I tried to collect the most frequently expressed arguments and criticisms that were put forward as reasons why the "missing Jews" did not want to be part of any Jewish community. Most people either began their explanations with, or hinted or inserted somewhere in the middle, a variation of the following information that would place them among the missing Jews:

> "Most of family members name is on Holocaust memorial walls"
> "I have Jewish roots..."
> "I have Jewish origins..."
> "I am not Jewish, but my grandparents were Israelites..."
> "I also had a Jewish grandparent..."
> "My family used to be Jewish/Israelite..."
> "All my cousins live in Israel they moved right after the war..."
> "I have a Jewish aunt..."

I can only assume that the reason why their Judaism and criticism of Israel begins with the confession of having a Jewish connection is to make sure that any antipathy they display towards Judaism or Israel is not perceived as coming from an antisemitic perspective. Occasionally respondents clearly state this logical connection, e.g. "I have Jewish origins, so I can't be accused of being an antisemite..." Their most common arguments for rejecting any participation in Jewish life were related to their perception of Jewish identity concepts, Jewish events, what they knew about Jewish religious traditions, and the current politics and policies of the state of Israel. Some have visited Jewish-themed events or programs before; some only heard about or experienced such events indirectly (they saw something in the news, or they saw the televised Hanukkah celebrations).

The only common element that groups this "non-group" together is that they do not want to belong to any Jewish community, organization, program or idea.

Below I selected only those frequently reoccurring opinions (from comments, blog posts, articles, online and offline debates, panel discussions) where the speaker or writer mentioned his or her connection to a Jewish background. Such critical thoughts are:
- Promoting Jewish pride slogans would be similar to the racial European nationalism that is at the core of antisemitism.
- Zionism might also lead to extreme nationalism. Zionism might include racism.
- Zionism combined with fundamentalist orthodoxy would maintain the conflict between Hungarian and Jewish nationalism within Hungarian citizens.
- Religious and often secular outreach language or missionary "buzzwords" that try to engage people with Jewish heritage into their activities is considered highly offensive towards Jews coming from mixed marriages or Jews married to non-Jews, or simply to many people that reject ethnic purity concepts.
- Jewish statements regarding integration into the Hungarian society are considered offensive and racist, when Jewish outreach warns against the "threat" of assimilation" and claims that intermarriage and assimilation have to be "combatted," "fought against" or "prevented," because they would lead to deterioration, loss, and fragmentation of the Jewish traditions and character.
- Jewish statements characterizing intermarriage as a "threat" to Jewish society, or an "issue," "problem," "concern," "worry," or "modern disease", are considered as offensive and unacceptable.
- Jewish organizations would more or less openly claim that intermarriage leads to divorce, lower birth rates, assimilation, disaffiliation, disappearance, the decline of Jewish communities, or a weakening of the Jewish character of the communities. These organizations also state that mixed marriages are a "devaluation" of Jewish values, that a person who marries a gentile becomes "vulnerable," and that the children of such a marriage will only be "half" Jewish (half-persons, or even dead to some communities).
- The interviewees reject the language and methodology of "Israel advocacy" or, as they often call it, "Israeli propaganda" (including Taglit/Masa/Hasbara).
- They also reject the ethnic or religious exceptionalism/exclusivity, or the emphatic declaration of Israel's "uniqueness."
- Many also state that in Israel, the special status of the Jewish population is an anomaly which poses significant and paradoxical issues to Israel's status as a democratic state.
- Israeli policies are often called "ethnocratic," "theocratic," "primordialist," "essentialist", and "exclusivist."

- The most criticized law is the principle of the law of return that is considered "racist," "ethnocentric," "undemocratic," "non-egalitarian," and against human rights (while many admit to understanding the security considerations behind the law).
- The most mocked Israeli law is that marriage between Jews is only legally accepted if it is approved by orthodox religious authorities. Therefore, marriage between those who identify themselves as Jews in Israel (and who are accepted as Jews by the State of Israel under the "law of return") can only take place if they meet the Halachic definition of "Jew" either through their maternal lineage or through orthodox conversion. There is no civil marriage in Israel, and many Hungarians who know about their Jewish heritage believe that the Halachic approach to defining who is a Jew is racist and "medieval."
- Many interviewees believe that Jewish religious laws are not only "ethnocentric" and divisive, but also "hopelessly homophobic."

From the above described opinions and attitudes, I concluded a few principles that this "non-group group" is opposing:
- Any essentialist "tribal" definitions, like "mixed," "hybrid," "half Jewish," "50% Jewish."
- Stereotyping, or collective cultural identity approaches that lead to positive or negative stereotypes, dangerous characterizations; collective descriptions can lead to collective "demonization."
- Any Jewish and Hungarian, Hungarian and Roma collective group distinctions within one nation.
- Extremely pronounced "Jewish content," like klezmer music and cholent nostalgia, i.e. "Fiddler on the Roof" sentimentality or "Jewish Kitsch", as they called it.

The respondents were also:
- Supportive of promoting equal rights for gays and lesbians, gay and lesbian marriage (almost regardless of age).
- Sensitive to discrimination against the Roma people (we checked and in some cases interviewed the initiators and supporters of several petitions).
- Sensitive to any kind of human rights violations.
- Against laws restricting people from sleeping in public spaces, or changes in the constitution that diminish gay rights.
- In favor of speaking out regarding hate speech and freedom of speech, and protesting against authoritarian actions by the government.

- Besides the obvious antisemitism of the extreme ethnic nationalists, the respondents sense and decode antisemitism and speak out when they sense coded hints and subtle differentiations between Jews and non-Jews (e.g. when Jews in Hungarian historical narratives are pointed out as Jews and not as Hungarian citizens), or any subtle trivialization of the Shoah.

Most of these opinions could be deduced in the person's biographical context as case studies. While I cannot provide that here, I must admit that I also have the impression that many of these opinions reflect interiorized criticisms against "the public image of Jewish outreach" (Festivals, Commemoration days, daily event advertisements) and the frequency of governmental Holocaust commemoration that causes harsh debates due to its inaccuracy (see Kovács 2014). The angry tone detected could be attributed to the fact that most have difficulty coming to terms with their family background.

Very few seemed to have an idea how diverse the Jewish identity continuum is in Europe or in Hungary, and how many options they might have. Most rely on certain perceptions that are also used in antisemitic arguments (Jewish networking, endogamy, Jewish exclusivity, in-grouping, Zionism as racism), and some back up these opinions with current Israeli policies (within a very similar ethnic nationalist approach that Hungary has), and limited level of religious knowledge.

While liberal, anti-nationalist, progressive civic values and extended egalitarian universalist human rights seem to be the underlying principle that creates a common basis for these groups, the survey results, opinions, interviews, and comments I have collected show similarities – except with regard to the question about the refugees. In this aspect, from extreme rejection to active acceptance, the spectrum of the opinion was wider and also divisive, both among the "missing" or "missed" Jews" and within the active Jewish community itself.

In conclusion, I found the strongest criticism came from people who know about their Jewish ancestors, but reject identifying themselves as Jewish. They do not want to affiliate or participate in any Jewish events, believe or have the perception that Jewish identity concepts are anachronistic, ethnocentric, racial, exclusivist, self-centered, and detached from modern social reality. The identity concept, based strictly on a fixed, essentialist, ethnic or religious background, seems to most people from the analyzed non-group like a dangerous road of extreme nationalism combined with fundamentalist religiosity. Since this concept distinguishes people, it cannot be other than non-egalitarian, exclusivist and undemocratic, ultimately leading to inter-group conflicts.

We could assume that certain "non-ethnic Hungarian nationalist values" could function as group border markers. Obviously, we are aware of many exceptions. For many Jewish Hungarians, non-ethnic nationalism does not apply when it comes to Israel's ethnic nationalist policies. Could we assume that the invisible, unreachable potential Jewish population (at least a proportion of them) are too liberal to be Jewish in Hungary? A country where neither state law nor the rules of MAZSIHISZ, the umbrella organization of Jewish faith-based communities, accepts the Jewish Reform/Progressive or the constructivist, masorti, or non-denominational movements like Aurora/Siraly clubs?

Perhaps the ethnocentric exclusiveness and tradition-bound nature of the officially state recognized, therefore highly publicized, Jewish community in Hungary pushes people away from participating in the activities of this community. At the same time, the Hungarian nationalist rhetoric also excludes them from the non-Jewish Hungarian majority. Some, without any family ties to Judaism, feel that they have become victims of antisemitic stereotypes. This leads to the question: what are those current or relatively new, upgraded stereotypes that make one believe that someone is Jewish? The Hungarian government's national cohesion campaign subtly implies that those that do not agree with the National Cooperation Plans or its ethnocentric concept of "the Hungarian nation" are "the others" who possess a "foreign heart" and are pursuing foreign interests. These "others" are considered non-patriots, sometimes even "the enemy of the nation," foreign in the corpus nationum in the sense of the current government's understanding of patriotism, which is deeply rooted in a 1930s concept of nationalism, which imagines a homogeneous ethnic nation based on Christian values and a 1,000-year history of ethnic national and political continuity. These "others" are all those who do not lament having lost Transylvania, do not fight for ethnic Hungarian communal interests, have dual loyalties to other ethnicities, cultures or nations, or no exclusive loyalty to the state.

It must be noted that while many of the previously described Neologue communities have shifted towards Jewish Orthodox practice with regard to defining Jewish identity, many formal and informal groups and congregations are actually increasingly inclusive, egalitarian, and constructivist (Dor Hadas, Moishe House). These groups currently are not recognized by the state, and therefore they do not receive any funds. Many observant religious Jewish leaders question even their Jewish character. Some of these "open" outreach organizations have liberalized their identity policy, attempting to learn inclusive techniques that celebrate diversity. They are increasingly receptive to the ongoing and unending changes that take place in society. The Israeli Cultural Center often organizes programs that not only accept but also celebrate the dynamics of a changing society and culture. Although the Reform movement has not been formally accepted in Hungary, it continues to provide many exceptions that provide examples of progressive identity concepts combining secularism, diversity, and respect for the interpretations of Jewish traditions. Some private Jewish high-schools, like the Lauder Foundation American school, have also changed their approach, opening up more and more to progressive values, attracting an increasing number of not only "simply Jewish" but people who are not Jewish, people who are uncertain or contextually Jewish, and many Budapest families that want to give their children a liberal education. The estimated number of families related to the Lauder school could reach 5,000 people.

As another example, we could mention that after four years of silence, the previously-mentioned Sirály concept was "revived" by the same people who had originally developed it, under the Marom banner at the end of 2014 in a new club called "Aurora" with a similar profile – supporting civil society, civil justice, promoting Tikun Olam (making the world we live in a better place), egalitarianism (within and outside Judaism), combining what are known to be Jewish values, e.g. from the Hebrew Bible, Talmud or rabbinical teaching, with universal values of social justice and human rights, while also offering lectures about Judaism and publicly celebrating Jewish holidays, mainly in a liberal, reconstructionist inclusive spirit. While cultural organizations have primarily been the ones that have publicly supported gay rights, in the last five years there are more and more religious (quasi reform) communities as well that support gay rights and hold special LGBT events.

The opinion of most people I interviewed emphasized the fact that the old infrastructure and administration of the Jewish Federation (MAZSIHISZ) could not get rid of its "communist legacy" – it would have remained authoritarian and corrupt for two and a half decades. However, since the new president of the Federation (András Heisler) was elected, the institution went through significant changes regarding its financial transparency, re-organized and revitalized activities. The Jewish Federation after 2010 has also changed its communication with the state, standing up more firmly for the Jewish community, speaking out against the coded antisemitism of the government or the misleading state-organized Holocaust commemorations while protecting the democratic, egalitarian, and progressive values with which non-religious "secular" Jews often identify. Still, the old, narrow spectrum of activity and identity which the traditional Hungarian Jewish cultural and religious infrastructure has promoted continues to appear to push away the majority of the imaginary "greater" Jewish population. Also, we must note again that social justice programs and informal civil organizations, think tanks or cultural centers – often especially those that are not explicitly and exclusively Jewish, or not marked Jewish at all – (Haver organization or the Mozaik Hub) especially Tikun Olam projects, or those semi-informal groups that engage in current anti-nationalist, anti-corruption, anti-centralization politics (like Aurora) nevertheless attract a growing number of people with Jewish heritage, despite not being explicitly identified as Jewish.

Having said that, the "missing Jews" occasionally do find these groups. We know that the revelation of "finding out" one's Jewish origins from the 1980s and 1990s is still an ongoing phenomenon: every month, we hear stories about people who mysteriously just found out that they are Jewish and want to act on it somehow.

While there does appear to be significant interest by people with some link to a Jewish heritage in programs and activities embodying progressive societal values often ascribed to the Jewish community, in proportion to the infrastructure, money, and effort invested in "Jewish life in Budapest," the commitment level and the number of active, non-professional (i.e., people that do not earn their living from Jewish organizations) participants indicates that most participants do not seem to feel the need to give back after "consuming" what they have been offered. We could assume that the producers are not matching their product to the needs of the consumers. The lack of general commitment to any community, the lack of stable organic communities, the lack of grassroots initiatives, the ad-hoc (often volunteer) leadership, and especially the constant re-invention of the "vibrant Jewish scene" from the euphoric decade of 1990s by the very same truly dedicated and creative group of friends would indicate that we cannot talk about "revival" or "renaissance" except in emic terms. This being said, from the perspective of the core of active Jews, primarily "professional Jews," (i.e., people that work as activists) there is an ongoing, undefeatable hope that they can still reach out and somehow attract the deeply "missed," "missing" Jews.

The Two Ends of the Phantom Pain

I started the paper with an analogy that the "missing Jews" feel like a phantom pain after amputation. This phantom pain, though, has two sides: it feels like the existing body is causing pain to the person, and it also feels like the missing body part, the amputated limb, is in pain. The body misses and longs for the lost body part. This analogy might shed light on a complex situation of the Hungarian Jewish community. While the "missing Jews" are themselves considered as being in a dire circumstance, by the "professional Jews" (religious or secular), "neither here nor there" as a 38-year-old community leadership trainer phrased it, the "lost" Jews are in pain. What is clearer is that the visible physical community is definitely in constant "pain" – longing and fixated on fixing some sort of hole. Reaching out to the missing Jews often comes across often as not feeling whole, not feeling independent, and not feeling strong enough.

In Hungary today, the effort to reach out, and the enormous investment in community building, are disproportionate compared to existing possibilities. From the number of participants and the quality of engagements, it seems like there are more organizations, initiatives, and programs than those which are really needed. However, the constant creation and re-creation of yet another new organization that will "this time, with this new concept, really" find the missing "90%" shows that active Jews believe in reaching out to the "greater Jewish Population" to attract them into Jewish communities. Although, from my research we could conclude that programs labeled specifically Jewish might push away most people, especially those who feel excluded from the Hungarian nation as well from its extreme ethnic and Christian nationalist rhetoric that sugar coats anti-otherness, there is also a reverse tendency in outreach that is committed to progressive liberal and socially sensitive values, civic loyalty, and the "Tikun Olam" idea which for many people in countries with growing authoritarianism in Europe becomes more and more a political mission.

Bibliography

Aviad, Janet. *Return to Judaism: Religious Renewal in Israel*. Chicago: University of Chicago Press, 1983.
Anderson, Benedict. *Imagined. Communities, Reflections on the Origin and Spread of Nationalism*. London-New York: Verso, 1983.
Azira, Régine. "A Typological Approach to French Jewry." In *New Jewish Identities: Contemporary Europe and Beyond*, edited by Zvi Gitelman, Barry Kosmin, and András Kovács, 61–73. Budapest: Central European University Press, 2003.
Balogh, Margit. "Az izraelita felekezet (1945–1989)" ["The Israelite denomination"]. *Tarsoly István (szerk.) Magyarország a XX. században. [Hungary in the XX. century.]* 2. Szekszárd: Babits Kiadó, 1997.
Barth, Frederik. *Ethnic Groups and Boundaries: The Social Organization of Cultural Difference*, 9–38. Oslo: Scandinavian University Press, 1969.
Bauman, Zygmunt. "From Pilgrim to Tourist—or a Short History of Identity." In *Questions of Cultural Identity*, edited by Stuart Hall and Paul du Gay, 18–36. London: Sage, 1996.
Benor, Bunin Sarah. *Becoming Frum: How Newcomers Learn the Language and Culture of Orthodox Judaism*. Rutgers University Press, 2012.
Boyd, Jonathan. "Jewish community 2.0: how the Internet may be changing the face of Jewish life." *JPR*. New Conceptions of Community, 2010.
Braham, Randolph L. *The Politics of Genocide: The Holocaust in Hungary*. 2 vols. New York: Columbia University Press, 1981.
Braham, Randolph L., and Brewster S. Chamberlain (eds.). *The Holocaust in Hungary: Sixty Years Later*. New York: Columbia University Press, 2006.

Caplan, Kimmy. "Israeli Haredi Society and the Repentance (Hazarah Biteshuva) Phenomenon." *Jewish Studies Quarterly* 8 (2001): 369–398.
Csoóri, Sándor. "Nappali Hold." [Moon in a Daylight]. *Hitel* 17, 18, 19 (1990): 4–7.
Danzger, Murray Herbert. *Returning to Tradition. The Contemporary Revival of Orthodox Judaism*. London: Yale University Press New Haven, 1989.
Davidman, Lynn. *Tradition in a Rootless World. Women Turn to Orthodox Judaism*. Berkeley, Los Angeles, Oxford: California Press, 1991.
DellaPergola, Sergio. "Hungary." In *The world Jewish population*, edited by Arnold Dashefsky, Sergio DellaPergola, and Ira Sheskin. Berman Institute, 2014.
Dencik, Lars. "Jewishness in Postmodernity: The case of Sweden." In *New Jewish Identities: Contemporary Europe and Beyond*, edited by Zvi Gitelman, Barry Kosmin, and András Kovács, 75–103. Budapest: Central European University Press, 2003.
Erős, Ferenc, András Kovács, and Katalin Lévai. "Hogyan jöttem rá, hogy zsidó vagyok?" ["How did I find out that I am Jewish?"] *Interjúk. Medvetánc* 2–3 (1985): 129–144.
Erős Ferenc, András Kovács, and Katalin Lévai. "How Did I Find Out I Was a Jew." *Soviet Jewish Affairs* 17, no. 3 (1987): 55–66.
Erős, Ferenc. *A válság szociálpszichológiája [Sociology of crisis]*. Budapest: T-Twins, 1993.
Erős, Ferenc. *A zsidó identitás Magyarországon a második világháború után. Holocaust emlékkönyv [Jewish identity after World War II. Holocaust memory book]*, 200–205. Budapest, 1994.
Gergely, Bárándy. *Centralizált Magyarország – Megtépázott Jogvédelem. A hatalommegosztás változásairól (2010–2014). [Centralized Hungary – The Destroyed Protective Rights; Changing the division of power between 2010–2014]*. Budapest: Scholar Publishing-house, 2014.
Gergely, Bárándy. *Centralizált Magyarország – Megtépázott Jogvédelem. A hatalommegosztás változásairól (2014–2018). [Centralized Hungary – The Destroyed Protective Rights; Changing the division of power between 2010–2018]*. Budapest: Scholar Publishing-house, 2018.
Gitelman, Zvi. "Reconstructing Jewish Communities and Jewish Identities in Post-Communist East Central Europe." *Jewish Studies at the Central European University* (2000): 35–50.
Gitelman, Zvi. "Becoming Jewish in Russia and Ukraine." In *New Jewish Identities: Contemporary Europe and Beyond*, edited by Zvi Gitelman and András Kovács, 105–37. Budapest: Central European University Press, 2003.
Gitelman, Zvi. *Jewish Identities in Postcommunist Russia and Ukraine: An Uncertain Ethnicity*. Cambridge: Cambridge University Press, 2012.
Gordon, Milton. "The Nature of Assimilation." *Assimilation in American Life: The Role of Race, Religion, and National Origins*, 60–83. New York: Oxford University Press, 1964.
Gudonis, Marius. "Particularizing the Universal: New Polish Jewish Identities and a New Framework Analysis." In *New Jewish Identities: Contemporary Europe and Beyond*, edited by Gitelman-Barry-Kovács, 243–261. Budapest: Central European University Press, 2003.
Gruber, Ruth Ellen. *Virtually Jewish: Reinventing Jewish Culture in Europe*. Berkeley: Berkeley University Press, 2000.
Gruber, Ruth Ellen. "A Virtual Jewish World." In *Jewish Studies at the Central European University, 1999–2001*, 3. Budapest: Central European University Press, 2002.
György, Péter. *Apám helyett. [Instead of my Father]*. Magvető, 2012.

Karády, Viktor. "Asszimiláció és társadalmi krízis. A magyar-zsidó társadalomtörténet konjunkturális vizsgálatához." ["Assimilation and social crisis. About the context of Hungarian-Jewish social history"] *Világosság* 3 (1993): 33–60.

Karády, Viktor. *Önazonosítás és sorsválasztás. A zsidó csoportazonosság történelmi alakváltozásai Magyarországon* [*Self-identification and Choice of Fate. Changes in Jewish group identifications in Hungary*]. Budapest: Új Mandátum Könyvkiadó, 2001.

Karády, Viktor. *Túlélők és újrakezdők. Fejezetek a magyar zsidóság szociológiájából 1945 után.* [*Survivors and Beginners. Additional chapters to the Sociology of Hungarian Jewry*]. Budapest: Múlt és Jövő, 2002.

Karády, Viktor. *The Jews of Europe in the modern era: a socio-historical outline*. Budapest: Central European University Press, 2004.

Kovács, M. Mária. *Zsidóság, identitás, történelem [Judaism, Identity, History]*, edited by Mária M. Kovács, M. Yitzhak, and Ferenc Erős. Budapest: T-Twins, 1992.

Kovács, M. Mária. "Rehabilitálnák azt az embert, aki a magyarokat hozzászoktatta az antiszemitizmushoz." ["The man that got Hungarians used to antisemitism, is about to be Rehabilitated."] In *Nol. Hu*. 2015. http://nol.hu/velemeny/homan-rehabilitalasarol-1534295. Accessed September 15, 2016.

Kovács, M. Mária. "Kettős beszéd: holokauszt-emlékév és Horthy-rehabilitáció." *Galamus*. March 7, 2014. http://www.galamus.hu/index.php?option=com_content&view=article&id=364077:kettos-beszed-holokauszt-emlekev-es-horthy-364077&catid=9:vendegek&Itemid=134. Accessed April 10, 2016.

Kovács, A., and I. Barna. "Identity à la carte: Research on Jewish identities, participation and affiliation in five European countries." In *Analysis of survey data*. Budapest: The American Joint Distribution Committee, 2010.

Kovács, A. "Jews and Jewry in Contemporary Hungary: Results of a Sociological Survey." *JPR Report No. 1*. London: JPR, 2004.

Kovács, András. "Zsidó csoportok és identitásstratégiák a mai Magyarországon." In *Zsidók a mai Magyarországon [Jewish groups and identity strategies in today's Hungary]*, edited by András Kovács, 9–40. Budapest: Múlt és Jövő, 2011.

Kovács, András, and Aletta Forrás-Biró. *Jewish Life in Hungary: Achievements, challenges and priorities since the collapse of communism*. London: Institute for Jewish Policy Research, 2011.

Kőbányai, János. *A halott arcán növekvő szakáll. A magyar zsidó történet vége?* [*Beard growing on a dead person. The end of Hungarian History?*] Budapest: Múlt és Jövő, 2001.

Liebman, Charles. "Jewish Identity in Transition: Transformation or Attenuation?" In *New Jewish Identities: Contemporary Europe and Beyond*, 341–350. Budapest: Central European University Press, 2003.

Mars, Leonard. "Cultural Aid and Jewish Identity in Post-Communist Hungary." *Journal of Contemporary Religion* 15, no. 1 (2000): 85–96.

Marsovszky, Magdolna. "Imre Kertész and Hungary Today." In *Imre Kertész and Holocaust Literature*, edited by Louise O. Vasvari and Tötösi de Zepetnek, 146–160. West Lafayette: Purdue University Press, 2005.

Miron, Guy. "Between 'Center' and 'East': The Special Way of Jewish Emancipation in Hungary." In *Central European University Jewish Studies Yearbook* 4, 111–138. Budapest: Central European Press, 2005.

Papp, Richárd. *Van-e zsidó reneszánsz?* [*Is there a Jewish renaissance?*] Budapest: Múlt és jövő, 2005.

Petresel, Alyssa. "Somehow I Am Different: Narratives of Searching and Belonging in Jewish Budapest." *Jewschool.com*. 2015. http://jewschool.com/2015/11/38479/getting-know-adam/.

Pinto, Diana. "The Third Pillar? Toward a European Jewish Identity." In *Jewish Studies at the CEU I. Yearbook* (Public Lectures), 177–199. 1996.

Pinto, Diana. "Are There Jewish Answers to Europe's Questions?" *European Judaism* 39, no. 2 (2006): 47–57.

Pinto, Diana. "Israel Poses a Serious Dilemma for Europe's Jews." *Haaretz*, February 14, 2013. http://www.haaretz.com/opinion/israel-poses-a-serious-dilemma-for-europe-s-jews.premium-1.503489.

Pinto, Diana. "I'm a European Jew—and No, I'm Not Leaving." *The New Republic*. March 27, 2015. https://newrepublic.com/article/121388/why-jews-arent-leaving-europe-contra-atlantics-jeffrey-goldberg. Accessed April 21, 2016.

Rosenson, A. Claire. "Polish Jewish Institutions in Transition: Personalities over Process." In *New Jewish Identities: Contemporary Europe and Beyond*, edited by Zvi Gitelman, Barry Kosmin, and András Kovács, 263–289. Budapest: Central European University Press, 2003.

Schepele, Kim Lane. "Hungary and the End of Politics. How Victor Orbán launched a constitutional coup and created a one-party state." *The Nation*, 2014. http://www.thenation.com/article/hungary-and-end-politics.

Shaffir, William. "The Recruitment of Baalei Tshvah in Jerusalem Yesihva." *Jewish Journal of Sociology* 25, no. 1 (1983): 33–46.

Stark, Tamás. *Zsidóság a vészkorszakban és a felszabadulás után 1933–1955* [*Jews during the Soah and after the Liberation 1933–1955*]. Budapest: MTA Történelemtudományi Intézet, 1995.

Stark, Tamás. "Kísérlet a zsidó népesség számának behatárolására" ["An attempt to limit the number of Jewish population in present day Hungary"]. In *Zsidók a mai Magyarországon* [*Jews in Contemporary Hungary*], edited by András Kovács, 101–127. Budapest: Múlt és Jövő. 2002.

Száraz, György. "Egy előítélet nyomában/ Egy interjú nyomában" ["The footprints of a prejudice. Interviews"]. In *Zsidókérdés-asszimiláció—antiszemitizmus. Tanulmányok a zsidókérdésről a huszadik századi Magyarországon. [The Jewish question – Assimilation – Antisemitism]*, edited by Peter Hanák, 295–365. Budapest: Gondolat Kiadó, 1984.

Tavares, Rui. "Committee on Civil Liberties, Justice and Home Affairs". http://www.europarl.europa.eu/sides/getDoc.do?type=REPORT&reference=A7-2013-0229&language=EN.

Vörös, Kati. "A Unique Contract: Interpretations of Modern Hungarian Jewish History." In *Central European University Jewish Studies Yearbook* 3, 229–255. Budapest: Central European University Press, 2002–2003.

Webber Jonathan. "Jews and Judaism in Contemporary Europe: Religion or Ethnic Group?" *Ethnic and Racial Studies* 20, no. 2 (April 1, 1997): 264.

Online Sources

János Gadó's interview with Andras Heisler, published in the monthly Jewish magazine
   Szombat. March 12, 2016. http://www.szombat.org/politika/a-zsido-felekezetek-kozott-a-mazsihisz-a-legdemokratikusabban-iranyitott-szervezet. Accessed September 15, 2016.
http://www.ksh.hu/docs/hun/xftp/idoszaki/nepsz2011/nepsz_10_2011.pdf, 150. Accessed
   March 15, 2015.
http://www.haaretz.com/jewish-world/jewish-world-features/1.642103. Accessed April 10,
   2016.
http://www.nytimes.com/2009/12/09/arts/09iht-GRUBER.html?_r=1&. Accessed April 10,
   2016.
http://marom.masortiolami.org/siraly-jewish-hangout-budapest-jta-video-article/. Accessed
   April 10, 2016.
Committee on Civil Liberties, Justice and Home Affairs, Rapporteur: Rui Tavares.
   http://www.europarl.europa.eu/sides/getDoc.do?
   type=REPORT&reference=A7-2013-0229&language=EN; Human rights in Hungary 2013.
   Accessed April 10, 2016.
http://www.csce.gov/index.cfm?FuseAction=ContentRecords.ViewDetail&ContentRecord_id=
   460&ContentType=S&ContentRecordType=S&IsTextOnly=True. Accessed April 10, 2016.
http://fra.europa.eu/en/publication/2013/jewish-peoples-experience-discrimination-and-hate-
   crime-european-union-member. Accessed April 15, 2016.
http://www.zsidokisebbseg.com/2005/09/29/beadvany-az-ovb-hez/. Accessed April 10, 2016.
http://regi.sofar.hu/hu/node/100547. Accessed April 10, 2016.
https://freedomhouse.org/report/nations-transit/2015/hungary. Accessed April 10, 2016.
https://www.politico.eu/article/second-brexit-referendum-george-soros-campaign-about-to-
   start/. Accessed May 28, 2018.

Avihu Ronen
Memories and Hopes: The Zionist Youth Movements and the Communist Regimes in Central Europe, 1944–1950

Introduction

This article deals with a short period in the life of the Central and Eastern European Jewish communities after the Holocaust and focuses on a unique chapter which appears to be relatively neglected at present: the relations between the new Communist regimes and the radical Zionist youth movements. The evolution of these complex relations through time was based on a common struggle against Nazism led by groups of Zionists and Communists during the war, and on a common belief in the future of Socialism, but one which was contradictory about how this would be implemented in Europe and in Palestine.

However, this short period, 1944 to 1950, ended abruptly with the fall of the "Iron Curtain" and the prohibition of Zionist activity. This was a tragic finale to the memory of a common past and to the illusion of "the world of tomorrow" which had formed the basis of these relations.

The article will focus on events characterizing three countries: Poland, Czechoslovakia, and Hungary.

The Anti-Fascist Struggle

During the Second World War, close cooperation existed between the members of the pioneering Zionist youth movements and the Communists in Central and Eastern Europe. During the Nazi occupation, even though their approaches to the Jewish national problem had separated the two groups until then and had led to no small amount of bitter rivalry, they succeeded in cooperating in the struggle against the Nazis. The ideological differences between the two were not forgotten, but they became secondary when considering the need for action against the Nazis. It should also be noted that Zionist and Communist activists usually came from the same social sector of the Jewish community, the educated middle class.

The cooperation was especially significant in the Polish ghettos, where the Zionist youth movements (actually, the pioneering movements, that is, the Zionist-Socialists) and underground Communist groups acted under joint organizational frameworks.

In Warsaw, cooperation began within the framework of the *Anti-Fascist Bloc*, established at the beginning of 1942, under the leadership of two Jewish Communists, Józef Lewartowski and Andrzej Szmidt, with the participation of the Zionist-Marxist movements Dror and Hashomer Hatzair. This bloc was short-lived and ceased to operate in June 1942. But the cooperation continued and in the new framework, Żydowska Organizacja Bojowa – ŻOB, four companies of Communists participated in the Warsaw ghetto uprising (April 1943). A certain level of cooperation also existed with the Communist underground outside of the ghetto which activated two groups in attempts that, in fact, were unsuccessful to help the fighters inside the ghetto.[1] Later, the survivors of the organization under the leadership of Zuckerman and Lubotkin took part in the fighting during the Warsaw uprising, under the command of the Communist *Armia Ludowa* – AL (August 1944).[2]

Cooperation between the Communists and the fighting organizations also existed in Vilna, where Yitzhak Wittenberg, a Communist, headed the United Partisan Organization (*Fareynigte Partizaner Organizatsye* – FPO) until he was handed over to the Germans. This cooperation also continued as partisans, despite the enduring conflicts about the existence of separate Jewish units.[3] In the Krakow ghetto, it was Gola Mira, a Communist and former Zionist, who initiated the cooperation between the Zionist *Hechalutz Halochem* and the pro-communist *Iskra* underground group.[4] The Communist Chaim Yelin was the head of the unified underground in Kovno (*Yidishe Algemayne Kamfs Organizatsie* – JFO) with 600 members[5] while Communist Daniel Moszkowicz was deputy to Zionist Mordechai Tenenbaum in the fighting organization of Bialystok, which led the uprising on August 15, 1943. Haika Grossman, a member of Hashomer Hatzair and the only one of the underground leadership in Bialystok to survive the up-

---

[1] Israel Gutman, *The Jews of Warsaw, 1939–1943: Ghetto Underground, Revolt*, translated from Hebrew by Ina Friedman (Bloomington: Indiana University Press, 1982).
[2] Yitzhak Zuckerman, *A Surplus of Memory: Chronicle of the Warsaw Ghetto Uprising* (Berkeley: University of California Press, 1993), 520–541.
[3] Yitzhak Arad, *Vilna Hayehudit Bekilion Vebamavak* [*The Struggle and Destruction of the Jews of Vilna*] (Jerusalem, Tel Aviv: Yad Vashem, Tel Aviv University, Sifriat Poalim, 1976).
[4] Yael Peled, *Krakov Hayehudit 1939–1946, Amida, Mahteret, Maavak* [*Jewish Krakow 1939–1943, Withstanding, Underground, Struggle*] (Tel Aviv: Ghetto Fighters Museum, Hakibbutz Hameuchad and Masuah, 1993).
[5] Avraham Zvi Brown and Dov Lewin, *Toldoteiha shel Mahteret: Hairgun Halohem shel Yehudei Kovna Bemilhemet Haolam Hashniya* [*A History of an Underground: The Fighting Organization of the Jews of Wilno during the Second World War*] (Jerusalem: Yad Vashem, 1962).

rising, served afterwards as a courier for the Soviet partisans until the liberation of Bialystok.⁶

In Hungary, where events did not culminate in an armed uprising, there was cooperation of another type. The orientation of the various Zionist pioneering underground groups was directed towards rescue, and the cooperation between them focused on that goal. The pioneering underground members, who specialized in forging documents, supplied thousands of counterfeit documents to their members, as well as hundreds of these false papers to the members of the Communist underground who needed them in order to hide out in the capital. The Communist underground groups, on their part, supplied the Zionist underground with hiding places, aided them in rescuing their members from the death march, and camouflaged units guarded the hiding places of Jews during the Szálasi regime. Among the leaders of these groups were Pal Demény, László Sólyom, and Iván Kádár.⁷

The peak of this cooperation between Zionists and Communists in Budapest was the establishment of the Megyeri unit, an underground pioneering group under the leadership of József Mayer-Megyeri, a Hashomer Hatzair member. The group accumulated arms and trained in preparation for an armed uprising, which, however, did not take place. The rescue of a group of about one hundred Zionist and Communist underground members from an Iron Cross prison just before the end of the war was another high point in their joint activity. Most of the contacts took place with members of Hashomer Hatzair, the more radical movement, but the joint activities also involved members of Dror and Maccabi Hatzair, the close partners with Hashomer Hatzair in the pioneering underground.⁸

In Czechoslovakia there was relatively little cooperation but what did exist had great symbolic significance. A small group of Hashomer Hatzair members decided to take an active part in the Slovak revolt and its members, led by Egon Roth, participated in battles, in cooperation with groups of Communists

---

6 Sara Bender, *The Jews of Bialystok during World War II and the Holocaust*, trans. Yaffa Murciano (Waltham, Mass.: Brandeis University Press; Hanover: University Press of New England, 2008); Haika Grossman, *The Underground Army: Fighters of the Bialystok Ghetto*, trans. Hebrew by Shmuel Beeri (New York: Holocaust Library, 1987), 367–376.

7 David Gur, *Brothers for Resistance and Rescue: The Underground Zionist Youth Movement in Hungary during World War II*, ed. Eli Netzer, trans. Pamela Segev and Avri Fischer (Jerusalem: Gefen, Society for the Research of the History of the Zionist Youth Movement in Hungary, 2007), 137.

8 Avihu Ronen, *Hakrav al Hachaim: Hashomer Hatzair Behungaria, 1944* [*The Battle for Life: Hashomer Hatzair's Underground in Hungary*] (Tel Aviv: Yad Yaari, 1994), 263–272. See also Avihu Ronen, *Harc az életért: Cionista (Somér) ellenállás Budapesten, 1944* (Budapest: Elvarosi Konyvkiado, 1998).

(August to October 1944). No less important was the arrival of five Israeli parachutists at Banská Bystrica during the revolt. The parachutists actually did not contribute much and most of the aid was to the Jewish community, still their deaths while the revolt was being quashed by the Germans and especially the death of Haviva Reik were perceived as a symbolic act of aid and cooperation between the members of the Zionist movements and the Communist fighters.[9]

Relations after the War

As the remnants of the Jews of Europe emerged from the ruins, the members of the pioneering underground found that some of their brothers in arms in the Communist underground had now become members of the new Eastern European governments. In Poland, immediately upon their release, Yitzhak Zuckerman and Avraham Berman, leaders of the ŻOB and the Jewish National Committee, met with Wladyslaw Gomulka, the former leader of the *Polska Partia Robotnicza – PPR* in the underground and now Secretary General of the Communist Party in Poland, and the three embraced warmly.[10] Even warmer relations existed between Zuckerman and General Marian Spychalski, the former commander of the *Armia Ludova – AL*, who had now become the Polish deputy Defense Minister. The Polish Foreign Minister was Yaacov Berman, a Jewish Communist who had spent the war years in the Soviet Union and was the brother of Avraham Berman.

Haika Grossman, the heroine of the Bialystok underground, acting inconsistently, joined the Polish Department of the NKVD after her release, which later became the *Urząd Bezpieczeństwa – UB* (Department of Security). She fell in love with and then married her supervisor in this department, Jan Kozlovski. "I will not deny", she later wrote, "that in the sea of loneliness, they were loyal to me in personal friendship. I avoided relationships with anyone who came into my surroundings from a different journey, not ours". Grossman, who later became an Israeli Knesset member, served in the UB until the end of June 1945 and only then did she return to Zionist activity.[11] It was not by chance that Yisrael Barzilai, a member of Grossman's movement, the Socialist Mapam Party, was appointed first Israeli Ambassador to Warsaw.

---

9 Akiva Nir, *Shvilim Bemaagal Haesh (Bamered Haslovaki)* [Trails in the Fire Circle (in the Slovak Revolt)] (Merhavia: Moreshet, 1967).
10 Zuckerman, *A Surplus of Memory*, 474–475.
11 Ziva Shalev, *Haika* (Tel Aviv: Moreshet, 2005), 216–222.

In Hungary, as well, members of Hashomer Hatzair had strong ties to a number of new government members. For example, Lazlo Sólyom, who had been provided with forged documents by the pioneering underground, became Chief of Police in Budapest and in 1948 was appointed Head of the Hungarian Army Special Staff. In addition, some of Yosef Meir's Communist friends in the resistance groups became members of the secret police under the command of Péter Gábor (Benjámin Eisenberger). Meir himself joined the police for about a year and primarily dealt with the identification of war criminals.[12] However, in Hungary, the disappointments were relatively not long in coming: Pál Demény, head of the resistance group with which the pioneering underground had been in warm contact, had already been judged and sentenced in 1945, while some of the members of the new government, who were mainly Moscow's men, related with relative indifference to the members of Hashomer Hatzair.[13]

In Czechoslovakia, contacts were very diversified, both due to the presence of many Jews in the higher levels of the Communist Party – some of them had been members of Zionist youth movements in their youth[14] – and members of other youth groups (primarily members of Hashomer Hatzair) crossing the lines, becoming Communists, during the war or a short time later, and now holding important positions in the new governments.[15] Hashomer Hatzair, which was the strongest of the movements in Czechoslovakia, had particular links to the Communists, and its members, who had voted for them, and even took an active part in the first election campaign in Czechoslovakia (May 1946). Yaakov (Benito) Rosenberg, an emissary of Hashomer Hatzair who was active in Bratislava from mid-1947 to the beginning of 1949, used to issue weekly reports about Zionist activities to Bedřich Geminder, Chief of the International Section of the Party Secretariat.[16] Later, his fellow kibbutz member Rafi Friedl (Benshalom) was appointed First Consul of the Israeli Embassy in Czechoslovakia.

---

12 Avihu Ronen, "Yoshka Meir: Agadat Horef Minahal Gaaton" ["Yoshka Meir: A Winter's Tale from the Gaaton Valley"], *Yalkut Moreshet* 57 (1994): 253–270.
13 Rafi Benshalom, *We Struggled for Life: The Hungarian Zionist Youth Resistance during the Nazi Era* (Jerusalem: Gefen, 2001), 111–112.
14 Rudolf Margolius (1913–1952), Deputy Minister of Foreign Trade; Bedřich Reicin (1911–1952), Deputy Minister of National Defense.
15 See Moshe Blumenfeld, *Du Kium Meahorei Hasorag* [*Coexistence Behind Bars*] (Merhavia, Tel Aviv: Sifriat Poalim, 1971).
16 Avihu Ronen, *Nidona Lechaim: Yomaneha Vechayeha shel Chajka Klinger* [*Condemned to Life: The Diaries and Life of Chajka Klinger*] (Tel Aviv: Yediot Sfarim and Haifa University Publications, 2011), 444.

The Political Climate

Certainly, the continuous connections between some of the functionaries of the new Communist regimes and the delegates of the left-wing Zionists were due also to the reconsideration of the "Jewish question" by USSR authorities. As Benjamin Pinkus has shown, the Russians were well informed about postwar Zionist activities in Central Europe (including the illegal immigration to Palestine), but they did not stop them because they had not yet made up their minds as to how to deal with this issue.[17]

On May 14, 1947, the reconsideration of the Jewish questions became an explicit dramatic turn in USSR policy. Andrei Gromyko, the permanent representative of the Soviet Union to the United Nations, declared that the Jews deserved a state of their own either in the framework of a bi-national state or as a second option, a Jewish state alongside an Arab state in a partitioned Palestine.[18]

Starting from the Gromyko speech, via the November 29, 1947 UN decision approving the partition of Palestine and the establishment of two states, the Czech-Israeli arms deal, and until the end of 1948, there was a "short honeymoon" of the Zionist movement and the state of Israel and the USSR and its allies. According to Pinkus, there were several reasons for this short positive turn of the USSR policy. The most important one was the motivation to get rid of British involvement in the Middle East. The second reason was that, in case of war between Arabs and Jews, the Russians could exploit the situation for their own benefit by sparking a dispute between the USA and Britain. In any case, besides these cynical considerations, there was also some sympathy for the Jewish suffering in the Holocaust which was more evident among the lower levels of Russian functionaries.[19]

This change in the global political climate was most welcomed in the left-wing movements in Eretz Israel (later: Israel)[20] and strengthened the legitimation

---

17 Benjamin Pinkus, *Yahasim Meyuhadim: Brit Hamoatzot vebaalot brita veyahasehen im haam hayehudi, hatzionut vemedint Isarael, 1939 – 1959* [*Special Relations: The USSR and its allies and their relationships with the Jewish nation, Zionism and Israel, 1939 – 1959*] (Beer Sheva: Machon Ben Gurion Leheker Israel vehazionut, Ben Gurion University 2007), 143–149.
18 Alyahu Eilat, *Hamavak al hamedinah. Washington 1945–1948* [*The fight for the state, Washington 1945–1948*] (Tel Aviv: Am Oved 1982, vol. I), 108; Pinkus, 167–168.
19 Pinkus, 172–173, 201–202.
20 Shaul Paz: "Hamahapecha sheboshesha lavo – Mapam besheelot hutz, 1945–1948" ["The revolution that was reluctant to come: The foreign policy of Mapam, 1948–1954"], *Lo yuchlu biladeinu* [*Couldn't do without us*], ed. Eli Tszur (Daliah: Yad Yaari and Yad Tebenkin, 2000), 187–189.

of the Zionist-Communist relationships in central European countries, which had been developed from below.

The Role of Memory

The memory of the Warsaw ghetto uprising played an important role in developing the complex relations between the Communists and the radical Zionists in Poland. The new Communist regime tended to minimize the memory of the Polish Warsaw uprising (August 1944) as much as possible, since it had been led by the *Armia Krajowa* nationalist underground which was declaredly anti-Soviet. However, the Warsaw ghetto uprising, in which leftist groups had taken an active role, well suited the Communist and Soviet outlook and was significantly highlighted.[21] Just as members of the *Armia Krajowa* were being pursued, charged, and deported, the fighters of the Warsaw ghetto were being recognized and honored. Fifty military decorations had already been awarded posthumously to Warsaw ghetto fighters on the first anniversary of the uprising.[22] During the following years, military decorations were also awarded to a number of ghetto fighters who had survived. National memorial ceremonies in the ruins of the ghetto, which included government representatives and members of the pioneering youth movements shaped this collective memory. Those ceremonies were crowned by the dedication in April 1948 of the Warsaw Ghetto Monument designed by Nathan Rapoport. The monument itself, which is an example of Soviet Realism, symbolized the cooperation of the two groups: most of the monument presented the anti-Fascist struggle, while the millions of victims were only partially represented.

In Czechoslovakia as well, the collective memory of the common struggle played a special role in forming relations with the new government. This was expressed particularly at the state funeral which took place on November 3, 1946, at the Jewish cemetery in Banská Bystrica for the Zionist victims of the Slovak uprising, Egon Roth and his fellow fighters.[23] The names of the two kibbutzim established by Hashomer Hatzair in Czechoslovakia after the war also represent-

---

[21] Krystyna Kersten, *The Establishment of Communist Rule in Poland, 1943–1948* (California, 1992), 220. James E. Young, "The Biography of a Memorial Icon: Nathan Rapoport's Warsaw Ghetto Monument", *Representations*, no. 26, (1989): 90–91.
[22] Beit Lohamei Hagetaot (Ghetto Fighter's House) Archives, 1796.
[23] Yosef Rav (ed.), *Shomrim Tamid: Hashomer Hatzair BeCzechoslovakia, 1920–1950* [*Guarding Always: Hashomer Hatzair in Czechoslovakia, 1920–1950*] (Givat Haviva: Hitahdut Yotzei Hashomer Hatzair Beczechoslovakia, Merkaz Teud Veheker, 1995), 235.

ed the legacy of the struggle: "Underground Fighters" (whose members later became the founders of Kibbutz Shomrat), and Lehavot Haviva, named after the parachutist, Haviva Reik (Reikova), whose members established a kibbutz with the same name in Israel. An act of special symbolic significance was the participation of the members of the pioneering youth movements (Hashomer Hatzair, Maccabi Hatzair, and Dror) in building the "youth railroad track" (Trať mládeže), acting as an independent "brigade" called the Haviva Reik Brigade (August 1948). The brigade composed an anthem to mark the event:

> We will turn the idea into deeds.
> We will break through the rock of ignorance.
> Arms are silenced,
> Here the Haviva Reik Brigade is laying tracks for comradeship.
> The palm of your hand will become a fist to honor the work.
> The march of work is already advancing
> The youth will lead the army of workers.
> Forward, forward – the railroad tracks of the youth.[24]

In Hungary, the direct memory of the past had a relatively weaker impact – because, as mentioned, the pioneering underground there mostly dealt with acts of rescue, and when grading acts of resistance, these were considered secondary, if they were at all included in the resistance classification. On the other hand, personal relations among ex-underground activists were very important, and especially those of Yoska Meir, who served in the secret police for almost a year.[25] Here too, as in Czechoslovakia, members of the pioneering movements took part in the May Day parade on May 1, 1945.

It should be noted that the memory of the cooperative anti-Fascist struggle overshadowed that of the Holocaust in general. Both in the new regimes and in the pioneering movements, the heroism of the fighters was emphasized far more than the suffering of the victims, which was required mostly during the trials against Nazi war criminals. But even in these cases, the victims were perceived not as Jews but rather as Poles, Hungarians or Czechoslovaks. "Suddenly the Jewish question no longer existed", wrote Rafi Benshalom as early as 1945. "Only the war of the Hungarian democracy and the victims of the Hungarian nation were spoken of."[26] But in fact, except for the important activities of the Jewish Historical Institute (Żydowski Instytut Historyczny – ŻIH) in Warsaw, even the

---

24 Giora Amir, *Kibbutz Ole min Hashoah: Sipur Meyasdei Lehavot Haviva* [*Kibbutz Emerges from the Holocaust: The Story of the Founders of Lehavot Haviva*] (Givat Haviva: Yad Yaari, 2009), 22.
25 Ronen. Yoshka Meyer.
26 Benshalom, *We Struggled for Life*, 111–112.

Jewish communities which were rehabilitating themselves did not deal with the memory of the Holocaust itself; everyone was turning towards the future.

Educational Activities and the "Escape" to Israel (Bricha)

Despite the many contacts and the common past, relations between the pioneering youth movements and the new Communist regimes should not be regarded as a kind of "honeymoon". These complex relationships involved various other and sometimes contradictory interests. For example, the relatively good relations between the governments and the youth movements were also the result of additional factors, and principally, from the attempt by the Communist regimes to present a seemingly democratic appearance. Thus, during those years, activities by non-Communist parties and organizations were also allowed and Zionist organizations were among them. Also at play was the maze of Soviet interests which had a dominant influence, reaching its peak with Soviet support of the UN Partition Plan and its recognition of the State of Israel (November 1947, May 1948). The activities of the Zionist organizations as well did not stem only from sympathy with the new regime, but also from the desire to acquire the greatest number of political connections in order to achieve advantages for the movements, whose activities were directed towards Israel rather than the nations within which they were acting. Furthermore, from the beginning, within the pioneering movements, several individuals (such as Abba Kovner and Mordechai Rosman) and organizations like the general Zionist youth movements viewed the immediate emigration from Europe and immigration to Israel as the main Zionist activity following the war.[27]

These years were also characterized by struggles between the Jewish Communists and the pioneers who, after all, were travelling along contradictory paths. In Poland the Jewish Communists under Szymon Zachriasz and Hirsch Smolar were leading the way towards rehabilitation of Jewish life and Jewish integration in Poland itself, while completely negating the pioneering way of life.[28]

---

[27] See Dina Porat, *The Fall of a Sparrow: The Life and Times of Abba Kovner*, trans. and ed. Elizabeth Yuval (Stanford, California: Stanford University Press, 2010).
[28] Hannah Shlomi, *Asufat Mehkarim Letoldot Seerit Hapleta Hayehudit Bepolin, 1944–1950* [Studies on the History of the Jewish Remnant in Poland, 1944–1950] (Tel Aviv: The Center for Polish Jewry, The Diaspora Research Institute, Tel Aviv University, 2001), 232–258.

Sharp disputes took place in Hungary as well, where the polemic took on a public character and was carried on openly in the newspapers and in public forums.[29] The arguments were, of course, even more bitter, when Jewish Communists tried to take the place of the Zionists in the new Jewish organizations, though sometimes these arguments took place with civility.

As an anecdote, we may note the story of Aryeh Yaari (Hunwald) who arrived in Hungary as an emissary from his young kibbutz (Ein Dor) in the Lower Galilee. He later related that one of his old friends had said to him: "Aryeh, why should you return to your far-away kibbutz in the desert? Stay here with us; you will be building a new Hungary and in a short while, you will become a minister…" Aryeh refused and at the end of his term as emissary he returned to his faraway kibbutz, while his friend ended his life in the next series of show trials.[30]

In any case, in this short period of legalization of Zionist activity, which continued from 1945 until 1949–1950, the members of the pioneering youth movements could develop wide ranging activities and for the first time they became a leading factor in the Jewish communities in Central Europe. Memories of the past, of course, influenced this development: in light of the failure of the traditional Jewish leadership during the Holocaust, these movements were the only ones who could appear to the survivors with their heads held high: as fighters, as partisans, as rescue activists. But there were also additional factors which involved the unique character of the Zionist youth movements.

Two central activities undertaken by the youth movements in central Europe had always been part of their goals – education and immigration to the land of Israel. After the Holocaust, these two activities took on new significance. At that time, education involved setting up Children's Houses to educate thousands of Jewish children, most of whom were orphans and many (especially in Poland) who had been gathered from orphanages, monasteries, and from Polish people who had hidden them during the war. Gathering and educating these children was conducted under the coordinating framework of the Zionist youth movements called the Zionist Coordination for the Redemption of Children in Poland. The Coordination was established with money from Jewish organizations in the United States (e.g., the World Jewish Congress and the Joint Distribution Committee), and it redeemed the children by paying the Polish families who had hidden them, but also accepted Jewish orphans who had arrived in the wave of repatriation from the Soviet Union. The children's houses of the Coordination were

---

[29] Yehuda Weiss, "Yeme Pricha LaShomer Hatzair BeHungaria" ["Days of Revival for Hashomer Hatzair in Hungary"], *Sefer Hashomer Hatzair* II, ed. Levi Dror (Merhavia: Sifriat Poalim, 1961), 373.
[30] Interview with Arie Yaari, Tel Aviv, January 1994.

administered in an educational format which included regular care and education along with informal movement educational methods, and some of these children's houses were defined as "children's kibbutzim". By mid-1946, about 50 children's houses had been established with about 3,300 children.[31]

In this mission, the members of the youth movements were in constant competition with children's houses run by the Jewish Communists and Bund members who controlled the Jewish Committee. While the Communists and Bund people intended to leave the children in Poland, the goal of the Coordination was to get them out of the country and to Palestine. However, there are no reports that the Communist regimes themselves impeded the youth movement members, and from the various testimonies it appears that they helped them, if not in money, then in places of residence and means of transportation.[32] A similar situation existed in Hungary, where youth movement activists established a number of children's houses,[33] while in Czechoslovakia, the youth movements collected the children into customary movement frameworks, leading them to immigrate to Palestine.[34] There too the movements received aid from the government.

The second aspect of youth movement activity in the first years after the war was problematic both for the regimes and for the Jewish community, and involved emigrating from those lands via the *Bricha* (escape), which directed immigration to Palestine. The activities of the *Bricha* began immediately following the end of the war in spring 1945, when a group of survivors from the Polish fighter groups were searching for a way to reach Palestine through Romania, Hungary, and Austria. When the first emissaries from Palestine reached them, this activity became more organized but most of its activists were and remained members of the youth movements from Central Europe. The *Bricha* from Poland sustained powerful momentum in summer 1946, after the pogrom in Kielce when

---

31 Nahum Bogner, *At the Mercy of Strangers: The Rescue of Jewish Children with Assumed Identities in Poland*, translated from the Hebrew by Ralph Mandel (Jerusalem: Yad Vashem, 2009), 193–206; see also Shlomi, *The Jewish Remnant*, 242.

32 Neomi Izhar, *Chasia Bornsteim Bielicka: Achat Mimeatim, Darka shel Lochemet Vemechanechet, 1939–1948* [*Chasia Bornstein Bielicka: One of the Few, The Story of a Fighter and an Educator, 1939–1948*] (Tel Aviv: Moreshet, 2003), 276–290.

33 See Yosef Ben Porat, *Zman Hatefer: Haodissea shel Ken Hatchia Behungaria Derech Streit Asher BeGarmania leEretz Israel, 1945–1948* [*Stitching Time: The Odyssey of the Tchia Branch in Hungary via Streit in Germany to Eretz Israel*] (Givat Haviva: Yad Yaari, 1993).

34 Yehoshua Bichler, "Yemey Hachraah" ["Days of Decision"], *Shomrim Tamid: Hashomer Hatzair BeCzechoslovakia, 1920–1950* [*Guarding Always: Hashomer Hatzair in Czechoslovakia, 1920–1950*], ed. Yosef Raz (Givat Haviva: Hitahdut Yotzei Hashomer Hatzair BeCzechoslovakia, Merkaz Teud Vehaker, 1994), 258–259.

most of the fugitives headed towards Czechoslovakia, and from there to Austria or Germany. In all, between 125,000 and 150,000 Jews had left Europe by 1948, and another 50,000 left for Israel in 1949–1950, before the Communist regimes finally closed their gates.[35]

This exodus, which was fundamentally illegal, obviously did not receive formal approval from the governments of the Communist regimes. According to many testimonies and documents the actions were known to them, and usually there was almost no intervention in the migration of the Jewish refugees, including their exit from the countries of children's houses of the Coordination. At the beginning of this wave of refugees from Poland, there was coordination between Yitzhak Zuckerman, one of the central *Bricha* activists, and General Spychalski, at that time the Polish Minister of Defense, about crossing the border between Poland and Czechoslovakia. Whenever there were mishaps by the border guards, Zuckerman could contact Spychalski directly.[36] Three years later, in 1949–50, a similar pattern emerged on the border between Hungary and Czechoslovakia. When Hungary forbade Zionist activity and legal immigration to Israel, secret cooperation took place between the Hungarian state police, the Czechoslovak security police, and the Israeli representatives, enabling about 20,000 Jews to cross the border from Hungary to Czechoslovakia.[37] In that case as well, the main activists of the *Bricha* were members of the pioneering movements while the Czechoslovak Defense Ministry and the Hungarian police were cooperating with people they already knew.[38]

Finally, it is also important to note the active military cooperation which involved the State of Israel. The Soviet vote in favor of the partition agreement and afterwards the arms deal between Israel and Czechoslovakia (at the beginning of 1948) deeply affected the political and military status of Israel and influenced its ability to win the War of Independence. This was followed by training of pilots and parachutists in Czechoslovakia, as well as pre-military training in Hungary and Poland, which enabled thousands of young people, many of them members of pioneering youth movements, to be directly inducted into the Israeli Defense Forces – IDF. David Gur, a member of the leadership of Hashomer Hatzair in

---

[35] Yehuda Bauer, *Flight and Rescue: Brichah* (New York: Random House, 1970).
[36] Zuckerman, *A Surplus of Memory*, 560–556; Bauer, *Flight and Rescue*, 212–213.
[37] Yehoshua Bichler, "Habrikha MeHungaria Minekudat Mabata shel Mishteret Habitachon HaCzechoslovakit, 1949" ["The Flight from Hungary from the Point of View of the Czechoslovak Security Police, 1949"], *Yalkut Moreshet* 54 (1994): 364; Moshe Elyihu Gonda, Yosef Cohen, and Yehuda Marton (eds.), *Yehudey Hungaria* [*The Jews of Hungary*] (Tel Aviv: Haaguda Leheker Toldot Yehudey Hungaria, 1980), 292.
[38] Bichler, "Yemey Hachraa", 264–266.

Hungary, talks about a meeting which took place at the beginning of 1948 at the headquarters of the Chief of Staff of the Hungarian army, László Sólyom, with a representative of the *Hagana* (the paramilitary Jewish organization which precede the IDF) in Hungary, Yaakov Solomon, in which the Hungarian Chief of Staff asked about the war going on in Palestine. A few months later, representatives of Sólyom were present at the graduation ceremony of the pre-military training course of about one hundred candidates for immigration to Israel.[39] In Poland as well, in Bolków, in Silesia, a training base of the Hagana was established which also served the Israel Defense Force from April to November 1948. It should be noted that in Poland also, the training in the use of arms took place with the knowledge of the government.[40] The high point of this cooperation was the establishment of the "Czech Brigade" in May 1948, trained by the Czechoslovak army. It numbered 1,334 men and was led by Jewish officers of the Czechoslovak army. About 600 of them reached Israel and fought in the Israeli War of Independence.[41]

The End of the Illusion

The relationships which have been described here, characterize an illusionary dimension among the Zionist Socialist youth movements and perhaps among some of the people in the new governments as well. They may have imagined that this cooperation would be continued and that the countries of Central Europe would establish a new model of Socialist society parallel to that which was meant to be created in Israel, whose social and political hegemony at that time was in the hands of Social Democratic and Marxist parties. This hope, which was partially the result of the utopian dreams of the former fighters, and partially a result of the delusions about Stalinist policy, was cruelly ended with the dramatic changes in Communist policies (even during 1948) and the show trials taking place in 1949–1953. At these trials, many Communists who had maintained contacts with Zionist movements, like Spychalski and Gomulka in Poland, Geminder and Blumenfeld in Czechoslovakia, and Sólyom and Gabor in Hungary, were charged. In some of the trials, and in particular in the Slánský trial in Czechoslovakia, there was also a clearly antisemitic dimension and the

---

39 Interview with David Gur, 5.10.2015.
40 The website of the Hagana, http://www.irgon-haagana.co.il (accessed 12.3.2016).
41 Yaakov Markovitzky, "Conspiratzia Comunistit o syua Leachim: Gius Habrigada Haczechit, 1948–1949" ["A Communist Conspiracy or Help to Brothers: The Recruitment of the Czechoslovak Brigade, 1948–1949"], *Iunim Betkumat Israel* 6 (2006): 190–201.

Jewish origin of many of the accused provided additional proof of their being Zionist spies. The contacts with the Zionist movements were presented at these trials as cooperation with "imperialist" forces, and as a plot to destroy the people's democracies. Also caught up in these trials were Zionist movement activists when the regimes were able to lay their hands on them. These included Aryeh Palgi in Hungary and the Israeli political figure Mordechai Oren whose trial in Czechoslovakia caused great turmoil in the Israeli workers' movements.[42] This was a tragic end to the complex relationships, which threatened the actual continuation of Jewish life in the Communist countries.

Conclusions

Sixty-five years after the events described here, they appear to be a far-off and distant story. The fall of Communism in Central and Eastern Europe and the privatization of kibbutzim in Israel blur the memory of this period in which, for a short historical moment, a new Jewish-Socialist identity seemed to be forming both in Europe and in Israel. The paradox was that, in fact, the attempt to form this identity subverted itself: the efforts of the Zionist movements led outside of Europe and thus caused the emigration of the best of the new forces to Israel, doing mortal harm to the Jewish communities, while the efforts of their Communist friends-foes led to the assimilation of Jewish Communists in their lands and to the loss of another types of Jewish identity. This can be exemplified by a personal note: my grandfather Moritz Rosenberg was the head of *Hevra Kadisha* in the Neolog community of Prešov between the two world wars. My father, Jacob, changed his family name to Ronen following his immigration to Eretz Israel. My uncle, Sandor, a Communist engineer in post-war Czechoslovakia, changed his to Rodak. So, nobody remained to follow the Rosenbergs.

As an epilogue to this historical story, it may be said that the rebirth of Jewish communities in the last two decades has also brought about a renewal of Hashomer Hatzair activity in Budapest and in Warsaw. In Budapest, activity resumed in 1990. There are about eighty active members of the group at present and it has been operating from an Israeli Cultural Center which opened in Budapest a few years ago. The educational orientation is mainly Jewish while most of the youngsters participate in an international Jewish camp during the summer.

---

42 See Karel Kaplan, *Report on the Murder of the General Secretary* (London: I.B. Tauris, 1990); Meir Kotik, *Mishpat Prague: Mishpat Haraava Haanti-tzioni Harishon Begush Hacommunisti* [*The Prague Trial: The First Anti-Zionist Trial in the Communist Bloc*] (Tel Aviv: Milo, 1985).

Activity in Warsaw resumed in December 2014. There are about three to four group leaders and about ten younger participants. A large number of the activists are youngsters who "discovered" their Jewishness only in recent years, after their grandparents or parents had told them about the family background. The content of activities is on a very basic level and deals with the essence of the movement as well as content about Judaism and about Israel.[43]

Bibliography

Amir, Giora. *Kibbutz Ole min Hashoah: Sipur Meyasdei Lehavot Haviva* [*Kibbutz Emerges from the Holocaust: The Story of the Founders of Lehavot Haviva*]. Givat Haviva: Yad Yaari, 2009.

Arad, Yitzhak. *Vilna Hayehudit Bekilion Vebamavak* [*The Struggle and Destruction of the Jews of Vilna*]. Jerusalem, Tel Aviv: Yad Vashem, Tel Aviv University, Sifriat Poalim, 1976.

Bauer, Yehuda. *Flight and Rescue: Brichah*. New York: Random House, 1970.

Bender, Sara. *The Jews of Bialystok during World War II and the Holocaust*. Translated by Yaffa Murciano. Waltham, Mass.: Brandeis University Press; Hanover: University Press of New England, 2008.

Benshalom, Rafi. *We Struggled for Life: The Hungarian Zionist Youth Resistance during the Nazi Era*. Jerusalem: Gefen, 2001.

Bichler, Yehoshua. "Yemey Hachraah" ["Days of Decision"]. In *Shomrim Tamid: Hashomer Hatzair BeCzechoslovakia, 1920–1950* [*Guarding Always: Hashomer Hatzair in Czechoslovakia, 1920–1950*], edited by Yosef Raz. Givat Haviva: Hitahdut Yotzei Hashomer Hatzair BeCzechoslovakia, Merkaz Teud Vehaker, 1994.

Bichler, Yehoshua. "Habrikha MeHungaria Minekudat Mabata shel Mishteret Habitachon HaCzechoslovakit, 1949" ["The Flight from Hungary from the Point of View of the Czechoslovak Security Police, 1949"]. *Yalkut Moreshet* 54 (1994).

Blumenfeld, Moshe. *Du Kium Meahorei Hasorag* [*Coexistence Behind Bars*]. Merhavia, Tel Aviv: Sifriat Poalim, 1971.

Bogner, Nahum. *At the Mercy of Strangers: The Rescue of Jewish Children with Assumed Identities in Poland*. Translated from Hebrew by Ralph Mandel. Jerusalem: Yad Vashem, 2009.

Brown, Avraham Zvi, and Don Lewin. *Toldoteiha shel Mahteret: Hairgun Halohem shel Yehudei Kovna Bemilhemet Haolam Hashniya* [*A History of an Underground: The Fighting Organization of the Jews of Wilno during the Second World War*]. Jerusalem: Yad Vashem, 1962.

Eilat, Alyahu. *Hamavak al hamedinah. Washington 1945–1948* [*The fight for the state, Washington 1945–1948*]. Tel Aviv: Am Oved 1982. Vol. I.

Elyihu Gonda, Moshe, Yosef Cohen, and Yehuda Marton. *Yehudey Hungaria* [*The Jews of Hungary*]. Tel Aviv: Haaguda Leheker Toldot Yehudey Hungaria, 1980.

---

43 Letter from Omer Khakim, the Head of the International Department of Hashomer Hatzair to the author, October 10, 2015.

Grossman, Haika. *The Underground Army: Fighters of the Bialystok Ghetto*. Translated from Hebrew by Shmuel Beeri. New York: Holocaust Library, 1987.

Gur, David. *Brothers for Resistance and Rescue: The Underground Zionist Youth Movement in Hungary during World War II*, edited by Eli Netzer. Translated by Pamela Segev and Avri Fischer. Jerusalem: Gefen, Society for the Research of the History of the Zionist Youth Movement in Hungary, 2007.

Gutman, Israel. *The Jews of Warsaw, 1939–1943: Ghetto Underground, Revolt*. Translated from Hebrew by Ina Friedman. Bloomington: Indiana University Press, 1982.

Izhar, Neomi. *Chasia Bornsteim Bielicka: Achat Mimeatim, Darka shel Lochemet Vemechanechet, 1939–1948* [*Chasia Bornstein Bielicka: One of the Few, The Story of a Fighter and an Educator, 1939–1948*]. Tel Aviv: Moreshet, 2003.

Kaplan, Karel. *Report on the Murder of the General Secretary*. London: I.B. Tauris, 1990.

Kotik, Meir. *Mishpat Prague: Mishpat Haraava Haanti-tzioni Harishon Begush Hacommunisti* [*The Prague Trial: The First Anti-Zionist Trial in the Communist Bloc*]. Tel Aviv: Milo, 1985.

Markovitzky, Yaakov. "Conspiratzia Comunistit o syua Leachim: Gius Habrigada Haczechit, 1948–1949" ["A Communist Conspiracy or Help to Brothers: The Recruitment of the Czechoslovak Brigade, 1948–1949"] *Iunim Betkumat Israel* 6 (2006).

Nir, Akiva. *Shvilim Bemaagal Haesh (Bamered Haslovaki)* [*Trails in the Fire Circle (in the Slovak Revolt)*]. Merhavia: Moreshet, 1967.

Paz, Shal. "Hamahapecha sheboshesha lavo – Mapam besheelot hutz, 1945–1948" ["The revolution that was reluctant to come: The foreign policy of Mapam, 1948–1954"]. In *Lo yuchlu biladeinu* [*Couldn't do without us*], edited by Eli Tszur, 187–189. Daliah: Yad Yaari and Yad Tebenkin, 2000.

Peled, Yael. *Krakov Hayehudit 1939–1946, Amida, Mahteret, Maavak* [*Jewish Krakow 1939–1943, Withstanding, Underground, Struggle*]. Tel Aviv: Ghetto Fighters Museum, Hakibbutz Hameuchad and Masuah, 1993.

Pinkus, Benjamin. *Yahaim Meyuhadim: Brit Hamoatzot vebaalot brita veyahasehen im haam hayehudi, hatzionut vemedint Isarael, 1939 – 1959* [*Special Relations: The USSR and its allies and their relationships with the Jewish nation, Zionism and Israel, 1939–1959*]. Beer Sheva: Machon Ben Gurion Leheker Israel vehazionut, Ben Gurion University, 2007.

Porat, Dina. *The Fall of a Sparrow: The Life and Times of Abba Kovner*. Translated and edited by Elizabeth Yuval. Stanford, California: Stanford University Press, 2010.

Porat, Yosef Ben. *Zman Hatefer: Haodissea shel Ken Hatchia Behungaria Derech Streit Asher BeGarmania leEretz Israel, 1945–1948* [*Stitching Time: The Odyssey of the Tchia Branch in Hungary via Streit in Germany to Eretz Israel*]. Givat Haviva: Yad Yaari, 1993.

Rav, Yosef (ed.). *Shomrim Tamid: Hashomer Hatzair BeCzechoslovakia, 1920–1950* [*Guarding Always: Hashomer Hatzair in Czechoslovakia, 1920–1950*]. Givat Haviva: Hitahdut Yotzei Hashomer Hatzair Beczechoslovakia, Merkaz Teud Veheker, 1995.

Ronen, Avihu. *Nidona Lechaim: Yomaneha Vechayeha shel Chajka Klinger* [*Condemned to Life: The Diaries and Life of Chajka Klinger*]. Tel Aviv: Yediot Sfarim and Haifa University Publications, 2011.

Ronen, Avihu. *Harc az életért: Cionista (Somér) ellenállás Budapesten, 1944* [*Fighting for Life: Zionist (Somér) Resistance in Budapest, 1944*]. Budapest: Elvarosi Konyvkiado, 1998.

Ronen, Avihu. *Hakrav al Hachaim: Hashomer Hatzair Behungaria, 1944* [*The Battle for Life: Hashomer Hatzair's Underground in Hungary*]. Tel Aviv: Yad Yaari, 1994.

Ronen, Avihu. "Yoshka Meir: Agadat Horef Minahal Gaaton" ["Yoshka Meir: A Winter's Tale from the Gaaton Valley"]. *Yalkut Moreshet* 57 (1994): 253–270.

Shalev, Ziva. *Haika*. Tel Aviv: Moreshet, 2005.

Shlomi, Hannah. *Asufat Mehkarim Letoldot Seerit Hapleta Hayehudit Bepolin, 1944–1950* [*Studies on the History of the Jewish Remnant in Poland, 1944–1950*]. Tel Aviv: The Center for Polish Jewry, The Diaspora Research Institute, Tel Aviv University, 2001.

Weiss, Yehuda. "Yeme Pricha LaShomer Hatzair BeHungaria" ["Days of Revival for Hashomer Hatzair in Hungary"]. In *Sefer Hashomer Hatzair* II, edited by Levi Dror. Merhavia: Sifriat Poalim, 1961.

Zuckerman, Yitzhak. *A Surplus of Memory: Chronicle of the Warsaw Ghetto Uprising*. Berkeley: University of California Press, 1993.

Section III: **Jewish Past and Present in the Czech Republic**

Jiří Holý
Jews and Jewishness in Cinema and Literature: The Case of the Czech Republic

I

The aim of this article is to briefly illuminate the Jewish topics presented in Czech cinema and literature after World War II. Depending on political and social developments, the images of Jews and Jewishness went through various phases within Czech art.[1] The changes of the Communist regime and its relation to the Jews played the defining role, possibly emphasizing the theme and the main argumentations of the article.

Immediately after World War II, literature primarily centered on documentary accounts of Nazi camp survivors. So *Továrna na smrt* (1946, *The Death Factory*) by Ota Kraus (1909–2001) and Erich Schön (1911–1995; later Erich Kulka) about Auschwitz was especially famous. It has been translated into the world's main languages. The designation of Auschwitz as "the death factory" became part and parcel of Holocaust terminology, even though Kraus and Schön were evidently not the first ones to use it. They concisely describe the link of racist ideology to the manner of segregation and liquidation of the Jews, emphasizing the impersonal role played by bureaucracy and the prevailing technology at the time.

Alfred Radok's (1914–1976) film *Daleká cesta* (1949, *The Distant Journey*) is in part a documentary. The movie begins with stylized documentary footage, and eventually also with a scene from Leni Riefenstahl's propaganda film *Triumph des Willens* (*Triumph of the Will*), celebrating Hitler and Nazi ideology. The main storyline centers on two Bohemian families: the Czech-Jewish Kaufmann family and the Czech Bureš family. The film demonstrates two parallel worlds – the reality of Jews stricken with Nazi persecution, and the reality of Czech "Aryans."

The film depicts the life of the Kaufmann family and of Toník Bureš, a Czech doctor, who falls in love with Hana and marries her. Hana's parents and younger brother are ordered to Terezín. Toník sets out to clandestinely visit the family, but

---

This article was supported by the programme Progres Q12 Literature and Performativity (In Czech: Literatura a performativita) at Charles University.

1 Cf. Jiří Holý, "The Jews and the Shoah in Czech Literature after World War II."

he does not get to Terezín in time; they have already been deported to the East. Later, Toník, as the husband of a Jewess, is interned in a labor camp. In the end, Hana, too, is interned in Terezín, where she as a doctor helps out during an outbreak of typhus. From her family, only she and Toník live to see the end of the war; in the final scene, they walk through the Terezín cemetery. Later, a similar style of contrast was used in the American television miniseries *The Holocaust* (1999).

Many scenes of this film are set in Terezín/Theresienstadt. The Terezín Ghetto resembles one of Kafka's labyrinths. The film bears the marks of Expressionism. Repeated shots of the gates of Terezín, from which people leave in one direction carrying coffins and masses of new arrivals stream in through the other direction, are probably an allusion to Fritz Lang's famous *Metropolis* (1927) and its shots of enslaved workers. It is paradoxical that this film was stigmatized in Czechoslovakia, while it met with great success abroad. The Communist regime prided itself on *The Distant Journey* at foreign film festivals. *The Distant Journey* was even screened in the USA, where it was compared to Orson Welles's famous *Citizen Kane* (1941).[2]

For ten years, *The Distant Journey* was the only Czechoslovak film dealing with the Holocaust. It significantly stimulated Czech and Slovak cinema in the sixties. Mainly two motifs of *The Distant Journey* appeared in later literary works and movies on this topic. The first one was the love between a Jewish woman and her gentile boyfriend or husband. The second was the complicity or indifference of many "ordinary" Czech/Slovak people in the fate of their Jewish fellow citizens.

After Radok had finished his film, Stalinist censorship was implemented in Czechoslovakia. *The Distant Journey* was banned from audiences and only shown abroad or in marginal cinemas. A similar case was with Jiří Weil's (1900–1959) novel *Život s hvězdou* (1949, *Life with a Star*). Like *The Distant Journey*, it not only presented the horrible brutality of the Shoah, but also its seemingly banal, even profane side. This novel is considered the most important work on this theme in Czech literature, and, along with *The Distant Journey*, inspired a multitude of other works.

Jiří Weil gained inspiration from modernist theories which shunned both traditional psychological analysis and structured plot. Instead, he re-established prose on facts and factual reality. The bearer of meaning is not the dramatic story of the hero or any authorial commentary, but much more often real-life sit-

---

2 See Šárka Sladovníková, *The Holocaust in Czechoslovak and Czech Feature Films*.

uations; a record of daily life which is described in a "dry," seemingly disinterested manner.

The protagonist Josef Roubíček is part of a mundane history, not a heroic history, as sought by Communist critics. Thus, this book does not resemble the majority of those novels about the Holocaust. However, Jiří Weil's art lay in purposefully choosing unheroic protagonists who went about living their daily lives, avoiding all that was important from an ideological point of view. Roubíček lives alone during the war in a dilapidated hut on the edge of Prague. Like the rest of the Jews, he goes to work and must regularly check in at the various bureaus. He dreams about a cup of coffee, which he cannot hope to ask for, and recalls how he went riding in the mountains with his girlfriend Růžena, to the cinema and to the cafés.

Life with a Star was in many respects founded on the author's personal experiences. It is a work primarily dealing with a person who, through no fault of his own, is subjected to humiliation, irrelevance, and anonymity – such as, for example, when he must sew on the yellow star with the lettering *Jude*.

The protagonist speaks with his friend Pavel about the loss of his human identity as they discuss the upcoming transport:

> [...] 'There's no other way but to become a number.'
> 'What do you mean?'
> 'A number – hanging around your neck, attached to your suitcase, glued to your rucksack. Then I'll load myself with fifty kilos and go. [...]' [3]

Jewish people undergo a gradual transformation into numbers, losing their humanity. It is a disastrous consequence of inhuman conditions in which they must live. This motif appears later often in Czech literature.

Throughout the entire narration, Roubíček almost never meets with either Germans or Nazis. The people which organized records of the Jews, confiscated money and property, prepared transports, etc. are office workers from the Jewish community.

The novel captures human degradation on a common level, yet made more conspicuous by Nazi despotism. This degeneracy is all the more convincing precisely because it was "normal" and rationalized. Its absurdity also lay in the fact that its victims accepted this fate. One of the most important scenes in the novel

---

[3] Jiří Weil, *Life with a Star*, trans. Rita Klímová and Roslyn Schloss (Evanston: Northwestern University Press, 1998), 99. Original: "[...] 'Nezbude mi nic jiného než být číslem.' 'Jak to?' 'Ano, číslem zavěšeným na krku, upevněným na kufru, nalepeným na batohu. Pak na sebe naložím padesát kilogramů a půjdu. [...].'" Jiří Weil, *Život s hvězdou*, 112.

describes an Aryan wife who forces her Jewish husband, Mr. Robitschek, to commit suicide, thereby "making the situation easier" for her and their daughter. He chooses to accept her argument.

> My wife divorced me and joined them, but she still finds it a nuisance that I'm alive. They blame her for it. She sent me a message that it would be best if I committed suicide. I wouldn't mind that, but my daughter wants it too. They wrote me that I'm a selfish bastard, that I never think of the family. [...] She was always a practical woman, and if now she thinks I should commit suicide, she must have good reasons. She wouldn't want it if it weren't necessary [...].[4]

The world depicted in this scene looks completely desperate and absurd. To think of one's family means to die, and to live means to be "a selfish bastard." The similarity of both names indicates that Robitschek is Roubíček's alter ego, his *doppelgänger*. Kohn and Roubíček are two favorite names in Czech Jewish anecdotes. In *Life with a Star*, Roubíček as well as Robitschek are tragicomic characters.

Later the narrator meets Robitschek again. Robitschek is in a Jewish hospital where he was transported after his suicide attempt.

> I had learned that his suicide attempt had been unsuccessful. He would have probably succeeded if he had had enough gas to inhale, but gas was scarce and had to be rationed. [...] 'I had bad luck', he [Robitschek] said. 'They saved me. I'm damn unlucky. I shouldn't have tried gas. I should have known what would happen with those economic measures. [...] But that isn't my only bad luck. The worst of it is that now I can't kill myself. They watch me here. They're for me, for putting me back in order for the transport. [...] My wife was here yesterday. [...] She was furious. She said I did it on purpose, that I didn't really want to kill myself. She complained that I'd made things very difficult for her, because she had gone to a lot of trouble to get a black dress and she'd had her best pair of stockings dyed black.'[5]

---

4 *Life with a Star*, 128 f.
5 Idem, *Life with a Star*, op. cit., 168 f. Original: "Dověděl jsem se, že se mu sebevražda nepovedla, bylo by se mu snad podařilo zemřít, kdyby se byl nadýchal dosti plynu, šetřili však plynem v oné době a přísně jej odměřovali. [...]
'Mám smůlu,' řekl, 'dostali mě z toho, mám prokletou smůlu. Neměl jsem to zkoušet s plynem, mohl jsem přece vědět, jak to dopadne, když s ním tak šetří. [...] Ale to není má jediná smůla. Horší je, že se teď nemohu zabít. Hlídají mě tady, ručí za mne, že mě odevzdají v pořádku do transportu. [...] Včera tady byla žena. [...] Hrozně zuřila. Říkala mi, že jsem to udělal schválně, že jsem se vůbec nechtěl zabít. Bědovala, že jsem ji přivedl do velkých nesnází, protože si opatřila s velkou námahou černé šaty a dala si přebarvit na černo své nejlepší punčochy.'" Idem, *Život s hvězdou*, op. cit., 151 f.

The absurdity of the world in which Robitschek lives leads to these situations. He calls survival after his suicide attempt "bad luck." He is kept under guard so he does not try to kill himself again, but instead to remain ready to join the transport. Robitschek does not have free choice to dispose of his life. His wife complains she had to buy a black dress, which apparently has much more value for her than the life of her former husband. And Robitschek accepts it as a legitimate statement. The description of these specific details in Weil's works are symptomatic of his writing.

Weil's novel is neither a description of the brutality of the Holocaust nor a satirical image of the Nazis. It is a Kafkaesque image of a man in a desperate situation, subjected to humiliation and anonymity. Accordingly, the novel uses literary procedures like self-deprecating, grotesque, as well as black humor.

II

The Stalinist regime, implemented in Czechoslovakia after the takeover in February 1948 and climaxed at the beginning of the 1950s, was marked by flagrant antisemitism, just as the other Communist regimes of that period. Publications on Jewish topics including the past of Czech Jews were quite exceptional. Descriptions of the Nazi regime and their concentration camps emphasized the Communists' heroic fight in all the Eastern-bloc countries, while the systematic extermination of the Jews was only mentioned in passing or not at all. So in the novel of the Czech prosaist, Norbert Frýd's (1913–1976) *Krabice živých* (1956, *A Box of Lives*), the Judaism of the main figures is suppressed or marginalized.

The filmed version of the well-known novella by Jan Otčenášek (1924–1979), *Romeo, Julie a tma* (1958, *Romeo, Juliet and the Darkness*), describes the tragic story of the love of one Czech boy and a Jewish girl during the German occupation in Prague. It was filmed by director Jiří Weiss, whose entire family was murdered in Auschwitz. In 1959, however, the film was censured by Václav Kopecký who was the Deputy Prime Minister and a party functionary at the time. The final scene was considered suggestive: the Czech inhabitants of the home where the young Jewish girl was hiding sent her directly into the hands of the Nazis, out of fear for their own lives.

The themes of Jewishness and the Shoah in Czech literature were increasingly developed only at the end of the 1950s, when the Stalinist system was being dismantled, and later as Czech culture underwent great liberalization – especially in the 1960s. The first prosaist to enter the scene in that period was Arnošt Lustig (1926–2011), who survived both Theresienstadt and Auschwitz. Thus, for him, the Shoah became the theme of his life's work. His first works are among his best: *Noc a naděje* (1958, *Night and Hope*) and *Démanty noci* (1958, *Diamonds in the Night*). These stories inspired two of the most exceptional films in the Czech "New Wave" movement of the 1960s – Zbyněk Brynych's *Transport z ráje* (1962, *Transport from Paradise*), based on the Theresienstadt stories from the first book, and Jan Němec's *Démanty noci* (1964, the filmed version of the story *Tma nemá stín* [*Darkness Casts No Shadow*]). Similarly to Jiří Weil, the author makes use of an intimate, inner perspective (even though he narrates in the third person) and works quite expressively with the subjective observation of time. His protagonists are not traditionally heroic figures but rather outsiders, children or old people. Lustig also distinguishes himself by his non-conventional picture of the war and the concentration camps usually depicted in terms of active resistance against the Nazis. Despite all the bleakness which these people must repeatedly undergo, donning an outer shell just to survive, the majority of them try to maintain basic moral values. For example, the story *Sousto* (*Morsel*) describes a boy who extracts his deceased father's golden teeth in order to exchange them for a lemon needed urgently by his ill sister.

Lustig's stories interweave several autobiographical experiences, such as his escape from the death transport in the fictionalized story *Darkness Casts No Shadow*. The author portrays exceptionally pivotal situations where life-and-death decisions must be made. His story *Druhé kolo* (*The Second Round*), for example, follows a boy who has three minutes to run up to a wagon, steal a loaf of bread, and return to his starving friends before the patrol makes its round once again. If he does not make it in time, the patrol will shoot him. The narrator details his inner consciousness during this situation for more than ten pages.

III

For some Czech authors, the Shoah becomes a metaphor for man caught in the machinery of the totalitarian regime and for the functioning of evil at large. Such is the case with Ladislav Fuks' (1923–1994) novel *Pan Theodor Mundstock* (1963, *Mr. Theodore Mundstock*), his collection of stories *Mí černovlasí bratři* (1964, *My Black-haired Brothers*), and his other works. Fuks belonged to a group of writers who had neither Jewish roots nor any personal experience from the Shoah, though they often wrote about it in the 1960s.

Based on elegant repetition and variation, Fuks' prose smoothly fuses realistic tableaux together with fantastic ones. The novel allows us to observe the main character in detail, however he himself lives at least half the time in his own head. In the beginning, for example, he converses with his imaginary *doppelgänger*, Mon, about items which the reader discovers in the second half of the book are only figments of his imagination. Otherwise, Mundstock is an unobtrusive character, previously employed as an office clerk before the war, and now lives in complete loneliness in an empty Prague flat waiting to be summoned for transport – similar to Weil's Roubíček. He tries to think up something which can save him. He comes up with an apparently faultless method of how to prepare himself for the concentration camp by training himself for the future hardships to be endured, including sleeping on uncomfortable wooden planks, being beaten, going hungry, etc. Based on his methodology, he dies in a tragicomic manner at the very moment they come to take him away for the transport: he must switch his suitcase from one hand to the other so that he will not become overtired. Yet as he does so while crossing the street, he does not notice the German car which runs him over.

Oftentimes in Fuks' prose, words and motifs which carry visual significance are repeated. Likewise, in the closing scene of the protagonist's death, the internal perspective which was dominant so far changes into an external one within the final few paragraphs:

> He heard a horrible noise. Glancing round he saw an enormous military truck bearing down on him. Everything went dark, some vast force tore his case from his hand, and he realized he had fallen into some dreadful trap... My God, what has happened? He heard the words shriek in his head; what were we doing, just practicing, we couldn't prepare ourselves for everything, it was all some terrible mistake I made [...].
> When the truck moved and they turned Mr. Mundstock over onto his back, the policeman, although he was no doctor, saw that this man who had so suddenly stopped still in the middle of the road would never get to the other side.
> It was a thin man with a graying face and motionless eyes, eyes turned beseechingly somewhere towards Heaven. The yellow Jewish star on his dark blue coat was covered in dust, but strange to say there was not a speck of blood on it.[6]

Symbols of a "star" as well as "dust" appear here that could be found often in the novel. The book was translated in many languages and became famous.

---

6 Ladislav Fuks, *Mr. Theodore Mundstock*, 213 f. Original: "Vtom slyší strašný hluk. Zahlédne, že se na něho řítí obrovské vojenské auto. Před očima se mu zatmí, jakási přeukrutná síla mu vyhodí kufr z ruky a tu pozná, že se dostal do nějaké strašlivé pasti... Bože, co se to stalo, vykřikne v jeho hlavě, co jsme to dělali, že jsme jen nacvičovali, vždyť jsme se snad opravdu nemohli na všechno připravit, vždyť to všechno asi byl nějaký můj omyl [...]. Když auto popojede a pana Mundstocka obrátí na záda, zjistí strážník, ačkoli vůbec není lékařem, že tenhle člověk, který se tak náhle zarazil uprostřed jízdní dráhy, už nepřejde. Je to jakýsi pohublý muž šedivých tváří a nehybných očí, obrácených kamsi prosebně k nebi. Žlutá židovská hvězda, kterou má na tmavomodrém kabátě, je celá od prachu zaprášená, ale kupodivu, není na ní ani stopy krve." Ladislav Fuks, *Pan Theodor Mundstock*, 180.

The story attracted the interest of such celebrated film directors as Charles Chaplin and Roman Polanski. Later, the novel was filmed in Poland by Waldemar Dziki under the title *Kartka z podroży* (1985, A Postcard from the Journey).[7]

Probably the most famous work of Ladislav Fuks was the novella *Spalovač mrtvol* (1967, *The Cremator*).[8] The main character of *The Cremator*, Karel Kopfrkingl, a crematorium employee, is even stranger than Fuks' previous characters. Kopfrkingl is also disturbed. Like Mundstock he believes that phantoms and mirages exist. In Mundstock's internal world his imagined *doppelgänger*, Mon, becomes a reality. In a similar way, Kopfrkingl believes that he has been visited by a messenger from Tibet and has been selected to be the new Dalai Lama. Unlike Mundstock, an outsider and an introvert, Kopfrkingl turns into an aggressor and a murder. The plot of the novella is concerned with his rise to power, in the crematorium and later by planning the extermination of Jews.

Kopfrkingl is therefore the opposite of Mundstock in terms of the hero's relation to the fictional world. *Mr. Theodore Mundstock* features aggression assailing the character unexpectedly from the outside versus *The Cremator* where aggression arrives just as unexpectedly, yet stems from the protagonist himself.

The system of names of the other figures is also unusual. The employees in the crematorium are: Vrána (Crow), Fenek (Fennec), Pelikán (Pelican), Lišková (Miss Fox), director Srnec (Roe Deer), Beran (Ram), Zajíc (Hare), Špaček (Starling), the dead persons Vlk (Wolf), Sýkorová (Chickadee), Daněk (Fallow Deer), and Piskoř (Mudfish). These names may also suggest the inhuman and bizarre nature of the entire fictional space.

The first scene in the novella, just as in the film, is situated in the Predator's House in the zoo. We can see in detail a leopard and a large a boa constrictor, both of which are in a cage. The boa constrictor is "hroznýš", literally meaning terrible animal, in Czech. Immediately afterwards the word "pouzdro" (casket) appears. The main character Kopfrkingl says: "... if one says the boa constrictor, it's quite clear. Everybody knows beforehand what to expect from a boa. [...] But a silver casket is a mystery. Nobody knows until the last moment what such a casket might contain until it's completely opened and examined."[9] This scene and these sentences anticipate Kopfrkingl's hidden perfidy and brutality.

---

7 Jan Poláček, *Příběh Spalovače mrtvol*.
8 Based on Juraj Herz's script (on which Fuks himself cooperated), an excellent film version was made starring Rudolf Hrušínský and Vlasta Chramostová in 1968. The film was premiered in the spring of 1969, but was quickly banned by the new neo-Stalinist regime which was implemented after the Soviet invasion of Czechoslovakia.
9 Idem, *The Cremator*, 10.

As is characteristic for the author, the reader finds factual reality mixed with the fantastic, and various visions within the fabric of the text. *Spalovač mrtvol* indeed verges on a horror story in terms of genre, while the narration of the story is unreliable. Thus, the fictional world is depicted in purely personal terms via the unstable perspective of the reality assumed by the protagonist.

However bizarre, psychologically disturbed, and insane, Kopfrkingl is not the embodiment of the traditional villain. In essence, he has a *petite bourgeois* mentality, works carefully, loves music (opera melodies waft from the crematorium), cares for his family, does not drink or smoke, and enjoys speaking in a flowery manner. Rudolf Hrušínský plays the role of Kopfrkingl with a smiling face and a seductive, monotonous sound tone of his voice. He calls his family "my angels," his profession "noble," his home "beautiful" and "blessed." Thus Kopfrkingl is "decent" as a person, yet is lacking personality and individuality, and thus is a complete conformist.

Characteristically, Kopfrkingl accepts the thoughts and empty phrases which he has previously heard. His hardness and aggression are lurking behind all these phrases. He often talks about his "blessed marriage," mentioning "I have intercourse only with my beloved wife," but in truth he visits the local brothel. He likes to manipulate his family and other people. He respects the law,[10] saying "peace, justice and happiness should reign" and "suffering is an evil we must be rid of, or at least alleviate." But he abuses all these words and becomes an informer and murderer: "We live in a great, revolutionary time, and we still have a lot of worries. We all lose. [...] All Jews in the German Reich are [...] excluded [...] it's the law, and as you know, we have to respect the law."[11] This rhetorical strategy is very similar to that of the Nazis, but also to Communist perpetrators. Kopfrkingl kills his wife, because she is a half Jew and would be an obstacle for his career. He kills his son and, in the film, also tries to kill his daughter.

In the film version, Kopfrkingl's perversity is evident in the dialogue with the Nazi boss (chief of the secret police SD in Prague). He has to prepare "the gas furnace of the future – equipment for incinerating as many people as possible." Kopfrkingl says, "If we had huge furnaces, to hold a hundred, five hundred, a thousand, it could be done in minutes. [...] We could quickly liberate all humanity, the whole world. [...] In such a huge hall, constantly in operation, once you entered you'd never come out alive." He bubbles over with the happi-

---

10 *The Cremator*, 8, 38.
11 Ibid., 173.

ness and loses his self-control. In his imagination, bodies of "the condemned" fall down to the abyss. It is reminiscent of the paintings of Hieronymus Bosch.

Fuks' novella skillfully and monstrously paints the fun-house atmosphere as solemnity, and stark ornamentation devolves into horror. The entire text is structured as a web of allusions, ciphers, and anticipations. The second chapter covers the Kopfrkingl family as they visit a carnival, including a fun-house with frightening scenes from the great plague of 1680 in Prague. The dying and killing which are demonstrated here (i.e., hanging, death by battering with a rod) foreshadow Kopfrkingl's own actions in chapters 13 and 14, when he hangs his wife in the bathroom and batters his son, Mili, with an iron rod in the crematorium. We can find an iron rod several times in the novella, as in the film, in advance of the murder. Right before the visit to the fun-house, the reader learns about Mr. Strauss who lost his wife. She died of "consumption of the throat" ("krční souchotě") and Strauss' son died from "scarlet fever"[12] ("na spálu").[13] Once again, this foreshadows both murders, since Kopfrkingl throws a noose around his wife's neck and burns the body of his son in the crematorium.

Fuks' novel, as well as Fuks, Herz's and Milota's film, are both very artfully constructed. They are not centered on testimony and documentary accounts, but on a picture of the human degradation on a "normal level" that is made more conspicuous by Nazi despotism. Kopfrkingl's effortless adaptation for the Nazi regime opens the problem of the responsibility of the "everyday Czech people"[14] for the persecution of Jews and for totalitarian regimes.

Ladislav Fuks understands the fate of the Jews as that of helpless people who have succumbed to both a fanatical, systematic hate as well as to their own personal fears. He purposefully describes the horror of a world which has lost its humaneness and has instead become a place of threat and persecution. Fuks' prose depicts the existential anguish of an individual surrounded by an incomprehensible, cold-blooded mechanism.

It is interesting that Arnošt Lustig used a similar approach in the mid-1960s. Previously, he opted to write using real-life situations as the basis. His most famous book remains the novel *Modlitba pro Kateřinu Horovitzovou* (1964, *A Prayer for Katerina Horovitzova*), which was made into a television movie by Antonín Moskalyk in 1965. The story was inspired by actual events which took place in Auschwitz in 1943 – the murder of a group of rich Jews whom the Nazis had promised safe passage across the border for a high price.[15] However, the prose

---

12 *The Cremator*, 15.
13 Idem, *Spalovač mrtvol*, 13.
14 Emphasis mine (JH).
15 Thomas Ammann and Stefan Aust (eds.), *Hitlers Menschenhändler*.

is structured in an extremely complicated way, foregrounding the intelligent Nazi Arthur Brenske who acts like the devilish Mephisto. He does not dominate his victims with brute force, but rather uses sophisticated double-edged talk and promises. For example, he talks about "the final solution" and "the gloom which will decide everything." The rich Jews gradually prepare to hand over their finances that were saved in American banks, hoping that this will help them survive:

> 'The final solution is at hand. You'll see for yourselves. Your worries will all go up in smoke and burn away like brush fire. From my own personal experience, I can testify that people often don't believe things which concern them in the most fundamental way until they feel them on their own skin. [...] We want to liquidate this exchange operation in the best possible way. You must listen to what I'm telling you. Until we've reached a destination which will be satisfactory to everybody and about which you'll have no fault to find, I'm doing my level best to make this trip as pleasant as possible for each and every one of you.'[16]

This manipulation of the people via language and the abuse of power is one of the most important themes in Czech literature from Karel Čapek and Karel Poláček to Václav Havel. The perspective of one protagonist (Mundstock, Kopfrkingl) dominates in Fuks' narrative. The perspective of narration in *A Prayer for Katerina Horovitzova* is also constructed so that the reader knows more than the characters in the story. In Lustig's novel, the rich Jews go into the gas chamber like sheep. One woman among them rebels: Kateřina Horovitzova grabs the SS-man Schillinger's gun and shoots him. Thus, Kateřina is likened to the Biblical character Judith.

IV

There are several Holocaust prose pieces which take place in Slovakia, but are written in the Czech language. After Czechoslovakia was dismantled, the Slovak Republic existed from 1939–1945 (excluding what would today be southern and

---

16 Arnošt Lustig, *Modlitba pro Kateřinu Horovitzovou. A Prayer for Katerina Horovitzova*, 218. Original: "'Konečné řešení už je na dosah ruky. Sami se přesvědčíte. Vaše starosti shoří jako seno. Ze zkušenosti mohu osobně dosvědčit, že lidé často nevěří některým věcem, které se jich nejbytostněji týkají, dokud se o nich nepřesvědčí právě na vlastní kůži. [...] Chceme odepsat navzájem likvidace této výměny z našich položek co nejspolehlivěji. Musíte poslouchat. Než budeme na místě, které nás všechny uspokojí a proti němuž, doufám, nebudete mít námitek, chci vynaložit nejlepší ze svých schopností, abych tuto cestu pro každého z vás učinil snesitelnou.'" Arnošt Lustig, *Modlitba pro Kateřinu Horovitzovou*, 98.

eastern Slovakia, which fell to Hungary). It was officially an independent country headed by the Catholic priest Jozef Tiso, however in reality it was only a satellite of Hitler's Germany. The Jews in Slovakia were also segregated, sent to holding camps and finally to the death camps in Poland.

The most well-known work on this topic is Ladislav Grosman's (1921–1981) *Obchod na korze* (1965, *The Shop on Main Street*). Grosman himself came from a Slovak Jewish family, and went through various labor camps during the war. He eventually escaped being transported to the extermination camp and thus had to remain in hiding illegally for the rest of the war. After the war, he studied in Prague and worked as an editor and screenwriter, publishing in both Slovak and Czech. His prose was made famous by the film directed by Ján Kadár and Elmar Klos, eventually winning an American Oscar in 1966 for Best Foreign Film. The screenplay and film were preceded by Grosman's story, *Past* (1962, *The Trap*), which was written in Slovak and published in Czech in the Prague journal *Plamen*. Arnošt Lustig referred Kadár and Klos to this work. The novel *The Shop on Main Street* was then created immediately after the screenplay (which Grosman cooperated on) and simultaneously with the film. Grosman's literary screenplay was published for the first time in the revue *Divadlo* in April 1964. It was edited for the second time posthumously in a rather different version in 1998. The novel was published for the first time as a serial story in the Prague magazine *Mladý svět* from August to December 1964. It was edited in the publishing house *Mladá fronta* in September 1965. The film premiered on October 8, 1965. Therefore, there are six different versions of this work.[17]

The temporal setting is the summer of 1942, when the first wave of Jewish transports from Slovakia was organized. The main figure, as very often occurs in Czech(oslovak) literature, is a small man who is not overly interested in politics or in public events – the rather naïve Tono Brtko, a joiner and carpenter. Brtko instinctively resents the Slovak Fascists and the war and keeps himself away from this ruling group. "Am I a parrot, to raise my arm and call to guard" he asks. But he is neither active nor brave enough to fight against the regime.

However, his brother-in-law is the leader of the local Hlinka Guard (the Slovak Fascist organization) and Brtko involuntarily becomes the so-called aryanizer. He acquires a small, worthless, and insolvent haberdashery store belonging to the old, almost deaf Jewish widow Rozália Lautmanová. Because he is a good person at heart and does not like conflicts, he helps the old Jewess to serve cus-

---

17 Jiří Holý, "The Six Versions of The Shop on Main Street."

tomers and to repair her furniture. He pretends to be her shop assistant, while at home he makes out to be a strict aryanizer.

Rozálie Lautmanová hopes the good old days of local community spirit when no problems existed between Jews and Slovaks will still continue. She knows nothing about the aryanization of Jewish property, not recognizing that Brtko is becoming the owner of her small shop. Her naivete is emphasized by the white color of her hair and her nightgown in the film. She can't understand that the times have changed and the shop is no longer hers.

In *The Shop on Main Street*, for a long time the story seems to present the idyllic life of a small town. But this lovely idyllic and relaxed mood comes to an end in the last chapter of the novel. This change is symbolized by a monstrous monolith of "Victory," for Brtko a "Babylonian tower" erected in the middle of the main street by the new power. The Jews from the city and neighboring regions receive summons and are lined up for transport. The moment that Brtko learns about the transport, the melody of Kaddish, a Jewish prayer known as a part of the mourning rituals, begins to play in the film for the first time. Brtko's friend Kucharský/Kuchár, who is involved in the resistance, is labelled as *White Jew* (someone who helps and protects Jews). He is brutally beaten, arrested, and deported. In the film, Kuchár is dumped into the street with the sign "White Jew." He is then tied up and left in the square for public ridicule.

Moreover, Brtko's wife Evelína is not satisfied with the money that Tono Brtko brings in. She knows the Jews should be transported and she wants more Jewish jewelry and gold.

> 'Where's her gold? Diamonds? What have you done with it? Where is the Jewess's jewellery? Go on, tell me!'
> 'Shut your mouth!'
> 'They're all going to be transported tomorrow, anyway.'
> [...] He seized her and shook her wildly. [...] He knocked her down onto the bed and severely beat her.[18]

After this scene, the desperate Brtko leaves for the pub. From now on, his dual existence cannot continue. In the pub he is drinking wine with old "Uncle Piti," the town crier, and they decide to save Mrs. Lautmanová and shelter her. However, Čarný, a fascist, starts shadowing Brtko, who finds himself under increasing pressure, paralyzed with alcohol and overcome by a terrible fear. The next

---

[18] Idem, *The Shop on Main Street*, op. cit., 104. Original: "'Ona má zlato! Drahokamy! Kde jsou? Kde jsou židovčiny šperky? Mluv!' 'Drž hubu!' 'Vždyť je zítra odvezou!' [...] Prudce ji odstrčil. [...] Svalil ji na postel a tloukl hlava-nehlava." Ladislav Grosman, *Obchod na korze*, 117 f.

morning Kolkotský manages the gathering of the Jews in the square close to Brtko's shop. The idyllic square becomes an *appelplatz*. Rozálie Lautmanová was forgotten due to a bureaucratic error (in the film, she does not wear a Jewish star, unlike other Jews in the town). Nevertheless, Brtko assumes that this is actually just a clever move on the part of his hated brother-in-law who will then also designate him a Jew lover and exponent and thus permanently get rid of him.

Thus, he convinces the old lady to go out to the square and join the deportees in transport. He argues that is the law: "Mrs. Lautman! The world is run this way now… there are special laws for Jews."[19] But then he regrets this and tries to save her. At first, Rozálie Lautmanová does not understand anything. In the film she even believes the police will protect her from Brtko. But then she realizes what is happening and is scared. Tono Brtko tries to hide her in a backroom by force and accidentally kills her while shoving her into her hiding place. He goes completely crazy and commits suicide by hanging himself.

The novel depicts the drama of a person who is roped into a dilemma, an oppressive, irresolvable situation through no fault of his own. A part of the text depicts not only dramatic and grotesque, but also imaginative scenes. On the last night before their death, Tono Brtko dreams that he and Rozálie, both youthful, happy, and carefree, are walking on the main street promenade. In this dream sequence, the idyllic mood returns, at least in Tono Brtko's mind. In the novel, this scene precedes the death of Lautmanová and the suicide of Brtko. At the end, it is only briefly noted, "so it's quite possible they are both up there now, promenading along the Main Street of heaven."[20]

In the film, this vision repeats twice. The second time it is a long scene at the very end of the film. Brtko finds out that he killed the old lady. He is completely confused. The camera follows his roving look. We can see a short shot of Rozálie's picture in Sunday clothes. After Brtko takes his own life, the door of the shop opens. Rosálie and Tono appear, both gracefully dressed in white – Rosálie in her Sunday clothes from the picture, and Tono in her husband's clothes. They smile at each other. They slowly walk down Main Street, and almost float and dance across the square. The fire brigade brass band is playing a waltz, the bandleader bows. It suggests relaxed, idyllic moments, like in heaven.

At this moment, Brtko addresses Mrs. Lautmanová using the informal form "ty" even though he was on formal terms with her in his real life. Therefore, these two characters are so close to each other and happy – but only in his

---

19 Ibid., 117.
20 Ibid., 122.

dream. By contrast, it is the harsh and absurd reality around them that separates them and leads them both to their deaths.

V

After the end of the 1960s, the Shoah did not play such a key role in Czech literature. The works which have been published continue in two main directions: genuine authentication and figurative stylization.

Arnošt Lustig's creations continue with these themes, with very few exceptions. Like many others, Lustig left Czechoslovakia after the invasion of the Warsaw Pact troops in August 1968 and lived in the United States from 1970. In the last years of his life, he shuttled between the U.S. and Prague. His later books, however, accentuate the more abrasive side of life in the camps (e. g., homosexual prostitution, lack of unity among the prisoners, etc.). He often recorded the stories of young Jewish girls and women, such as in *Dita Saxová* and *A Prayer for Katarina Horovitzova*. Their beauty and youth form a moving contrast to the horrors of the Shoah. Lustig oftentimes reworked his older prose, thus producing new, extended versions which are occasionally renamed. This may have been the author's attempt to reach out to the American public. Most critics, however, agree that his original, more laconic prose versions are stronger.

Viktor Fischl (1912–2006) was another author who also frequently addressed Jewish themes. He was a Zionist who came from a Czech Jewish family. Although the Czech lands were already occupied, he successfully immigrated to Britain in 1939 where he worked as a clerk for the Czechoslovak emigration authority. After the Communists came to power, he left for Israel, took up the name Avigdor Dagan, and worked in the diplomatic corps. Fischl began his literary career before the war, though most of his prose was written much later in his life after retirement. In addition to his literary endeavors, he was also the chief editor of the exhaustive, three-volume publication *The Jews in Czechoslovakia: Historical Studies and Surveys* (1968, 1971, 1984). The Shoah appears in the background of several books by Fischl, but it takes center stage in his novel *Dvorní šašci* (*The Court Jesters*; originally written in 1982, first in Hebrew and eventually published in Czech in 1990). The "court jesters" were in fact four Jewish prisoners in an extermination camp where the Nazi camp commander spared their lives for the sake of his nightly entertainment.

Czech literature has also concentrated on the Shoah after 1989. Arnošt Goldflam's (b. 1946) play *Sladký Theresienstadt* (*Sweet Theresienstadt*, premiere in 1996, in book format 2001) serves as an example. Harkening to his Jewish roots, he originally recorded a famous episode from the Theresienstadt ghetto in the propaganda film *Theresienstadt*, better known as "Vůdce daroval Židům město" (*The Führer Gave the Town to the Jews*). The film starred the famous German actor and Theresienstadt prisoner Kurt Gerron (named Gerroldt in Goldflam's play). In this "documentary," people state how peaceful life is in Theresienstadt, how they play football, visit various cafés, libraries, etc. After the filming, the director and the actors in the "documentary" were sent on the transport to Auschwitz to die. Within his play, Goldflam used comic, tragic, and grotesque elements as well.

Jewish topics can also be found in other Goldflam plays and short stories. In one of his later plays, *Z Hitlerovy kuchyně* [*From Hitler's Kitchen*, 2007], six ministories linked by the character Adolf Hitler add up to a slightly unorthodox perspective of Hitler. Goldflam's grotesque reconstructions of Hitler's life remove any demonic qualities and present him as a completely private, bookish, and slightly bizarre person. In the first scene, Hitler and Stalin meet (by coincidence sometime before World War I) at a train station in Brno. They hope the trains will take them away to meet their dreams. At the same station, a little Jewish boy from Hungary named Georg Tabori is lost. Tabori was a Hungarian-Jewish writer and dramatist who survived World War II in exile and was famous for his provocative plays about the Shoah.

Goldflam's interest in the life of the Jewish community in Czechoslovakia was expressed in documentary films shot for Czech TV in the mid-nineties: *Ztracený domov* (*Home Lost*) and *Domov nalezený* (*Home Found*). These two documentaries contain Goldflam's interviews with Czech, Slovak, and German Jews who emigrated from Czechoslovakia to Israel. Older interviewees recalled their lives before World War II and the Nazi persecution. They also described Czech and especially Slovak antisemitism. Among them are celebrities such as the writers, journalists, and researchers Viktor Fischl, Erich Kulka, Ruth Bondy, and Joab H. Rektor.

Perhaps the most remarkable attempt at a new approach to the Shoah is the "Auschwitz" chapter of Jáchym Topol's (b. 1962) novel *Sestra* (1994, *The Sister*, in English as *City Sister Silver*), entitled "I Had a Dream." This text, written as a frightening vision, fusing horror, vulgarity, the grotesque, and banality, may thus be understood as a call against conventional Holocaust iconography and it blasphemes these images. The narrator describes a dream he had wherein he found himself, together with his friends, on a type of flying carpet which lands in a sea of ash and bones in Auschwitz:

> It was the ashes of cremated people, my brothers, the ashes of cremated Jews. Any last hope we had that maybe there'd been a mix-up, and at least we were in some slightly cosmopolitan wicked old gulag, was lost. And the ashes stirred up by our landing stuck to our shoes

and clothes, and made it hard for us to walk. And where there weren't ashes, brothers, there were bones, human bones, an endless ghastly sea of bones. Then we saw towers in the distance and so we started walking... using one of the taller towers as our point of orientation... and we were afraid because the skulls were watching us, looking at us, and we asked ourselves: Why are we here? Why us? Why did it happen for me? And some of the skulls seemed to answer: Why not? Some of them lay there softly, jaws set in a knowing smile, but more, far more, just peered out blankly at us, what was left of the jaws twisted into a grimace of pain, because they'd got these the hard way, brothers, and heavy-duty, alive. There was a sea of them, an ocean. And this comparison occurred to us when we couldn't walk anymore because we kept plunging into the bones and so we tried to swim our way through, we tried to move and crawl and shove our way through with our arms. [...] And we inched along toward the towers, trying not to catch the skulls' empty glances so we wouldn't go insane. There were children's skulls, my brothers, and there were piles of skulls smashed to bits, and there were skulls shot full of holes, and skulls that looked like they'd been crushed in a press, and skulls with small holes mended shut with barbwire, and one of us, O knights and skippers, cracked another joke: Guess that's what you'd call his-and-hers skulls, ho! Ho! But then he started to vomit. And the one creeping in front of him didn't hear him because he was weeping, and the one crawling behind him didn't hear him because he was praying out loud. And, friends and brothers of mine, it wasn't hell we were going through but whatever it is that comes after it.[21]

The figures meet with the live skeleton of the Czech Josef Novák, who did not end up in Auschwitz for being a Jew or for having taken part in the resistance, but

---

21 Jáchym Topol, *City Sister Silver*, 101–102. Original: "'Byl to popel ze spálenejch lidí, bratři moji, ze židů. Poslední naděje, že snad došlo k mýlce a že sme aspoň, dyž už, tak v ňákým lehce kosmopolitním starým zlým gulagu, vzala za svý. A ten popel, kterej sme svým dopadem zvířili, se nám začal lepit na boty a šaty a těžce se nám šlo. A tam, kde nebyl popel, byly kostry, lidský kostry, moře, nekonečný příšerný moře kostí, bratři. Šli jsme všichni, celá partička, k nějakými městu, jehož věže sme zahlídli v dálce... jedna byla vysoká a sloužila nám jako orientační bod... a báli sme se, protože ty lebky naší cestu sledovaly, dívaly se na nás a my byli plný hrůzy a říkali jsme si: Proč jsme tady? Proč my? Proč se stalo zrovna mně? A některý ty lebky jako by odpovídaly: A proč ne? Některý lebky ležely mezi ostatníma lebkama lehce, sledovaly nás čelistma a jakoby vědoucím úsměvem, ale víc, daleko víc jich na nás hledělo jen prázdnými očními důlky, protože to, co zbylo z čelistí, bylo zkroucený v šklebu bolesti, protože tihle to dostali natvrdo a těžce zaživa, bratří moji. A bylo jich moře, oceán. A tohle přirovnání nás napadlo, když už jsme nemohli jít, protože jsme se v hromadách kostí propadávali, tak jsme, bratří moji, v těch kostech zkoušeli plavat, zkoušeli sme se pohybovat a lézt a odstrkovat se rukama. [...] A tak sme si to šinuli k věžím a snažili se nechytat ty mrtvý pohledy z očních důlků lebek, abychom nezešíleli. Byly tam dětský lebky, bratří moji, a byly tam hromady lebek roztříštěnejch napadrť a byly tam rozstřílený lebky a lebky jakoby drcený v lisu a lebky s malejma otvorama svázaný ostnatým drátem a jeden z nás, rytíři a kapitáni, zas zavtipkoval: To budou asi takzvaný manželský lebky, ho! ho!, ale pak začal zvracet. A ten, co se plazil před ním, ho neslyšel, protože plakal, a ten, co lezl za ním, ho taky neslyšel, protože se hlasitě modlil. A my jsme, přátelé a bratři moji, nešli peklem, my jsme šli něčím, co bylo po pekle.'" Jáchym Topol, *Sestra*, 88 f.

instead because of an illegal store selling food products. This strange guide – the prototypical "small-scale Czech person"[22] speaks in a lower-class type of Prague slang (the remainder of Topol's entire novel uses normal, oral Czech speech in a non-traditional manner). To horrified listeners, he recounts drastic scenes of torture and murder in which he, himself, partially participated in as the "capo." This alternates with gallows humor. Despite all this, he and the famous Dr. Mengele (who apparently atoned for his sins after the war by devotedly treating the Indians in South America) go to heaven after their deaths. His very arrival in heaven recalls the selection process in Auschwitz: "...that was Mr. God, my boys, and as the lines went past he'd just smile and go 'Rechts! and Links!' with this like white cane, and in the line I was in, the angels took care our wounds (...) but the debbils tore inda that other line wid pitchforks an whips..."[23]

Topol's text may be explained not only as an evocation of the Shoah's horror, but also as a reminder of the "Czech's responsibilily"[24] in the Jews' extermination. At the end of his dream, the narrator meets with the lofty Face (of God) from whom he learns that the future Messiah died as a young Jewish child in Auschwitz. The hundreds of thousands of skulls in the bone fields cry out to him and to his friends, "Our blood on you and your children!"[25] It is an evident allusion to the cry of the Jews in the Gospel of Matthew (Mt 27, 25). This quotation from the New Testament was interpreted as an inspiration to anti-Judaism. Here, in the text of Topol, it looks more like the complicity of the Czech people on the fate on Jews. Also, the words "I had a dream" seem to be an indirect polemic with the famous speech by Martin Luther King in Washington in August 1963. While M. L. King's speech was a call for tolerance, Topols' Auschwitz chapter means the end of hope and humanism, which were buried with the unknown Jewish child – the Messiah in Auschwitz.

Conclusion

After World War II, Czech cinema and literature featuring Jewish themes were confronted with the Holocaust. About 80,000 Czech and Moravian Jews were murdered during the Nazi persecution. Clearly, portrayals of the persecution and mass murder of these people and millions of European Jews cannot be seen as a purely aesthetic affair. On the other hand, it is necessary to assess lit-

---

22 Emphasis mine (JH).
23 *City Sister Silver*, 116.
24 Emphasis mine (JH).
25 *City Sister Silver*, 123.

erature and film depicting the Shoah not just using ethical yardsticks, but literary and cinematographic yardsticks as well.

Immediately after the war, two excellent works were produced, Alfred Radok's *Distant Journey* and Jiří Weil's *Life with a Star*. They not only presented the horrible brutality of oppression and murder, but also a mundane history: the seemingly banal, profane side of the Shoah. Radok's film depicted two motifs that were later often used: love between a Jewish woman and her gentile boyfriend/husband, and indifference or even the complicity of many "ordinary" Czech people in the fate of their Jewish fellow citizens. Weil's novel described a "small man," the typical figure of Czech cinema and literature with Jewish topics. It used grotesque, black humor and absurdity.

The reflection of the Shoah in Czech literature always depended upon the political situation in Czechoslovak (Czech) society. During the latent and sometimes even open antisemitism of the Stalinist period of the early 1950s, and the neo-Stalinist period of the 1970s, Jewish themes hardly came to public light at all.

The theme of Jewishness was developed mainly in the 1960s. Many Czech and Slovak writers and film directors were inspired by the Holocaust: the survivors Josef Bor, Leopold Lahola, Erich Kulka, Arnošt Lustig, Ladislav Grosman, J. R. Pick, Ján Kadár, and Juraj Herz, as well as Ladislav Fuks, Josef Škvorecký, Ladislav Mňačko, Hana Bělohradská, Zbyněk Brynych, and Jan Němec, who don't have any Jewish roots. For some authors, the Shoah was a metaphor for man caught in the machinery of the totalitarian regime and for the functioning of evil at large. This is the case with the novels *Mr. Theodore Mundstock* and *A Prayer for Katerina Horovitzova*, as well as with the films *Diamonds in the Night*, *The Shop on Main Street*, and *The Cremator*.

After the end of the 1960s, the theme of Jewishness did not play such a key role in Czech cinema and literature. For the generations which never experienced the Holocaust, they primarily understood it in terms of set images and stark iconography (e. g., Hebrew religious ceremonies, transports, lonely children, gas chambers). Some authors attempted to push these boundaries – mainly Arnošt Goldflam in his plays *Sweet Theresienstadt* and *From Hitler's Kitchen*, and Jáchym Topol in his novels *City Silver Sister* and *The Devil's Workshop*.

This article presented two distinct approaches. On one hand, there is a "closed narrative" which portrayed the Shoah as a monumental narrative with authentic details (Norbert Frýd, Jan Otčenášek, Arnošt Lustig's later works).[26] On the other hand, there is an "open narrative," which presented the persecution

---

26 See Jiří Holý and Hana Nichtburgerová, "Jurek Becker: Jakob der Lügner."

of Jews and the war without any heroism, pathos, and sentiment, and featured objective reports, but also with farcical and tragicomic elements (Jiří Weil's novel *Life with a Star* and Alfred Radok's film *Distant Journey*, both exceptional works for their time, initiated this approach in the post-war years). They inspired several of the masterpieces of the 1960s (Ladislav Fuks, Josef Škvorecký, Ladislav Grosman, Lustig's early works) and, after the Velvet Revolution of 1989, the works of Arnošt Goldflam, and Jáchym Topol.

Bibliography

Ammann, Thomas, and Stefan Aust. *Hitlers Menschenhändler: Das Schicksal der "Austauschjuden."* Berlin: Rotbuch Verlag, 2013.

Fuks, Ladislav. *Pan Theodor Mundstock*. 2nd ed. Prague: Československý spisovatel, 1969.

Fuks, Ladislav. *The Cremator*. Translated by Eva M. Kandler. London: Marion Boyars, 1984.

Fuks, Ladislav. *Mr. Theodore Mundstock*. Translated by Iris Urwin. New York: Four Walls Eight Windows, 1991.

Fuks, Ladislav. *Spalovač mrtvol* [*The Cremator*]. Prague: Odeon, 2003.

Grosman, Ladislav. *Obchod na korze* [*The Shop on Main Street*]. 2nd ed. Prague: Mladá fronta, 1966.

Grosman, Ladislav. *The Shop on Main Street*. Translated by Iris Urwin. New York: Doubleday & Company, 1970.

Holý, Jiří. "The Jews and the Shoah in Czech Literature after World War II." *Russian Literature* 77, no. 1 (2015): 35–53.

Holý, Jiří. "The Six Versions of The Shop on Main Street." In *The Aspects of Genres in the Holocaust Literatures in Central Europe*, edited by Jiří Holý, 97–113. Prague: Akropolis, 2015.

Holý, Jiří, and Hana Nichtburgerová. "Jurek Becker: Jakob der Lügner (1969)." *Holocaust. Zeugnis. Literatur. 20 Werke wieder gelesen*, edited by Markus Roth and Sascha Feuchert, 152–168. Göttingen: Wallstein Verlag, 2018.

Lustig, Arnošt. *Modlitba pro Kateřinu Horovitzovou*. 7th ed. Prague: Andrej Šťastný, 2003.

Lustig, Arnošt. *Modlitba pro Kateřinu Horovitzovou. A Prayer for Katerina Horovitzova*. Anonymous translation. Prague: Nakladatelství Franze Kafky, 2008.

Poláček, Jan. *Příběh Spalovače mrtvol. Dvojportrét Ladislava Fukse* [*The Story of The Cremator. A Double Portrait of Ladislav Fuks*]. Prague: Plus, 2013.

Sladovníková, Šárka. *The Holocaust in Czechoslovak and Czech Feature Films*. Stuttgart: ibidem, 2018.

Topol, Jáchym. *Sestra* [*Sister*]. Brno: Atlantis, 1994.

Topol, Jáchym. *City Sister Silver*. Translated by Alex Zucker. North Haven: Catbird Press, 2000.

Weil, Jiří. *Life with a Star*. Translated by Rita Klímová and Roslyn Schloss. Evanston: Northwestern University Press, 1998.

Weil, Jiří. *Život s hvězdou. Na střeše je Mendelssohn. Žalozpěv za 77 297 obětí* [*Life with a Star. Mendelssohn Is on the Roof. Elegy for 77,297 Victims*]. Prague: Nakladatelství Lidové noviny, 1999.

Marcela Menachem Zoufalá
Ethno-religious Othering as a Reason Behind the Central European* Jewish Distancing from Israel

Introduction

What image of Israel do Czech Jews have across generations? Is Israel seen as a distant beloved homeland or rather as an ambivalent bond or even an uncomfortable social burden? The following contribution is dedicated to an analysis of the contemporary attitudes of the Czech Jews towards Israel and the Israelis. This study will argue that Czech Jews are experiencing an often-described phenomenon of distancing from Israel. Nevertheless, Czech Jews are partially distancing themselves for quite different reasons than, for example, American Jews. It accrues from long-term qualitative research results that Czech Jews are besides others significantly discomforted by the changing ethnic composition of the Israeli population. In particular, it is the culture of Mizrahi Jews that is increasingly coming into the foreground. Secondly, a correlative theme of particular prominence indicates that Czech Jews are caught up in nostalgic, idealized views of the past and relate to contemporary Israel in some aspects through the mental representation of a once-flourishing, then forever lost Central Europe created in a large scope by Jewish intellectuals. To a certain extent, this identification with the imagined pre-war Central Europe contributes to the distancing of Czech Jews from Israel and the Israelis.

---

* The notion of "Central Europe" appears frequently in the testimonies of the respondents, even though there was no direct question related to that term posed by the interviewer. This topic was brought up exclusively by the respondents following an expressed interest in the nature of their relation to Israel. This is also the main reason why the title of this paper, "Ethno-religious Othering as a reason behind the Central European Jewish distancing from Israel", includes the notion of "Central Europe" rather than solely "Czech Republic".

https://doi.org/10.1515/9783110582369-010

Methodology

The presented findings[1] are based on long-term qualitative anthropological research on Czech Jewish identity that mainly took place from 2014 to 2018. The conceptual framework of the study involves a hermeneutic-narrative approach using an emic perspective focusing primarily on the respondents' own interpretations of the world and reality surrounding them. The main methods applied to gathering data are participant observation (from 2010 for almost a decade), interview, and follow-up socio-cultural analysis of accounts and exploration of general printed and online Czech media sources with a special focus on Jewish community. The interviews were processed in line with the demands of methodological reflexivity. Maximum emphasis was placed on respondents' anonymity and emotional security. Altogether, twenty-nine episodic interviews were conducted and analyzed.[2]

Czech Jewish Population: Current Figures

In 2018, according to demographer Sergio DellaPergola, the Czech core Jewish population numbered 3,900 and enlarged Jewish population 6,500 people out of an overall population of 10,600,000.[3]

Except for Prague, where the majority of Czech Jews live, there are nine smaller-sized communities in the cities of Brno, Plzeň, Karlovy Vary, Olomouc, Liberec, Děčín, Ostrava, Ústí nad Labem, and Teplice.[4] There are ca. 3,000 registered members in ten local Jewish communities and more ca. 2,000 additional members of other Jewish associations.[5]

According to the Federation of Jewish Communities, it is estimated that 15,000 to 20,000 Jews live in the Czech Republic. However, most of them are not listed.[6]

---

1 The presented findings dedicated to contemporary attitudes of the Czech Jews towards Israel and the Israelis are part of a larger study, a work in progress, which focuses on specifics of Czech Jewish identity, perception of antisemitism among Czech Jews, and relations between the majority of the Czech society and the Jewish minority.
2 The interviews were conducted and analyzed in the years 2016–2019.
3 Sergio DellaPergola, "World Jewish Population, 2018," Arnold Dashefsky and Ira M. Sheskin (eds.), *The American Jewish Year Book* 118 (Dordrecht: Springer, 2018): 361–452 at 52.
4 "Jewish communities", website of the Federation of Jewish Communities in the Czech Republic. Accessed August 2019.
5 "Statistics", website of the Federation of Jewish Communities in the Czech Republic. Accessed August 2019.
6 Ibid.

Phenomenon of Distancing

Much of the current research on relations between Israel and diaspora pays particular attention to the phenomenon of distancing, also known as disaffection or disengagement, which in the given context refers to the Jewish diaspora that distances itself from Israel. This almost global trend is being observed among the especially younger and more secular generation which expresses not only increasingly resolute critical views and discontent, but often also indifference towards Israel.

Distancing in the U.S.

A high number of currently conducted studies, e. g. "70 Years of Israel-Diaspora Relations: The Next Generation", "A Portrait of Jewish Americans.", "Beyond Distancing: Young Adult American Jews and Their Alienation from Israel" or "Survey of American Jewish Opinion", are aimed at the American Jewish community which represents the second largest Jewish population worldwide.[7] This is the principal reason why the distancing discourse will be demonstrated firstly on the example of the U.S. Jewish diaspora. From the often-quoted study "A Portrait of Jewish Americans"[8] which, inter alia, looks at attitudes and connection with Israel, we have learned that more than half of American Jews (fifty-three percent) aged sixty-five and older claim that "caring about Israel is essential for their Jewish identity".[9] The same view is shared by "only" thirty-two percent of Jewish adults under the age of thirty.[10]

Frequent arguments, presented by a significant number of American Jews for their growing disengagement from Israel, are serious objections against Israel's approach to the peace process[11] and the continued building of settlements.[12]

---

[7] DellaPergola, "World Jewish Population, 2018," 20: "Israel (with 6,558,100; i.e., 45% of total Jewish Population) and the United States (with 5,700,000; i.e., 39% of total Jewish Population) represents countries with two of the largest core Jewish populations worldwide."
[8] See "A Portrait of Jewish Americans," Pew Research Center, Washington, D.C., October 1, 2013, accessed August 2019.
[9] "A Portrait of Jewish Americans", 84.
[10] Ibid.
[11] Only 38% believe the Israeli government is "making a sincere effort to establish peace with the Palestinians" (while in comparison "12% – think Palestinian leaders are sincerely seeking peace with Israel"). Ibid., 89–90.
[12] "A 44% plurality of American Jews say the continued building of Jewish settlements in the West Bank hurts the security of Israel." Ibid., 91.

An extensive report with the title "70 Years of Israel-Diaspora Relations: The Next Generation"[13] points to how the "distancing"[14] discourse is increasing.[15] Besides the aforementioned Israeli policies[16] on peace and war, the report refers to another focal points of the crises, among others, religious and political issues as well-known Western Wall crises,[17] conversion problematics, the "Who is a Jew" legislation and the status of non-Orthodox Judaism in Israel.[18] Even the latest studies reconfirm such a development, showing a significant degree of indifference. In the "Survey of American Jewish Opinion",[19] launched in June 2019, a quarter of respondents disagree that a "thriving State of Israel is vital for the long-term future of the Jewish people".[20]

In a similar vein, the 2007 study "Beyond Distancing: Young Adult American Jews and Their Alienation from Israel"[21] brought even higher numbers, declaring that fifty percent of American Jews under the age of thirty-five would not perceive "Israel's destruction [as] a personal tragedy".[22]

The nature of the relationship of American Jews towards Israeli Jews was illustrated by the following findings. Based on the metaphor of a family, American Jews consider Israeli Jews their "Siblings (thirteen percent), First Cousins (fifteen percent), Extended family (forty-three percent) [or] Not part of [their] family (twenty-eight percent)".[23]

Distancing in the Czech Republic

As indicated above, the crux of the negative delimitation of Czech Jews partly differs from the stances of the American Diaspora. According to the presented research, Czech Jews do not necessarily perceive Israel's government official line

---

[13] Shmuel Rosner and John Ruskay, "70 Years of Israel-Diaspora Relations: The Next Generation" in Barry Geltman and Rami Tal (eds.), *The Jewish People Policy Institute (JPPI)* (2018). Accessed August 2019.
[14] Ibid., 9.
[15] Ibid.
[16] "70 Years of Israel-Diaspora Relations: The Next Generation", 30.
[17] Ibid., 9.
[18] Ibid., 30.
[19] "Survey of American Jewish Opinion", AJC, June 2, 2019, accessed August 2019.
[20] Ibid.
[21] Ari Y. Kelman and Steven M. Cohen, "Beyond Distancing: Young Adult American Jews and Their Alienation from Israel" (Andrea and Charles Bronfman Philanthropies, 2007).
[22] "Beyond Distancing: Young Adult American Jews and Their Alienation from Israel", 9.
[23] "Survey of American Jewish Opinion".

as problematic, even though these issues do randomly appear in the testimonies and public discourse, predominantly among the younger generation.

One of these rather sporadic cases illustrates a new initiative called "Jewish Voice of Solidarity" (JVS) which has been established in the Czech Republic in May 2019. In the introductory formula of JVS the distancing discourse rings out quite loudly, pointing out "our"[24] different views and "our" distinct experiences, concluding with clear demarcation: "We don't speak on behalf of Israel, but we also need that Israel doesn't speak on our behalf."[25]

In the mission statement, the signatories delimit themselves against currently ruling Israeli establishment "approving undemocratic laws",[26] However, the overall tone of the statement is rather conciliatory. In a commentary for a Czech newspaper, one of the platform's founding members clarifies that:

> Most of such initiatives, [e. g. Jewish Voice for Peace (JVP)], deal mainly with affairs in Israel and Palestine. We don't want to neglect this topic however, we would like to extend our interest also to Jewish culture and art, in general, mainly to its forms created out of the State of Israel.[27]

Comparing this definition to much more radical expressions being used by the mentioned U.S. based activist organization JVP, openly calling for boycott of Israel,[28] Czech Jewish initiative endorse more neutral stances and hold a broader perspective on Israel, not focusing dominantly on Israeli-Palestinian conflict.[29]

Considering that the JVS represents a very rare if not the only publicly and for media formulated critique on behalf of Czech Jews towards Israeli current politics, it implies that Israeli-Palestinian conflict and surrounding Israeli government policies are not entirely in the focus of the Czech Jewish community

---

24 "Our" views mean "Czech Jewish Diaspora" views versus Israeli.
25 Original: "My nemluvíme za Izrael, ale také potřebujeme, aby Izrael nemluvil za nás." Statement of the platform of Jewish Voice of Solidarity, June 6, 2019, accessed August 2019.
26 Ibid.
27 Markéta Hrbková, "Solidarita jako jedno z důležitých témat židovských dějin" ["Solidarity as one of the important themes of Jewish history"]. *Deník Referendum*, May 28, 2019, accessed August 2019. Author's translation.
28 See the website of Jewish Voice for Peace, https://jewishvoiceforpeace.org/jvp-supports-the-bds-movement/.
29 Nevertheless, on August 1, 2019, just a few days before a submission of this article, an open letter to the Czech minister of foreign affairs was published. This letter was protesting against the demolition of Palestinian houses. The letter was signed, among others, by the JVS. This points out a possible future strengthening of the mentioned emerging tendency. See http://ism-czech.org/2019/08/01/otevreny-dopis-ministru-zahranicnich-veci-cr-tomasi-petrickovi-ve-veci-nelegalniho-bourani-palestinskych-domu-izraelskou-okupacni-spravou/#more-10088.

mainstream debate. This consideration is in accordance with the latest findings of current research, investigating general attitudes towards Israel among Visegrad countries, stating that there is rather lower awareness of Israeli-Palestinian conflict and the topic is even being marginalized in Central/Eastern European political and public debates. Joanna Dyduch even argues that "the conflict between Israel and Palestine does not play an important role in [Visegrad countries] foreign policy or internal social debate".[30]

Research Question

Working with the assumption that the aforementioned Israeli policies may not be of significant importance in the eyes of the majority of the Czech Jewish community, in contrast to the American one, the key research question of this study strives to analyze the main patterns of status quo and reasons why the distancing discourse appears among the Czech Jews anyway.

In general, the outcomes of presented study testify that Czech Jews across generations have currently a rather ambivalent relationship towards Israel. Two divergent and often conflicting patterns emerged. On the one hand, an overwhelming number of respondents reflected on Israel with affection, like about their distant homeland they feel connected to and "will always fight for"[31], and on the other hand, there is obviously also a high degree of mixed attitudes and negative delimitations.

The most striking result which emerged from the analysis was that Czech Jews are besides others significantly discomforted by the changing ethnic composition of the Israeli population. It is concretely about the culture of Mizrahi Jews that is increasingly coming into the foreground. This finding was unexpected and suggests that Czech Jews might be still strongly rooted in an ethnically homogenous Central Europe and despite the influence of globalization; they are not too confronted by the ethnically diverse groups of population that are present for example in Western Europe. As a result of decreased contact with other cultures that is familiar to Central and especially Eastern Europe, Czech Jews enclose themselves to their own ethnic shell and become estranged after meeting the "Oriental" Israel.

---

30 Dyduch, Joanna. "Die Visegrád-Staaten und Israel Dimensionen und Funktionen einer Sonderbeziehung", Osteuropa, 9–11/2019, s. 361–362. Original quote: "Weder in ihrer Außenpolitik noch in der internen gesellschaftlichen Debatte spielt der Konflikt zwischen Israel und Palästina eine wichtige Rolle; keiner der Visegrád-Staaten verfügt über eine klare Nahost-Agenda."
31 Respondent no. 1 – male; age group 30–35, (first part).

Concept of Mizrahim in Israeli Society

> In a world where Jewish is synonymous with Central and Eastern European, where North African/Middle Eastern is synonymous with Arab Muslim, where "of color" is synonymous with "not Jewish", and where communities are generally represented through their men, our mere existence threatens to destroy the foundation of numerous identity constructs as society knows them today.[32]

The personal observation above, from one of the first English-language anthologies dedicated to the voices of Mizrahi Jewish women, aptly summarizes the lasting ethnic cleavages within Jewish society which accompany State of Israel from its outset. It is an inarguable fact that Israeli elite, especially in the political, economic, and cultural dimensions, has been mainly formed by the Ashkenazim,[33] while the Mizrahim[34] have represented the rather social "periphery".[35]

Even though the Mizrahim today constitute almost half of the Israeli Jewish population,[36] Mizrahi identity is in Israel perceived as a minority identity.[37]

Marginalization and exclusion of the Mizrahim began immediately upon their arrival into the country. The conditions and integration efforts were extremely insufficient and they were often omitted during the distribution of sources and goods, both material and symbolical. The Israeli historian Tom Segev in his challenging book *1949: The First Israelis* captured a number of statements of representatives of government or press, whose attitudes towards Jews from Arab lands were quite questionable, clearly possessing patterns of orientalism.

For example, at a meeting with Jewish intellectuals and writers, the Israeli prime minister David Ben-Gurion noted that:

---

[32] Loolwa Khazzoom, *The Flying Camel: Essays on Identity by Women of North African and Middle Eastern Jewish Heritage* (New York: Seal Press, 2003), xi.
[33] Ashkenazi – Jews who originally came from European or largely Christian countries.
[34] Mizrahim – Jews who originally came from Arab speaking or largely Muslim countries.
[35] Ruth Halperin-Kaddari and Yaacov Yadgar, *Religion, Politics and Gender Equality among Jews in Israel* (United Nations Research Institute for Social Development (UNRISD) and Heinrich-Böll-Stiftung, 2010), 13.
[36] Immigrants from Asia and Africa (of first and second generation) constitute forty-seven percent of the Jewish population. Israel Central Bureau of Statistics (ICBS), *Statistical Abstract* (Jerusalem: Israel Central Bureau of Statistics, 2017).
[37] *Religion, Politics and Gender Equality among Jews in Israel*, 13.

> Even the immigrant from North Africa, who looks like a savage, who has never read a book in his life, not even a religious one, and doesn't even know how to say his prayers, either wittingly or unwittingly has behind him a spiritual heritage of thousands of years.[38]

However, some years later Ben-Gurion shared an even more severe observation:

> The ancient spirit left the Jews of the East and their role in the Jewish nation receded or disappeared entirely. In the past few hundred of years the Jews of Europe have led the nation, in both quantity and quality.[39]

Since Ben-Gurion was undoubtedly one of the most significant figures of the modern State of Israel, his views were and still are highly influential and widely shared. Statements along these lines were entirely not unique in the beginning of the state, their impact has persisted and is shaping Israeli public discourse till today. The image of Mizrahi Jews created through Euro-centrist lenses as lazy, wild, and primitive savages is often reactivated by different agents of Israeli society in order to achieve political, media or even commercial success. As concluded by Segev, Ben-Gurion sanctified a view that the "[…] house of a Jewish rag-merchant in Plonsk, Poland […] was endowed with the 'ancient spirit' while that of, say, a Sorbonne-trained Jewish physician in Algeria, was not."[40]

Shlomo Fischer argues that in reaction to their disqualification from Israeli-Jewish society, the Mizrahim gradually developed a "semi-traditional ethno-religious Jewish collectivity, […] *a counter-collectivity*",[41] as a tool for how to, in the opposite direction, exclude Ashkenazim and to question an authenticity of their Jewishness.[42]

The fact that the Mizrahim are more religiously observant[43] and prefer to affiliate rather with Jewish[44] than Israeli ancestry implies they are still not feeling

---

38 Ben-Gurion's meeting with Writers, 10.11 1949, *Divrei Sofrim*, State Archives, Quoted acc. Tom Segev, *1949: The First Israelis* (New York: Owl Books by Henry Holt and Company, 1998), 156.
39 David Ben-Gurion, *Netsah Israel* (Jerusalem: Shnaton Hamenshala, 1954), 17. Quoted acc. Tom Segev, *1949: The First Israelis* (New York: Owl Books by Henry Holt and Company, 1998), 156.
40 Ibid., 9. Quoted acc. Tom Segev, *1949: The First Israelis* (New York: Owl Books by Henry Holt and Company, 1998), 156.
41 Shlomo Fischer, "Two Patterns of Modernization: An Analysis of the Ethnic Issue in Israel", *Israel Studies Review* 31, no. 1 (2016): 66–85 at 66.
42 Ibid.
43 *Religion, Politics and Gender Equality among Jews in Israel*, 12.
44 Fifty-two percent of third generation Mizrahim prefer Jewish over Israeli ancestry, compared to thirty-one percent of Ashkenazi. Noah Lewin-Epstein and Yinon Cohen, "Ethnic origin and identity in the Jewish population of Israel", *Journal of Ethnic and Migration Studies* (2018): 9.

part of the originally European, Zionist project and strive to belong through adherence to Judaism.[45]

The Israeli psychologist Dan Bar-On, among many others, argues that in contrast to the dominating Zionist identity, several different "Others" have emerged, divided between external Others (Nazi and Arabs) and internal Others (Diaspora Jews, Holocaust survivors, Orthodox, and "Oriental" Mizrahi Jews).[46] However, due to their certain proximity to Arab societies and cultures, the Mizrahim were often seen as they would originate from the external Arab Other with who Israel has often been in a mortal combat.[47] This fluid merging of "mistaken" identity of Jews from Arab lands with Arabs has stemmed not only from certain cultural closeness but undoubtedly also from a similar appearance. Indicated categories of "Otherness" (Alterity) are almost fully represented in a tense biased depiction of the Israeli population during the Eichmann trial in 1961 in Jerusalem shared by German-Jewish American philosopher Hannah Arendt in her infamous private letter to Karl Jaspers, clearly written from standpoint of Central European Jewish diaspora:

> My first impression: On top, the judges, the best of German Jewry. Below them, the prosecuting attorneys, Galicians, but still Europeans. Everything is organized by a police force that gives me the creeps, speaks only Hebrew, and looks Arabic. Some downright brutal types among them. They would obey any order. And outside the doors, the Oriental mob, as if one were in Istanbul or some other half-Asiatic country. In addition, and very visible in Jerusalem, the peies (sidelocks) and caftan Jews, who make life impossible for all reasonable people here.[48]

The inter-ethnic gaps influence not only how Jewish tradition is perceived and carried out but also the forming of the national identity,[49] and are broadly seen as an unsuccess, so far, to decrease cultural and socioeconomic divisive attributes within a Jewish population with different geographical backgrounds.[50]

However, in the recent period so-called Mizrahization is being mentioned in the context of ongoing demographical and cultural changes of Israeli society that is, due to certain observations, gradually becoming more "Oriental", visibly

---

45 Ibid., 16.
46 See Dan Bar-On, *The Others Within Us: Constructing Jewish-Israeli Identity* (Cambridge: CUP, 2008).
47 Ibid., 18.
48 Lotte Köhler and Hans Saner (eds.), *Hannah Arendt and Karl Jaspers' Correspondence, 1926– 1969* (Harcourt Brace Jovanovich, 1992), 435, letter number 285.
49 *Religion, Politics and Gender Equality among Jews in Israel*, 12.
50 "Ethnic origin and identity in the Jewish population of Israel", 1.

reflecting its Middle Eastern geographical setting. This transformation of societal morphology is often portrayed together with growing haredization, nationalism, and right-wing views which are seen as the main interconnected characteristics of the Mizrahi world.

In the 2017 article called "Israel on the Road to the Orient?: The Cultural and Political Rise of the Mizrahim", Lidia Averbukh describes the "Orientalisation" or "Mizrahization" as a long-term development in Israeli politics and society and even attributes to it an importance of a paradigm shift.[51] However, Averbukh perceives this recently stabilized trend of appreciating Mizrahi roots at the expense of pro-European focus as rather an intra-societal phenomenon limited for the time being to Israeli domestic environment.[52] She sees no significant implication of Mizrahization for foreign policy in general and for political and economic relations with Europe in particular, and she even observes that in the past "European cultural identification with the Ashkenazi leadership was no guarantee of political consensus."[53] Nevertheless, Averbukh does not specifically elaborate on European Jewish communities voices and their potential impact on the relationship between EU and Israel.

Analysis of Interviews and Literary Excerpts

Turning now to the experimental part of the study, it is worth noting that all the following statements come from respondents who openly endorse their Jewishness and/or are somehow involved with Jewish organizations; it can be said that Jewishness is one of the central topics of their identity.

Respondent no. 1: male; age group 30–35[54], (first part).

Motto: "I will always fight for Israel…"
Question of interviewer (further only "Q."):
"What is your relation to Israel?"

---

[51] Lidia Averbukh, "Israel on the Road to the Orient?: The Cultural and Political Rise of the Mizrahim", in *Stiftung Wissenschaft und Politik* (German Institute for International and Security Affairs, SWP Comments 9, April 2017), 6.
[52] Ibid., 1.
[53] Ibid., 7.
[54] Due to given emphasis on respondents' anonymity, only an age group is stated (in the time of conducting the interview) and not their exact age.

Answer of interviewee (furthermore only "A."):
"Well, I lived there once...I love Israel and I have connections there through my ancestors [...] I will always fight for Israel because I know this is right [...] the State is not perfect...but any state is perfect [...]."[55]

Motto: "...His faith is alive and real"
Q: "Would you choose Israel for your life?"

A: "God, they would drive me insane, half of them are 'Zulukafern'. You don't know that word? That's an expression from [our][56] synagogue: 'hetzi[57] Zulu, hetzi Morocco'".

The mentioned word "Zulukafern" has an interesting etymological connotations. The first part of the word, "Zulu", denominates an African indigenous ethnic group; the second part "kafern" is derived from an Arabic word "kāfir" that is usually translated into English as "non-believer", i.e. a non-Muslim. However, the contemporary usage serves mainly as an ethnic slur.[58]

The most probable reasons for the use of this term for indicating the Mizrahim Jews, besides its obvious racist context, is also an attempt to doubt their Jewish heritage. As mentioned earlier, the Jews who came originally from Arab countries have been often interchanged with Arabs.

The second component of expression includes "hetzi Morocco". Moroccan Jews have gained this stereotypical reputation of the most problematic among other Mizrahim. All the negative "oriental" characteristics attributed to the Mizrahim apply in Jewish Israeli popular imagination primarily to Jews from Morocco. This is how, for example, the Israeli Hebrew expression "Marokai Sakinai"[59] [Moroccan with knife] was born.

Even more particular is the local context, i.e. Czech usage and connotations of the term. The word "Zulukafer" has in the Czech daily lexicon quite a rare occurrence, however it is not entirely unusual in literature. For example, the prom-

---

55 Translation of interviews and literary excerpts from Czech into English was made by Anna Hupcejová, unless otherwise stated.
56 One of Prague's currently functioning synagogues; the name is not given due to anonymity measures.
57 "Hetzi" means "Half" in Hebrew.
58 Definition of "Kaffir (ethnic slur)" from *Academic Dictionaries and Encyclopedias*, accessed August 2019.
59 Daniel Mayer, "Maročtí Židé" ["Moroccan Jews"], *Maskil* 8/9 (2013): 4, accessed August 2019.

inent Prague writer Gustav Meyrink[60] introduces "Zulukafr" to the reader as a black man and dark magician in his esoteric novel *The Green Face*. The following image describes Zulukafer's essence in the story: "If he is a murderer?! – Who is to judge? [...] He is a savage and has his own faith; God forbid that many people have such an awful faith as him, but his faith is alive and real."[61]

Motto: "...except they were both black..."

The same respondent reveals identical views in the different part of interview, explaining further why he would not live in Israel:

> It is often seen even here [describing environment of Prague Jewish quarter] – yesterday I saw two [Jews] who were dressed as if they were from somewhere like Krakow, except they were both black...

On a fluid category of "blackness" as a tool of othering, exclusion, and dehumanization in different Jewish contexts, Esther Benbassa ruminates in her contribution: "The Jews were black for Western non-Jews; the Sepharadim were black for the Ashkenazim; for the Jews, the Arabs too were black. The color black removed you from the realm of the human in order to submerge you in a vision, a look, an imagination that classified things and nature as it saw fit."[62]

Motto: "We have defended the community against Hitler and Stalin..."

In the following excerpt our respondent describes the religious event from Spring 2017 of bringing new *sefer Torah* into the Old-New synagogue. The Torah scroll was donated to the Prague Jewish community by an Israeli businessman living in Prague. Our respondent felt highly resentful about the whole event:

> [...] and that was seen again in our carrying in of the Torah... it had at the start an Ablauf,[63] which first was kept and then not anymore... XY hosted it... he invited individual agents of the community to add a letter... first of all, there was almost no one from the Prague Jewish

---

[60] Gustav Meyrink, 1868–1932, Viennese born writer living for twenty years in Prague. A large part of his works was influenced by the romanticized and enigmatic image of Prague's Jewish Ghetto.

[61] Gustav Meyrink, *Zelená tvář* [The Green Face] (Prague: Volvox Globator, 1991), 149.

[62] Esther Benbassa, "Otherness, Openness and Rejection in Jewish Context", *Journal of the Interdisciplinary Study of Monotheistic Religions (JISMOR)* 5 (February 2010): 23, accessed August 2019.

[63] "Ablauf" is a German word which means "course" in English.

community… that was like the invasion of the Israelis… whom I've never seen before […] suddenly the Israelis stole the microphone… nobody knows who they were, how did they got here, what title they were speaking from – one of them just took that microphone and began moderating alone in Hebrew, thereby the entire structure fell apart and the messieurs completely took over…

On the same occasion, another respondent **(no. 2, male; age group 60–65, first part)** expressed his views in a similar fashion, mentioning that "One of the Israelis was singing as a Muezzin". Then he added: "We are not those who basically came from Morocco, those are Moroccan Israelis with Torahs, that's not our tradition." Even *Community newspaper*, the official periodical of the Jewish community, informed about the event, though in a somewhat different way:

> The ceremonious bringing of two new Torah scrolls on Sunday 19th of March was an event that surely nobody who was present will ever forget. The opening remarks by the supreme Prague rabbi David Peter and the vice-chairman of the Jewish Community of Prague František Bányai were followed by the writing of the final thirty letters of the Torah. Donators and those who contributed the most to the acquisition of the new script and our community's activity had the honour of adding these last letters.
> […] Today is the cherry on top of the cake in terms of achieved successes. After many years when Torah scrolls were destroyed and burned, after many years when their dispensability was demonstrated and after many years of other priorities, the community with its rabbi celebrates today a new Torah scroll. It is perhaps the best expression of the Prague Jewish Community's development.[64]

The presentation of the event as the hoped-for climax of community life as portrayed in public media discourse has obviously a very different tone than our respondents' testimonies. With certain bitterness, the respondent **(no. 1 male; age group 30–35, second part)** lets himself be heard: "We have defended the community against Hitler and Stalin – and now it should be given to some guys from Bnei Brak?!"

Motto: "…absolutely crazy Chasid-Tayman[65] discotheque…"
Then he resumes in an emotional retelling of the bringing in of the Torah that seems to have left a quite strong impression on him:

---

[64] František Bányai, "Slavnost vnášení tóry" ["Celebratory bringing of the torah"], *Obecní noviny* 14, no. 7 (April 2017): 7.
[65] In contemporary usage in Israel, Tayman is identified with Yemen.

> And then it continued with that absolutely crazy Chasid-Tayman discotheque, which really doesn't fit here at all, that would be acceptable only on a Goa trans in the dessert, but really not here... in the rhythm of camel caravans and some crazy Yiddish somewhere from Belarus... such gibberish that they speak in their enclaves, where you're scared to go if you're not veiled...

Here the entire image gets even more complicated. It is like if today's expressively religious Israelis are evoking in the reservoir of the collective memory of Czech Jews the remembrance of Jewish refugees from the East who were looking for haven in Czech lands in different time periods and were viewed very contradictorily by the local settled community. On the one hand, there were many who tried to support the refugees and improve their living conditions and on the other, the local Jewish minority feared discreditation in the eyes of the majority population and attempted to distance themselves from the refugees.

In terms of imagology, an Eastern Jew ("Ostjude"[66]) represents an example of hetero-image, a counterpart of a Jew from Central/Western Europe, who considered themselves more culturally mature.[67]

The stereotype of the Eastern Jew is in the Czech environment personified by famous depiction of writer Jiří Mordechai Langer by his brother František. František Langer thus describes the personal change and religious fervor of Jiří and presentiments and emotional reactions of the strongly assimilated Prague Jewish family at the onset of the twentieth century:

> Father looked almost horrified when he informed me that Jiří had returned. I understood his horrification when I saw my brother. He stood before me in a torn overcoat cut like a kaftan, ranging from his chin all the way to the floor and on his head he wore a circular wide hat from black plush, shoved deep into his nape. He was hunched, his whole cheeks and also chin were overgrown with a gingerish beard and from the front of his ears down to his shoulders hung curled hair and payots. [...] My brother did not depart home, to civilization, from Belz, he brought Belz with him... [...] It is understood that this religious or whatever kind of exhibitionism was deeply embarrassing to everyone at our home. Like the entire Jewish society at that time, our family has completely assimilated to all external signs and habits of our surroundings – then did not Jiří's appearance now convict everyone of pretence and hypocrisy? I do not know how our neighbours interpreted it, they probably were compassionately tapping their foreheads. But I think that my brother's disguise has

---

[66] On this topic see Magdaléna Fottová, "Images of Eastern Jews in Czech-Jewish and Czech language journals 1910–1925", in Jiří Holý (ed.), *Close and Foreign: Jews, Literature, and Culture in the Czech Lands in the Twentieth Century Prague* (Prague: Akropolis, 2016).

[67] On this topic see Steven E. Aschheim, *Brothers and Strangers, The East European Jew in German and German Jewish Consciousness, 1800–1923* (Madison, Wisconsin: University of Wisconsin Press, 1982).

shaken up father and his peers in also other ways. It infringed the feeling of security and stability, maybe evoking memories of already forgotten stories of the times and constraints of the ghetto, of life without rights and freedom, of the emotional states filled with indignity and injustice. [...] Here among us walked the ghost of the past, someone who awoke from the dead and was a warning. I can imagine all such feelings – feelings which were felt by millions of Jews a quarter of a century later, who were making their way into gas chambers with a yellow star sewn onto their coats.[68]

Taking into consideration that in the imagination of Czech Jews, the two discussed stereotypes, "Oriental" Jew" and Eastern Jew/"Ostjude", have possibly become interchangeable and merged into one hetero-image, their striking presence and displaying "religious or whatever kind of exhibitionism", as Langer put it above, might be, in both cases, "deeply embarrassing" for their local counterparts.

It is as if Czech Jews still today prefer to maintain a lower visibility in a public space and their "Oriental" guests are violating "a silent agreement" and "particular laws" of coexistence between Jews and "[Czech] nation" from which "the principal is perhaps inconspicuousness. These laws would have been broken if the state had opened itself to the influx of very strangely appearing Jews from the East."[69] The mentioned Jewish discreetness was recommended by Ferdinand Peroutka, the both admired and highly controversial icon of Czech journalism, a little less than two years before he was arrested and imprisoned in a concentration camp.

Patterns of distancing springing out from a sense of different – disturbing, perhaps even uncanny – ethnicity also appeared in other respondents' accounts, after they were requested to reflect on their relationship to Israel. Even though these were expressed more moderately.

---

68 Jiří Langer, *Devět bran: Chasidů tajemství*. [*Nine Gates to the Chassidic mysteries*] (Prague: Československý spisovatel 1965). Foreword by František Langer, 14–16.
69 Ferdinand Peroutka, "Něco o českém národu a o židech" ["Something about the Czech nation and the Jews"], *Přítomnost* XV, no. 19/1 (1938): 34. Full quote: "I demand that our Jews do not invoke too loudly the laws of hospitality in this matter. The coexistence of Jews with our nation is taking place according to particular laws, from which the principal is perhaps inconspicuousness. These laws would have been broken if the state had opened itself to the influx of very strangely appearing Jews from the East." It was issued in reaction to the Czechoslovak borders closure, to prevent the entry of fleeing Jews from Romania.

Respondent no. 3: male; age group 40–45 (first part).

Motto: "...isolate you from *the others*..."
Q: "How would you describe your relation to Israel? Have you ever considered Aliyah?"

A: "I think that a Jew has the right to not see Israel as his society. I think that it is right for a Jew to see Israel as his nation in some respects, but not as his society. Because that society is from a large number Polish and from a huge part Arab in the sense of Jewish-Arabic and why would a person from Prague feel like this is his country, when it isn't. You can somehow conform or swim into it, you can live in a certain bubble. If you're going to go to university, then you will still feel like the intellectual DNA of the German Central Europe, which is still there. So you can live in your own bubble that will isolate you from the others, but the Israeliness, the most typical Israeliness isn't something that a normal Czech could consider as his own culture. Most probably not."

Respondent no. 2: male; age group 60–65 (second part).

Motto: "...but at the same time, they are still Jews..."
Q: "How would you describe your relation to Israel? Have you ever considered Aliyah?"

A: "Once, Israel truly looked like a nation of Einsteins. People had the feeling that even though they were in an exotic country, they could build it like their home. That is essentially like Bauhaus, those German emigrants who left and wanted to build Germany there. It was not only a case of buildings, but entire society. And that is happening also today with people from the Arabic countries, Sephardic Jews, million people from Russia. That is again completely different from the initial vision. I don't mean the aim, the aim is fulfilled, but the vision was different at the start. And I don't know if I would feel good there, I would feel well in the former European Israel.

However, I do like it there, one gets used to it quite fast. Initially I was thinking about how come that Arabs can be vendors at the markets, but then I noticed that they all wore kippahs on their heads. It is such a diverse community in all aspects. It reminds me slightly of America, something "multi-cultural", but at the same time they are still Jews. I'm not sure if a European who has roots here from several hundreds of years back would be able to adapt there. And I know that people from Germany and also Czechoslovakia moved there and weren't happy."

Respondent no. 4: male; age group 40 – 45 (first part).

Motto: "…things between the Orient and Jewishness…"
Q: "Have you ever considered Aliyah?"

A: "During puberty, but luckily, I was a Central European punk – first I came here [to Israel], loved it, but… then I started understanding differences between Israel and Jewishness, I liked it, but didn't want to live here. There are many things between the Orient and Jewishness, many things that aren't Jewish. A number of things that are archetypically done to make the Jews Jewish in Central Europe died out here [in Israel] …"

Respondent no. 3: male; age group 40 – 45 (second part).

Motto: "Israeliness […] like the antipode of Jewishness…"
Q: "How would you describe your relation to Israel?"

A: "Israeliness is something like the antipode of Jewishness, somehow elemental, some strange vitality to aggression. And that is something different from our Central European, specifically Prague image of Jewishness where we see a gloomy Kafka wandering through the streets around the Old Town Square. Question is if this is some form of Jewishness, like a long-term phenomenon. Rather not – why would the symbol of Jewishness not be something like Kafka's father, who had exactly that "Gründer" [Founding Father] enthusiasm, nothing was a problem for him and he was always feisty. We have decided that we will consider Jewishness to be the mentality of one specific generation of Prague Jews."

Motto: "…Israel [as] the true heart of Europe…"
Except of perceived "Orientalization" of Israeli population, emerging as a one of the motives behind negative delimitation from Israel, there was another interconnected issue of particular prominence in the testimonies.

A recurrent theme in the interviews was a sense of loss referring to the Central Europe. The conception, however, does not relate to the Central Europe of today, but rather to mediated vision of what could be called Kunderian Central Europe, i.e., "not a state: it is a culture or a fate. Its borders are imaginary and must be drawn and redrawn with each new historical situation."[70] It was the place where "the Jews in the twentieth century were the principal cosmopolitan, integrating element [...] a condensed version of its spirit, creators of its spiritual unity."[71] For Kundera, "in destiny [of Jewish people] the fate of Central Europe seems to be concentrated, reflected, and to have found its symbolic image."[72]

The German-speaking intellectuals who created groundbreaking works in the Central European space and had their upswing cruelly ended by the Second World War. This Central Europe stands in the imagination of contemporary Czech Jews for some kind of mythical haven – a lost home that they desire and whose way of Jewishness they identify.

The idealized image of the fading Ashkenazi Israel in the minds and hearts of Czech Jews melts with the mental imagery of pre-war Central Europe and its fervent intellectual climate which was partly formed by Jews; retrospectively this climate has, created one of the most influential generations of Jewish thinkers of all time. It is of particular interest that this Central European nostalgia carries in some measure onto their relationship with Israel and, by extension, the Israelis.

This theme appears for example in following excerpts:

R3[73]: "Israeliness is something like the antipode of Jewishness, somehow elemental, some strange vitality to aggression. And that is something different from our Central European, specifically Prague image of Jewishness where we see a gloomy Kafka wandering through the streets around the Old Town Square."

R2: "Once, Israel truly looked like a nation of Einsteins. People had the feeling that even though they were in an exotic country, they could build it like their home. That is essentially like Bauhaus, those German emigrants who left and wanted to build Germany there. It was not only a case of buildings, but entire society."

R3: "If you're going to go to university, then you will still feel like the intellectual DNA of the German Central Europe, which is still there."

R4: "A number of things that are archetypically done to make the Jews Jewish in Central Europe died out here [in Israel]."

---

[70] Milan Kundera, "The Tragedy of Central Europe", *New York Review of Books* 31, no. 7 (April 26, 1984): 6.
[71] Ibid., 7.
[72] Ibid.
[73] "R3" stands for Respondent no. 3 and so on.

R2: "I would feel well in the former, European Israel."

Milan Kundera: "'Even after Europe so tragically failed them, the Jews nonetheless kept faith with that European cosmopolitanism; and Israel, their little homeland, finally regained, seems to me, the true heart of Europe – a heart strangely located outside the body."[74]

It is as if the disturbing remark made in 1985, during the famous Czech-French writer's oration when accepting the Jerusalem Prize for Literature, deeply resonated with the above expressed contemporary nostalgy for the vanished Central Europe.

Conclusion

The discourse of distancing, apparent in Jewish diasporas almost all over the world, has found its way of being expressed even in the Czech Republic. However, to a certain extent, it has adapted to specific local conditions. The present study strived to analyze the main patterns of its occurrence and roots among Czech Jews today.

One of the original indisputable presumptions was the distinct patterns of the motivation of American Jews. It was assumed that the crux of their distancing from Israel stems mainly from reservations regarding the peace process, the building of settlements, egalitarian prayers, etc.

In parallel, we argued that our respondents do not necessarily perceive Israel's government official line as problematic. This assumption is in a certain way mirroring, e. g., above-mentioned lower awareness of Israeli-Palestinian conflict in mainstream discourse in Visegrad countries.

Returning to the research question posed at the beginning of this paper, it is now possible to state that ambivalent attitudes of Czech Jews towards Israel might be partly discussed in the framework of ongoing changes within Israeli society. In this context, specifically, so-called "Mizrahization" or "Orientalization" of Israel should be considered as a potential reason behind the growing discourse of distancing. At this point, it is necessary to emphasize the sensitivity of the presented findings. The given claims could be easily understood as threatening for positive self-perception of both; those who would identify themselves

---

[74] Nan Robertson, "Kundera accepts Jerusalem Prize", *New York Times*, Section C (May 10, 1985), 28.

with the Czech Jews and equally for those representing Israelis. The nature of this study is qualitative, i.e., under any circumstances, expressed observations cannot be generalized and applied to the Czech Jewish community as a whole. It is rather an effort to determine possible, often unconscious, motivations behind described stances and conduct. In other words, these results should be undoubtedly interpreted with caution.

In conclusion, it is worth mentioning that there is ample room for further progress in determining similar, perhaps latent, patterns in American Jewish diaspora's[75] attitude towards Israel. Surprisingly, the discourse on ethnicity as a reason behind distancing has not been so far extensively investigated in contemporary research. For example, the above-quoted significant latest study "70 Years of Israel-Diaspora Relations: The Next Generation" was occupied in detail with the mutual relationship between the Jewish diaspora and Israel mentioned it rather marginally and touched on this topic only twice and very carefully on 130 pages of extent.[76] The report is composed of academic research and political analyses, which is focused on avoiding the distancing discourse. The report also warns that "a discourse that concentrates on distancing may itself generate distancing."[77] We could speculate that lack of attention dedicated to the impact of inter-ethnical cleavages on the Israel-diaspora relation may also stem from the similar supposition; i.e., efforts not to deepen the feelings of estrangement while Mizrahi Jews and their culture are, at least in reduced conventional perception, increasingly representing the external image of Israel. Future studies on the current topic are therefore recommended.

---

[75] Our comparison consistently supports that the ethnic homogeneity of the Czech Jewish population is comparable with the American one. The overwhelming majority of American Jews (ninety-four percent) consider themselves as "white non-Hispanics" (comparing to sixty-six percent of U.S. general public), "A Portrait of Jewish Americans", 46.

[76] Firstly, the frame of section "Recommendations for Diaspora Conduct and Objectives" suggests "Encouraging Jewish pluralism in Israel through social and educational means (…) and *honoring the composition and character of Israel's Jewish population*", "70 Years of Israel-Diaspora Relations: The Next Generation", 12. Secondly, while describing changes which have occurred in Israel: "the transition from a small and intimate society to a large population encompassing subgroups that all have their own social and ideological agendas; high birthrates and rapid demographic change; military might and political power; economic growth and the development of a Western-style society of abundance; the dominance of a political right based on religious and traditional voters, many of them *Mizrahim*", "70 Years of Israel-Diaspora Relations: The Next Generation", 18.

[77] Ibid., 7.

Bibliography

Averbukh, Lidia. "Israel on the Road to the Orient?: The Cultural and Political Rise of the Mizrahim". In *Stiftung Wissenschaft und Politik*. German Institute for International and Security Affairs, SWP Comments 9. April 2017. https://www.swp-berlin.org/fileadmin/contents/products/comments/2017C09_avk.pdf. Accessed August 5, 2019.

Bányai, František. "Slavnost vnášení tóry" ["Celebratory bringing of the torah"]. *Obecní noviny* 14, no. 7 (2017).

Bar-On, Dan. *The Others Within Us: Constructing Jewish-Israeli Identity*. Cambridge: CUP, 2008.

Benbassa, Esther. "Otherness, Openness and Rejection in Jewish Context". *Journal of the Interdisciplinary Study of Monotheistic Religions (JISMOR)* 5. February 2010. http://www.cismor.jp/en/series/jismor/. Accessed August 2019.

DellaPergola, Sergio. "World Jewish Population, 2018." *The American Jewish Year Book*, edited by Arnold Dashefsky and Ira M. Sheskin, 361–452. Dordrecht: Springer, 2018.

Dyduch, Joanna. "Die Visegrád-Staaten und Israel Dimensionen und Funktionen einer Sonderbeziehung", Osteuropa, 9–11/2019, s. 351–367. https://www.zeitschrift-osteuropa.de/hefte/2019/9-11/die-visegrad-staaten-und-israel/ Accessed April 2020.

Fischer, Shlomo. "Two Patterns of Modernization: An Analysis of the Ethnic Issue in Israel". *Israel Studies Review* 31 (2016): 66–85.

Halperin-Kaddari, Ruth, and Yaacov Yadgar. *Religion, Politics and Gender Equality among Jews in Israel*. United Nations Research Institute for Social Development (UNRISD) and Heinrich-Böll-Stiftung, 2010.

Hrbková, Markéta. "Solidarita jako jedno z důležitých témat židovských dějin" ["Solidarity as one of the important themes of Jewish history"]. *Deník Referendum*. May 28, 2019. http://denikreferendum.cz/clanek/29646-solidarita-jako-jedno-z-dulezitych-temat-zidovskych-dejin. Accessed August 2019.

Israel Central Bureau of Statistics (ICBS). *Statistical Abstract*. Jerusalem: Israel Central Bureau of Statistics, 2017.

"Jewish communities". Website of the Federation of Jewish Communities in the Czech Republic. www.fzo.cz/zidovske-obce. Accessed August 2019.

"Kaffir (ethnic slur)". *Academic Dictionaries and Encyclopedias*. https://enacademic.com/dic.nsf/enwiki/698681. Accessed August 2019.

Kelman, Ari Y., and Steven M. Cohen. "Beyond Distancing: Young Adult American Jews and Their Alienation from Israel". Andrea and Charles Bronfman Philanthropies, 2007.

Khazzoom, Loolwa. *The Flying Camel: Essays on Identity by Women of North African and Middle Eastern Jewish Heritage*. New York: Seal Press, 2003.

Köhler, Lotte, and Hans Saner. *Hannah and Karl Jaspers' Correspondence, 1926–1969*. Harcourt Brace Jovanovich, 1992.

Kundera, Milan. "The Tragedy of Central Europe". *New York Review of Books* 31, no. 7 (April 26, 1984).

Lewin-Epstein, Noah, and Yinon Cohen. "Ethnic origin and identity in the Jewish population of Israel". *Journal of Ethnic and Migration Studies*, 2018. https://doi.org/10.1080/1369183X.2018.1492370. Accessed August 2019.

Mayer, Daniel. "Maročtí Židé" ["Moroccan Jews"]. *Maskil* 8/9 (2013). https://www.maskil.cz/5773/12.pdf. Accessed August 2019.

Meyrink, Gustav. *Zelená tvář* [*Green Face*]. Prague: Volvox Globator, 1991. "A Portrait of Jewish Americans." Pew Research Center Washington, D.C. October 1, 2013. http://www.pewforum.org/2013/10/01/jewish-american-beliefs-attitudes-culture-survey/. Accessed August 2019.

Peroutka, Ferdinand. "Něco o českém národu a o židech" ["Something about the Czech nation and the Jews"]. *Přítomnost* XV, no. 19/1 (1938).

Robertson, Nan. "Kundera accepts Jerusalem Prize". *New York Times*, Section C. May 10, 1985.

Rosner, Shmuel Rosner, and John Ruskay. "70 Years of Israel-Diaspora Relations: The Next Generation". In *The Jewish People Policy Institute (JPPI)*, edited by Barry Geltman and Rami Tal, 2018. http://jppi.org.il/new/en/article/dialogue-at70/#.XSxnAugzaUk. Accessed August 2019.

Segev, Tom. *1949: The First Israelis*. New York: Owl Books by Henry Holt and Company, 1998.

Statement of the platform of Jewish Voice of Solidarity. June 6, 2019. https://www.facebook.com/pg/%C5%BDidovsk%C3%BD-hlas-solidarity-2280561515605510/about/?ref=page_internal. Accessed August 2019.

"Statistics". Website of the Federation of Jewish Communities in the Czech Republic. http://www.fzo.cz/o-nas/statistika. Accessed August 2019.

"Survey of American Jewish Opinion", AJC, June 2, 2019. https://www.ajc.org/news/survey2019. Accessed August 2019.

Unquoted recommended sources

Aschheim, Steven E. *Brothers and Strangers, The East European Jew in German and German Jewish Consciousness, 1800–1923*. Madison, Wisconsin: University of Wisconsin Press, 1982.

Fottová, Magdaléna. "Images of Eastern Jews in Czech-Jewish and Czech language journals 1910–1925". In *Close and Foreign: Jews, Literature, and Culture in the Czech Lands in the Twentieth Century Prague*, edited by Jiří Holý. Prague: Akropolis, 2016.

Website of the Jewish Voice for Peace. https://jewishvoiceforpeace.org/jvp-supports-the-bds-movement/. Accessed August 2019.

Zbyněk Tarant
Jews and Muslims in the Czech Republic – Demography, Communal Institutions, Mutual Relations

Introduction

It is not easy being Jewish, as the old saying goes, and it is certainly not easy being a Muslim these days. Until now, the Czech Republic seems to have been protected from both the eruptions of antisemitism that we are currently witnessing in Western Europe and the violent jihadist extremism that has engulfed Muslim society. Yet both the Jewish and Muslim communities in the Czech Republic are facing several serious challenges, most of which do not seem to be directly connected to the contemporary crises. While some of these challenges bring them closer together than even the two communities would be ready to admit, others place them on opposite sides of the barricade. This chapter provides a basic introduction to the current state of the Czech Muslim and Jewish communities, their demography, some of the challenges they face, and the mutual relations they experience, if any. While some literature on the contemporary situation of the Jewish and Muslim communities exist in the Czech Republic,[1] the sources in English are very limited. Therefore, the following chapter should address this problem and provide a basic introduction for foreign readers.

Demography – the Jews

Estimating the demography of religious and national minorities presents a serious methodological as well as ethical challenge, especially in the case of the

---

This research was made possible thanks to the generous support by the Jan Hus Educational Foundation (Stipendium vzdělávací nadace mistra Jana Husa 2015).

[1] For the Jewish communities, see Alena Heitlingerová, *Ve stínu holocaustu a komunismu – čeští a slovenští židé po roce 1945*. [*In the shadow of the Holocaust and Communism – Czech and Slovak Jews after 1945*] (Prague: G Plus G, 2007). For the Muslim communities see Miloš Mendel, Bronislav Ostřanský, and Tomáš Rataj, *Islám v srdci Evropy* [*Islam in the Heart of Europe*] (Prague: Academia, 2008); or Lukáš Lhoťan, *Islám a islamismus v České Republice* [*Islam and islamism in the Czech Republic*] (Prague: Self-published, 2011).

https://doi.org/10.1515/9783110582369-011

Jews. The collection of such data by official authorities was outlawed in most European countries after World War II, because census results and community membership lists were previously abused as part of the plan to carry out a systematic extermination program. In today's Czech Republic, religious affiliation or ethnicity are declared intimate personal details and it is not allowed to systematically collect and store such data without explicit and informed consent. Any data about the population and the demography of religious minorities in the European Union are thus only qualified estimates that are more or less accurate. The simple question of how many Jews and Muslims are currently living in the Czech Republic has, in fact, multiple potential answers depending on the methodology employed in data collection or estimation. As Table 1 shows, different sources yield different results:

Table 1: Comparison of Jewish and Muslim population estimates in the Czech Republic according to various sources.

	Jews	Muslims
Census Results	345 (521)*	1,921
Community members (stated by the community)	3,000	13,000
Qualified estimates	3,900[2]	19,000[3]
Estimates that commonly appear in the media (without source data)	15,000–20,000	10,000–20,000

* Jews by "nationality"

Table 1 requires additional explanation. First, we should consider the national census. The forms for the Czech national census, the last one being carried out by the Czech Statistical Office in 2011, contain voluntary fields for "religion" and "nationality". However, the majority of the Czech population leaves these non-obligatory fields blank or fills them in with disingenuous answers (e.g. 15,055 claimed to be Jedi Knights in one census). If we were to accept the census results literally, then there would be only 345 Jews by religion and 521 by nationality. The fact that hundreds of respondents claimed to be Jewish by nationality but not by religion is indeed noteworthy and could be connected to the phenomenon of "Jews by descent," i.e. those individuals who claim to be Jewish and even have Jewish parents, yet do not maintain any Jewish traditions or customs

---

[2] Sergio DellaPergola, "World Jewish Population." *Berman Jewish DataBank* 2013. Appendix A, accessed February 20, 2016.

[3] "Cizinci podle státního občanství k 12. 31. 2014". ["Foreigners according to nationality"]. *Czech Statistical Office*, accessed February 20, 2016.

at all. According to the same census, there would be only 1,921 Muslims by religion.[4] These are suspiciously small numbers, even by Czech standards. The numbers are still valuable, however. One can read them as an "index of trust" – how many people actually trust the state enough to share their "nationality" and "religion" with its official authorities.

The second source of data are the numbers provided by the community institutions themselves. Both the Jewish and Muslim communities publicly state the approximate size of their membership base. The *Federation of the Jewish Communities in the Czech Republic*, which is an umbrella organization of Czech Jews, reports having 3,000 members.[5] The number seems to be close to the one given by the leading expert on Jewish demography, Sergio DellaPergola, who estimates the Czech Jewish population to be 3,900. However, he accompanies this with a strong disclaimer that the number is simply a "base estimate derived from less recent sources and/or unsatisfactory or partial coverage of a country's Jewish population; updated on the basis of demographic information illustrative of regional demographic trends."[6]

With the Jewish communities, the old question "Who is a Jew?" becomes a serious methodological problem as some of the communities and organizations only include *halachic* Jews, while others may also include those who are eligible for the *aliyah*,[7] but are not Jewish according to the *halacha* (see the painful issue of "Fatherjews", "tatínkovci" in Czech). In the end, the Federation of the Jewish Communities estimates that while it has 3,000 members, there might actually be between 15,000 and 20,000 individuals[8] who could be considered as either *halachic* Jews or at least eligible for the *aliyah* according to the Israeli Law of Return. However, most of them are assimilated and non-religious, and many are unaware of being Jewish in the first place. The notoriously high degree of agnosticism in Czech society affects its Jewish community as well, and one should not be surprised to meet community leaders who do not know how to utter the Hebrew blessing over bread. On the other hand, the community allows foreigners to register as members, which means that not all of these 3,000 members are Czech Jews. For example, there are sizable groups of American Jews who have found a partner or do their business in the country and who participate in communal life.

---

4 "Obyvatelstvo podle věku, náboženské víry a podle pohlaví". ["Population according to age, religion and gender"]. *Czech Statistical Office*, accessed February 20, 2016.
5 Statistics. *Federation of the Jewish Communities*, accessed February 20, 2016.
6 DellaPergola, "World Jewish Population".
7 For example Immigration to Israel according to the Law of Return.
8 Statistics. *Federation of the Jewish Communities*. Accessed February 20, 2016.

There are also about 900 Israeli citizens living in the Republic,[9] the majority of whom are likely to be Jewish, although Israeli Arabs or Messianic Jews do actually live in the Republic as well and congregate in their own respective communities.

The demographic profile of the Czech Jewish community seems to be very unfavorable. Tomáš Jelínek, a former chairman of the Prague Jewish community, made a quick demographic survey in 2014. Six of the ten regional communities sent information in response to his request, including the largest ones in Prague and Brno, thus covering a sample of about 2,500 Czech Jews. Jelínek found that the mean age of Czech Jews is between fifty-two and sixty-six[10] and that some communities, such as Ostrava, have almost no members younger than thirty years of age. According to Jelínek, the situation is somewhat better in Prague, which records about sixteen percent of its members as being either thirty years old or below.[11] If Jelínek's data are valid, then it points to a severe level of population aging. Without a "demographic miracle", some of the regional communities could well disappear over the next twenty to forty years.

Let us quickly discard the beliefs of Czech antisemites, who claim that there are 200,000 Jews living in the country.[12] This absurd number makes sense only to the Holocaust deniers – if there has been no Holocaust, then the 146,000 Jews living in Bohemia and Moravia before World War II must still be living somewhere. The difference between the actual number of Jews living of the country and the numbers that are believed to be living there by antisemites is an interesting example of so-called "antisemitism without Jews". This difference between myth and reality can be also partially responsible for the obsession of Czech antisemites with the production of "lists of Jews", containing the names of individuals who are considered to be Jewish, even if they are not.[13]

---

9 "T14 Cizinci podle kategorií pobytu, pohlaví a občanství k 12. 31. 2014." [English title: "T14 Foreigners according to residency category, gender and citizenship"]. *Czech Statistical Office*. Accessed February 20, 2016.
10 In December 2014, the mean age of the Czech Republic ranged between forty-one and forty-two. See "Průměrný věk obyvatel v krajích České republiky k 12. 31. 2014." ["Mean age of the population in regions of the Czech Republic"] *Český statistický úřad*. Accessed February 20, 2016.
11 I would like to thank Dr. Jelínek for sharing his yet unpublished data. The data is valid to 2014.
12 See the introduction for the Czech translation of Traian Romanescu's pamphlet "Masters and Slaves of the 20th Century."
13 Such lists were published in print by *Týdeník Politika* (96 (1992): 2) and online by neo-Nazi website *Národně-vzdělávací institut*. Available only through WebArchive, accessed February 20, 2016, http://www.vzdelavaci-institut.com/_files/index.php?option=com_content&task=view&

Demography – Muslims

The Muslim communities are surprisingly small in the Czech Republic when compared to Western Europe. The Centre for Muslim Communities, which claims to be an umbrella organization for Czech Muslims (see below), also claims to represent about 13,000 members. Again, similar problems arise with such numbers. First, the method used in calculating this estimate is unclear. Second, if the number comes from the membership lists of the Centre for Muslim Communities, then it represents only the organized members. If there is a non-practicing Muslim, or a practicing one who, for whatever reason, refuses to become a member of this particular organization, he/she does not appear in the statistics. After all, although the Centre for Muslim Communities claims responsibility for over 13,000 members, in the mid-2000s it was unable to collect the 10,000 signatures needed for the so-called "second level" of registration that would grant it the full rights of an organized religious community.[14]

Another way of estimating the Muslim population in the Czech Republic is to count foreign nationals from "Muslim" countries who are currently living in the Czech Republic. This is possible due to the fact that foreigners represent a significant majority of the Czech Muslim population.[15] Only a small fraction (several hundreds) are Czech converts to Islam. According to the data provided by the Czech Foreign Police from December 2014, there were 19,345 foreigners from countries that have dominant Muslim majorities and/or Islam as the state religion.[16] The statistics include all kinds of residence status, from asylum seekers

---

id=327&Itemid=57. The latest list is being compiled by Adam B. Bartoš on his blog, https://cechycechum.wordpress.com.

14 Miloš Mrázek, "Muslimská unie." ["Muslim Union"]. *Dingir* 1/2006, 23, accessed February 20, 2016. So far, it has managed to achieve only the "first level", which grants it only legal subjectivity (it can act as an organization), and due to the "Islamosceptic" nature of Czech society later attempts have been met with fierce opposition.

15 Daniel Topinka et al., "Muslimové imigranti v České Republice: Etablování na veřejnosti" ["Muslim immigrants in the Czech Republic: Public Emancipation"], *Fenomén moci a sociálne nerovnosti – nultý ročník konferencie pre doktorandov a mladých vedeckých pracovníkov*, ed. Lukáš Bomba, Estera Kövérová, and Martin Smrek (Univerzita Komenského v Bratislave, 2014), 244, accessed February 20, 2016. See also Jiří Bečka and Miloš Mendel, *Islám a České země* [*Islam and the Czech Lands*] (Olomouc: Votobia, 1998).

16 This would include the following countries: Afghanistan, Albania, Algeria, Azerbaijan, Bahrain, Bangladesh, Burkina Faso, Chad, Djibouti, Egypt, Gambia, Guinea, Indonesia, pre-war Iraq, Iran, pre-war Yemen, Jordan, Kazakhstan, Kosovo, Kuwait, Kyrgyzstan, Lebanon, Libya, Malaysia, Maldives, Mali, Morocco, Mauritania, Niger, Nigeria, Pakistan, The West Bank and

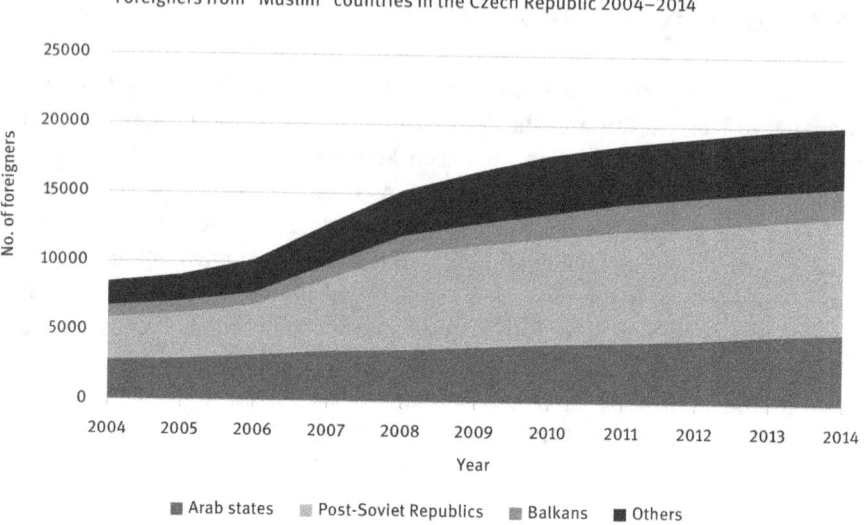

Figure 1: Source: Czech Statistical Office, Public Database, Foreigners in the Czech Republic.

to long-term residence permit holders. Of course, this number is not final as there are non-Muslim countries with significant Muslim minorities (Russia, Israel, France, India etc.) and there are also Muslim countries with non-Muslim minorities (Egypt, Lebanon, pre-war Syria). But even with a disclaimer about the potential inaccuracy of information, along with taking into consideration the several hundred Czech Muslim converts, we are able to estimate that there might be approximately 20,000 Muslims currently living in the Republic. Topinka et al. arrived at a similar estimate in 2009. By using a more sophisticated methodology that took into account the issue of minorities, they recorded a number of 19,097.[17] The timeline of migration statistics reveals that the number of Muslim migrants in the Czech Republic has nearly doubled since the mid-2000s, from about 10,000 to today's 20,000. However, Muslims still represent only about four percent of the migrant population in the Czech Republic and about 0.0016 percent of the total Czech population.[18] This recent change has also influenced the inter-

---

Gaza in pre-1967 borders, Saudi Arabia, Senegal, Sierra Leone, Somalia, United Arab Emirates, Sudan, pre-war Syria, Tajikistan, Tunisia, Turkey, Turkmenistan, and Uzbekistan.

[17] Daniel Topinka et al., "Muslimové imigranti v České Republice: Etablování na veřejnosti," 244.

[18] "Cizinci podle státního občanství k 12.31.2014" ["Citizens according to nationality"]. *Český statistický úřad*. Accessed February 20, 2016.

nal structure of the Muslim community, an issue that will be discussed below. Unfortunately, there are no reliable data from which we are able to derive the demographic profile (the age groups) of the community. The charts that appear on Wikipedia, for example, are derived from 2011 Census data, which may be biased as the majority of the Czech Muslims appears to have left the field "religion" blank.

Jewish Community Institutions

The umbrella organization of the Czech Jews is The Federation of the Jewish Communities in the Czech Republic (Federace židovských obcí v ČR). In addition to the largest Jewish Community in Prague and its regional branches (see below), the Federation includes several "collective members," namely the Jewish Liberal Union, which represents the Liberal stream, the Conservative movement, *Bejt Praha*, as well as *Bejt Simcha*, which unites the Reform Jews. From a legal standpoint, these representatives of Liberal, Conservative and Reform streams are not registered as "religious organizations." They are mere "civic associations" with collective membership within the Federation, having no voting rights in the established Jewish communities.

The Federation also includes other institutions and organizations such as the Theresienstadt Initiative, The Jewish Museum in Prague, Ha-Koach Sports Club, The Union of Jewish Youth, The Association of Jewish Academics etc.[19] It runs a kosher dining hall for its members as well as a care center for the elderly, most of whom are Holocaust survivors who have no relatives. There is a Jewish kindergarten, a basic school, and a high-school in Prague, established thanks to the support of the Ronald S. Lauder Foundation.

Several Jewish periodicals are published on a regular basis, namely the *Roš Chodeš* (a mouthpiece of the Prague Jewish Community, formerly known as *Věstník židovských náboženských obcí*), *Židovské listy* (which provides critical opposition to the leadership of the Federation), *Maskil*, published by the Reform community, *Bejt Simcha*, or *Šaliach*, published by the community in Teplice. There are several news sites for Czech Jews, including the *Židovský tiskový informační servis*, *Blog Židovských listů*, the liberal website *Shekel*, the pro-Israel website *Eretz.cz*, and others. Some of these "Jewish" websites are actually run by non-Jews. Czech Radio broadcasts a special short program for Czech Jews, enti-

---

[19] "Federace židovských obcí – organizace" ["Federation of the Jewish Communities – About us"]. Accessed February 20, 2016.

tled *Šalom alejchem*, which is presented by the chairman of the Jewish Museum in Prague, Leo Pavlát.

It can sometimes be difficult to label the Czech Jewish Communities according to their rites as there can be differences between theory and daily practice. For example, while the leadership and rabbinate of the Prague Jewish Community follows the orthodox stream, the majority of its members are either non-practicing "Jews by descent" or descendants of a very diverse variety of streams, including Conservative, Liberal and Reform, which in itself is a source of bitter personal conflicts and controversies within the community. The Czech Jewish communities face a typical dilemma, which is well summarized by Peter Salner: "tradition without people, or people without tradition?"[20] After the Velvet Revolution, the largest community in Prague became divided by tensions between the orthodox approach of the post-1993 chief rabbinate and voices that called for a more liberal, "open" approach. While the former chief rabbi, Karol Sidon, had viewed the orthodox stream as a way of preserving Jewish traditions against assimilation, the more liberal faction had been worried that its strictness actually weakened the community.[21] The quarrels then included topics such as community membership conditions, restituted property etc.

The conflicts inside the Prague community erupted into a series of swift, abrupt changes in leadership in the mid-2000s.[22] Ten years later, the bitterness from these past skirmishes and the polarization of the community around the two main conflicting personalities (Karol Sidon and Tomáš Jelínek) is still palpable, making it difficult for an outsider to navigate the minefield of personal animosities. In that sense, the Prague Jewish Community confirms the classic Jewish anecdote that if there are two remaining Jews on a deserted island, they will always establish at least three synagogues. The new Chief Rabbi of Prague, David Peter, a young, fresh graduate of the orthodox rabbinical school in Israel, has a difficult mission in reuniting the divided community, keeping pace with the as-

---

[20] Peter Salner, *Židia na Slovensku medzi tradíciou a asimiláciou* [Jews in Slovakia between tradition and assimilation] (Bratislava: Zing Print, 2002), 103. Alena Heitlingerová, *Ve stínu holocaustu a komunismu – čeští a slovenští židé po roce 1945* [In the shadow of the Holocaust and communism – Czech and Slovak Jews after 1945] (Prague: G Plus G, 2007).

[21] See L. Arava-Novotná, "Židovská tradice mezi pokračováním a mytologizací. Pražský judaismus a jeho předávání v letech 1945 – 2005" ["Jewish tradition between continuation and mythologization. Prague Judaism and its tradition between 1945 – 2005"]. *Mýtus – "Realita" – Identita. Národní metropole v čase "Návratu do Evropy"*, ed. Blanka Soukupová and Andrzej Stawarz, Urbánní studie, sv. 9. FHS UK v Praze (Prague, 2015), 45 – 48.

[22] For a full explanation of the roots, causes, and aftermath of the disputes, see A. Heitlingerová's *Ve stínu holocaustu a komunismu – čeští a slovenští židé po roce 1945*.

sertive Chabad and continuing the crucial process of transforming a "caretaker community" into the spiritual center of Jewish life that it once was.

Rabbi David Peter himself leads an orthodox congregation in the Jerusalem Synagogue. The Spanish Synagogue maintains Conservative rites and is visited mostly by foreign Jewish visitors and tourists as the domestic Conservative community is rather small. The Reform Jews congregated in the Jerusalem Synagogue until 1994, but they have had no regular place of prayer since then, and their *minyan* meets either in the rabbinical courtroom at the Jewish town-hall in Maiselova Street, in one of the regional communities, or in the Pinkas synagogue, where they organize common holiday services, sometimes together with the Jewish Liberal Union. Outside these joint services, the JLU itself usually meets in the Tall Synagogue for Kabbalat Shabbat.

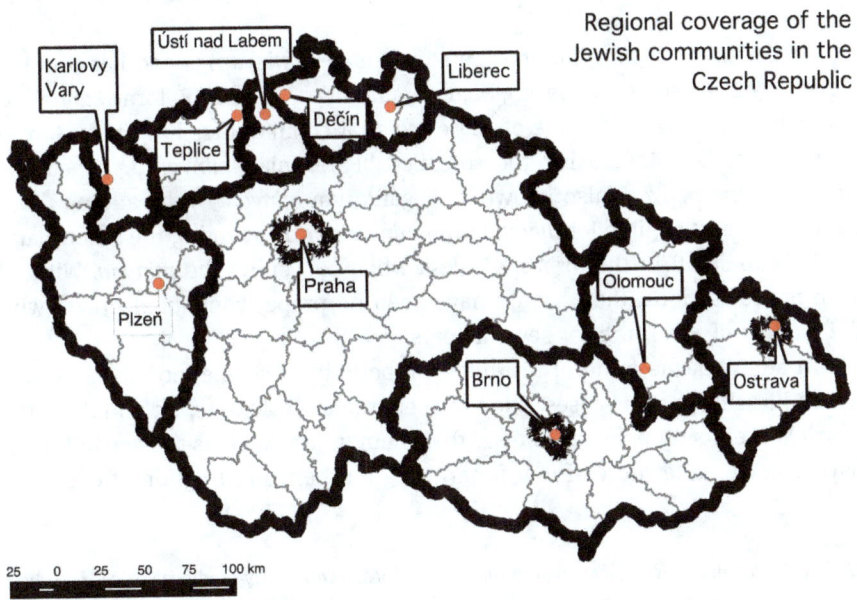

Figure 2: Source: Map created by the author, based on data published by the Federation of the Jewish Communities.

Outside Prague, there are regular services on Saturdays and most holidays are observed in the Brno, Olomouc, Liberec, Karlovy Vary, and Teplice communities. The regional communities find themselves in an even more difficult situation than the ones in Prague: they sometimes rely on one or two leading personalities, whose unplanned indisposition can effectively paralyze all activities and projects. This includes even the more active communities such as in Teplice, which is dependent on several families. Others, for example in Děčín, organize shabbat services only once or twice a month. One community, in Ústí nad Labem, is currently under threat of being formally dissolved due to inactivity.

The phenomenon of the so-called "flying rabbis" (i.e. one rabbi or *chazan* providing religious services to multiple communities) is very common in the Czech Republic. Sometimes, due to the lack of qualified spiritual leaders, the community cannot choose a rabbi or *chazan* who follows its own rites, but has to temporarily adjust its rites according to the persuasion of the particular rabbi or *chazan* who is currently available. This is the experience of Ostrava, where an orthodox rabbi from Olomouc leads Shabbat prayers once a month for its mainly Reform members. The replacement of Kabbalat Shabbat by an academic lecture delivered by a non-Jewish scholar occasionally takes place in Pilsen and Ostrava. Both communities currently have no permanent rabbi or *chazan*.

Achieving a proper quorum of ten adult males (*minyan*) is another continuing struggle for some of the regional communities. As one Reform *kantor* explained in an interview, some of these regional communities have been forced to improvise by either using the so-called "Palestinian *minyan*",[23] by adding women to the *minyan* (also known as "egalitarian *minyan*"[24]), by adding those who are at least eligible for *aliyah* to the *minyan*, or by counting the second, unused Torah scroll in the *aron ha-kodesh* when convening the *minyan*. Without such improvisations, there would have been no proper communal prayer with the reading of the Torah for several years.

In addition, the issue of restituted property has become more of a burden than a blessing in some cases, due to the enormous costs of repair and the bitter quarrels over its proper use. Some of the communities, such as the one in Pilsen, have become quite successful at fundraising for the reconstruction of their syn-

---

23 This is in analogy to a local custom (*minhag*) recorded from early-medieval Palestine, where only seven men were required for *kaddish* and *brachot*. The issue of minyan is discussed in Reform Judaism, yet not even Reform Judaism is willing to make significant changes to this practice and tries to find alternative solutions as to what to do if a *minyan* is not present. See for example "The Need for a Minyan" in *Teshuvot for the Nineties: Reform Judaism's Answers to Today's Dilemmas*, ed. W. Gunther Plaut and Mark Washofsky (New York: Central Conference of American Rabbis, 1997), 23–28. The author would also like to thank the Reform chazan, Michal Foršt, for sharing his personal experiences in relation to the practice.

24 While the practice of "egalitarian *minyan*" is most typically identified with the Reform stream of Judaism, one can find egalitarian *minyanim* in other streams as well. In the orthodox stream, for example, there is a well-known synagogue, *Shira chadasha*, in Jerusalem. Recently, the Israeli Supreme Court ordered the creation of a space for egalitarian minyan at the Western Wall in Jerusalem.

agogues, but their religious life is limited and their reconstructed monumental synagogues are often converted into tourist landmarks or rented out for secular purposes, such as concerts or exhibitions, while the small community meets on Saturday evenings for tea, cake, and a lecture in the office of the *kehila*.

Muslim Community Institutions

The Czech Muslim community can hardly be described as a "community" as it is divided along multiple lines into sub-communities. One division relates to ethnic boundaries: Arabic, Turkic, Bosnian, Albanian, Chechen, Turkmen, Iranian, Pashtuni etc. But groups are also divided according to the Sunni-Shiite sectarian divide, or according to their actual attitudes to religion as such – ranging from "secular" Marxists and Baathists on the one hand to highly devout, conservative Muslims on the other. The influx of Muslim migrants from post-Soviet republics has further changed the ethnic and national map of the Czech Muslim community, with the various Turkic and Caucasian nations replacing the Arabs as the strongest group. This also means that the largest group of Czech Muslims today stems from a Turkic and Sunni background, followed by Sunni Arabs.

At first glance, the organization of the Muslim community superficially resembles the Jewish one. However, its structure is much more complex. There are two parallel networks of regionalized Muslim community institutions ("Muslim Communities" and "Islamic Foundations"), and several smaller organizations with a national coverage. First, there is an umbrella organization, *The Centre for Muslim Communities* (Ústředí muslimských obcí, ÚMO) that unites the Muslim Communities of Prague, Brno, Hradec Králové, Karlovy Vary, and Teplice. The establishment of a new one is rumored to be taking place in Plzeň. The Centre for Muslim Communities is supposed to represent the Muslims vis-á-vis the state as only this institution has the legal status of a religious organization. Its regional branches lack legal autonomy.[25]

Parallel to these "Muslim Communities" are the so-called Islamic Foundations, namely the *Islamic Foundation in Prague* (Islámská nadace v Praze), the *Islamic Foundation in Brno* (Islámská nadace v Brně), and the *Islamic Foundation in Teplice* (Islámská nadace v Teplicích). These foundations run some of the most important mosques and prayer rooms. In simple terms, the difference can be explained by stating that the "Muslim Communities" unite and organize people in general, while the "Islamic Foundations" have been established by the

---

25 They do not have a Company Registration Number (IČO) of their own.

"Communities" for specific purposes, such as the establishment and operation of mosques. The Islamic Foundation in Teplice and Karlovy Vary mostly provide religious services for visiting Arab patients at the local spas. The Islamic Foundation in Brno covers the Moravian regions of Jihomoravský, Vysočina, Olomoucký, Moravskoslezský, and Zlínský. Its *imam*,[26] Muneeb Hasan al-Rawi, also chairs the umbrella organization, the Centre for Muslim Communities. However, the structure is confusing, with the institutions tending to overlap in relation to both their activities and their membership base. Furthermore, they do not publicly comment on one another and it is not surprising to meet a Czech Muslim who does not understand the difference between the "Foundations" and the "Communities".

Miroslav Mareš pointed out in 2015 that the trend towards the fragmentation of the Czech Muslim communities continues to increase.[27] Smaller, more specialized or dissenting organizations with national coverage are in the process of building their own parallel structures, such as the *General Union of Muslim Students* (Všeobecný svaz muslimských studentů), or the recently revived offshoot, *Muslim Union* (Muslimská unie), led by Muhammad Abbas, notorious for his extremist attitudes. Some Arab businessmen have attempted to establish a parallel network of prayer rooms, independent of the Centre, as in the case of the *Alfirdaus* language schools, which are also being used for prayer gatherings. Some of the Muslims, such as the Turks or the small groups of Shiites, are loosely organized around their particular place of prayer or around the Lebanese and Iranian Embassies, without having established an official institutional basis. There are also organizations and institutions of a non-religious character, which are nonetheless important in the lives of Arabs and Muslims in the country, such as the *Czech-Arab Society* and the *Dar Ibn Rushd* publishing house, founded by the former Syrian political dissident, Charif Bahbouh.

According to one directory,[28] there are five places of prayer for Muslims in Prague, two run by the Islamic Foundation in Prague (the first has been moved to the *Národní třída* in Prague, the second is a large complex in the Černý Most neighborhood), one prayer room being run by Turkish Muslims, one by the Muslim Union, one being located at the student dormitories in Trója, and the last one existing only semi-officially under the banner of Alfirdaus

---

26 For example, Muslim religious leader, lit. "the one sitting in front during prayer".
27 Miroslav Mareš, "Muslimská politika v ČR na rozcestí" ["Muslim politics in Czechia at the crossroads"]. *Naši politici*, January 20, 2015, accessed February 20, 2016.
28 "Adresář modliteben" ["Directory of prayer rooms"] *Infomuslim.cz*. Accessed February 20, 2016.

Language School.[29] The largest of these is the complex at Černý Most, which is *de facto* a mosque. It belongs to the Islamic Foundation in Prague and does not have the legal status of mosque, although it is commonly referred to as such ("Pražská mešita"). It contains a large prayer room with a *mihrab* and is fully equipped for regular prayers, Friday prayers, and communal activities, including the provision of special programs during Ramadan and assistance in the preparation of halal food.

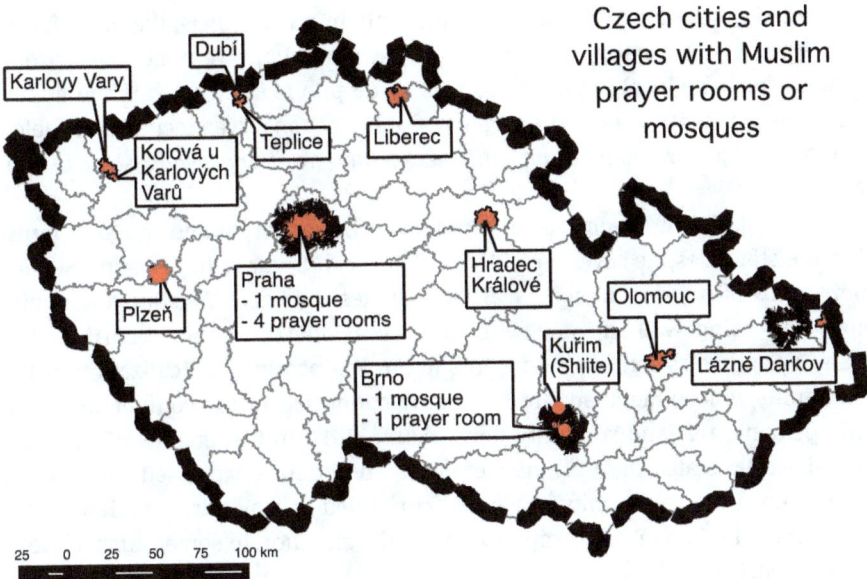

Figure 3: Source: map compiled from the directory of mosques and prayer rooms. ("Adresář modliteben" ["Directory of prayer rooms"]. Infomuslim.cz, accessed February 20, 2016)

In the legal sense of the term, there is only one official mosque in the Czech Republic and this is the one in Brno, run by the Islamic Foundation in Brno. It was established in 1998,[30] following a painful public debate, as a small single-story structure without a *minaret* and being only plainly and modestly decorated. Attempts to establish additional mosques have failed, either due to the un-

---

29 Lukáš Lhoťan, "V Praze je mešita, kterou muslimové vydávají jen za jazykovou školu. Podvedli u toho úřady." ["There is a mosque in Prague which the Muslims present only as a language school, deceiving the authorities."]. *Reflex.cz*, June 20, 2014, accessed February 20, 2016.
30 "Historie brněnské mešity" ["History of the Brno mosque"]. *Islámská nadace v Brně* [*Islamic foundation in Brno*], September 27, 2008, accessed February 20, 2016.

realistic nature of such projects (Orlová) or due to the negative public response (Teplice, Hradec Králové).[31] Arab patients in Teplice spas can choose between two prayer rooms, one in the town center and the second in the nearby town of Dubí. There are also about a dozen smaller prayer rooms in regional cities, run by the Union of Muslim Students which are very often located in student dormitory blocks as most of the Muslims in these regional cities are foreign students from Muslim countries. The Shiites have one small prayer room in Kuřim, established in 2012 by the Ahlu al-Bayt society, led by Abdulrahman Adday.[32] While it had already existed there for several years without any conflicts, the plans for its transfer to a more central location in Brno-Královo Pole were met with strong hostility from the local populace, as was made public in 2016 when the migration crisis was at its peak.[33] The negative public response was incited and fueled by the nativist movement *Islám v ČR nechceme* [*We do not want Islam in Czechia*].[34]

The majority of Czech Muslims appear to be moderate and on good terms with the state, despite several incidents. The Czech Security Information Service (Bezpečnostní informační služba, BIS) evaluates the Czech Muslim community mostly as being well-adjusted with only occasional incidents occurring. The Arab Spring has raised some concerns about the possible radicalization of the community, whose members are frequently connected to the conflict zones via their personal ties and solidarity networks. In 2011, in the shadow of the Arab Uprisings, BIS stated that: "There were isolated opinions expressed with radical elements, yet the community in general continued to maintain a moderate religious character, namely in comparison with the situation in some states of Western Europe."[35]

Most concerns have been raised in connection with the Salafist and Jihadist propaganda that appears on the web as it could potentially target the domestic Muslim community in the Czech Republic as well. This is especially the

---

31 Z. Vojtíšek, "Český boj o mešity." ["The Czech Mosque disputes"] in *Dingir* 1/2006, 19–21, accessed February 20, 2016.
32 Zuzana Taušová, "Kuřim má netušené prvenství. První centrum šíitských muslimů v zemi." ["Unexpected primacy for Kuřim – A first Shiite religious center in the country"]. *Idnes.cz*, August 8, 2012, accessed February 20, 2016.
33 Lukáš Valášek, "V Králově Poli vzniká Islámská čtvrť, vyhrožují brněnské letáky." ["An Islamic neighborhood is being created in Královo pole, Brno flyers warn."]. *iDnes.cz*, February 9, 2016, accessed February 20, 2016.
34 "AKTUALIZACE: Brno se bouří proti nové islámské modlitebně." ["UPDATE: Brno rises up against the new islamic prayer room"]. *IVČRN*, February 4, 2016, accessed February 20, 2016.
35 "Výroční zpráva Bezpečnostní informační služby za rok 2011." ["Annual report of the Security Information Service for 2011"]. *Bezpečnostní informační služba*, accessed February 20, 2016.

case as the community lacks qualified religious leaders and has to improvise by either asking under-qualified enthusiasts to serve as *imams* or by inviting *imams* from abroad, which involves a degree of risk. The security services make it publicly clear that they follow and monitor the activities of jihadist propaganda or suspicious persons attempting to establish contact with the community. In its reports, which are written meticulously in order to provide an overview of its activities, but which do not give away any concrete classified details, the BIS summarizes:

> In 2014, the BIS investigated several cases of foreigners who were active on Czech territory and about whom there was a reasonable degree of suspicion that they had travelled to Syria to join the battle. Apart from the presence and activities of these individuals, the BIS has also focused on those who sympathize with I[slamic] S[tate] or other ideologically similar terrorist organizations. From this perspective, the controversial converts and some native-born, mostly young Muslims were the most dangerous group. However, the condemnation of IS activities dominated the Czech Muslim community in 2014.[36]

This summary points to the fact that while the community in general is moderate and loyal to the Republic, there could potentially be several radical individuals, both from inside the Czech community and among its foreign visitors, whose activities should be closely watched.

There are concerns that foreign powers, namely Saudi Arabia, might use their funding activities as a means of leverage for the export of Wahhabism or Salafism. The extent to which the Czech Muslim community is really threatened by this phenomenon is still not clear. Some evidence exists, but more is needed. The activities of the so-called Third World Relief Agency (the foundation suspected of financing terrorism) were investigated in the mid-2000s[37] in the Czech Republic and it is possible that the agency might have been involved in the funding of the Brno mosque.[38] Multiple sources suggest that Jihadist propaganda may be active in Czech mosques and that several Jihadi-linked organizations have used the Czech territory for their activities, namely Jama'at shari'a, Tablighi jama'at, Hizballah-linked Al-Aqsa' foundation, and others. Activities of such foundations suspected of having links to Jihadism were investigated in several

---

36 "Výroční zpráva Bezpečnostní informační služby za rok 2011." ["Annual report of the Security Information Service for 2011"]. *Bezpečnostní informační služba*, accessed February 20, 2016.
37 "Výroční zpráva Bezpečnostní informační služby za rok 2014." ["Annual report of the Security Information Service for 2014"]. *Bezpečnostní informační služba*, accessed February 20, 2016.
38 Zdeněk Vojtíšek, "Posouzení 'Ústředí muslimských obcí' vzhledem k řízení o registraci podle Zákona 3/2002 sb." ["Evaluation of the 'Center of the Muslim Communities' in regard to the process of registration according to the Act no. 3/2002 sb."], accessed February 20, 2016.

cases.³⁹ The Czech Security Information Service (BIS) has commented on the "repeated discrepancies in the incomes and expenditures of the islamic organizations active within the Czech territory".⁴⁰ Lukáš Lhoťan (see below) has recently published a document that he claims proves the direct financing of the Brno mosque from Saudi sources.⁴¹ However, the document contains graphical glitches and is unsigned, so it is difficult to verify its authenticity.

There have been several serious revelations and leaks attributed to the former member of the community, Lukáš Lhoťan – a former Czech Muslim convert, founder of an anti-Zionist website *Bezcenzury.org* (now defunct),⁴² and member of the community in Brno. He later converted to Christianity and started leaking the community's internal documents, videos, and recordings. Lhoťan has accused the Czech security services of underestimating the jihadist activities in Czech mosques and claims that the police ignore radical sermons and slogans in Arabic, due to a lack of qualified translators.⁴³ Some of Lhoťan's revelations have directly affected Muslim-Jewish relations in the Czech Republic as they contain a recording of a sermon delivered by a radical Czech convert, Lukáš Větrovec, that makes reference to medieval anti-Jewish texts as the justification for Middle-Eastern violence.⁴⁴ This particular case has caused a degree of damage in the relationship between the communities (see below).⁴⁵ Lhoťan has published a book, entitled *Islam and Islamism in the Czech Republic*, which claims that the Czech Muslim community has been taken over by radicals who are funded from abroad, and who have also managed to push out the more moderate

---

39 Miroslav Mareš, "Islamismus jako bezpečnostní hrozba pro Českou republiku" ["Islamism as a security threat for the Czech Republic"]. *Vojenské rozhledy* 2010/4, 118–128, accessed February 20, 2016.
40 "Důvodová zpráva k návrhu zákona, kterým se mění zákon č. 153/1994 Sb., o zpravodajských službách České republiky, ve znění pozdějších předpisů." ["Explanatory report to the proposal of an act that mends the the Act no. 153/1994 sb."]. *Bezpečnostní informační služba*, accessed February 20, 2016.
41 Lukáš Lhoťan, "Jak čeští islamisté "nepřijímají" peníze ze Saúdské Arábie" ["How the Czech Islamists are 'not receiving' money from Saudi Arabia"]. *Eurabia.cz* February 28, 2016, accessed February 20, 2016.
42 *Bezcenzury.org*. Available only via WebArchive, accessed February 20, 2016.
43 Lukáš Lhoťan, "Islám a islamismus v České Republice" ["Islam and Islamism in the Czech Republic"]. Prague, 2011.
44 See the video on Lukáš Lhoťan's YouTube account. *YouTube* 11.10.2011, accessed February 20, 2016.
45 "Prohlášení Federace židovských obcí v ČR, Židovské obce v Praze a Židovského muzea v Praze." ["Proclamation of the Federation of the Jewish Communities in the Czech Republic, Jewish Community of Prague and Jewish Museum in Prague"]. *Federace židovských obcí*. 12.12.2011, accessed February 20, 2016.

Muslim communities. In addition, the umbrella organization Centre for Muslim Communities, funded from suspicious sources, represents only a minority of Czech Muslim voices.[46]

Lhoťan repeatedly criticizes and attacks his former co-believers. One of his claims, that an extremist book was being distributed in Czech Muslim centers, led to quite a controversial police crackdown on the Prague Mosque.[47] The book in question was Bilal Philips's *Fundamentals of Tawheed*, which contains quotations from several anti-Jewish medieval texts. Only one person was indicted and the police operation was criticized for being inappropriate, both in terms of its use of force (sending a heavily armed anti-terrorist unit to indict a single person in relation to a translated book) as well as in relation to its timing, coinciding as it did with the Friday sermon.[48] It should be noted that Bilal Philips is a controversial personality, being banned from several Western countries, and that the choice of his books from among the dozens of better and more moderate works on the concept of Divine unity (*tawheed*) in Islam certainly does not leave us with a good impression of the Czech Muslim community, which itself does not appear to act strongly enough to curb extremist propaganda among its members. In this sense, Lhoťan's criticism is fully justified.

Lhoťan continues to provoke others, but he himself proves to be a complicated personality, to say the least. In 2011, he published pictures showing the presence of masked men with guns during a sermon in the Brno mosque.[49] Interestingly enough, as one Jewish liberal website pointed out, the guns were actually air-soft replicas and, judging from the unique paint stains on his camouflage uniform, one of the masked men might well have been Lhoťan himself, who was an air-soft enthusiast at the time.[50] It does not change the fact that such weapons, even if they are non-lethal, are not supposed to be present in Mosque grounds, nor does it change the fact that the preacher, Lukáš Větrovec, allowed this and let himself be photographed. Neither is Větrovec alone in relation to his

---

[46] Lhoťan, *Islám a islamismus v České Republice*.
[47] "Zátah v mešitě. Policisté chtěli, abychom měli hlavu dole, říká svědek." ["Mosque crackdown: The Police demanded that we keep our heads down, witness recounts"] *Lidovky.cz* April 25, 2014, accessed February 20, 2016.
[48] Zuzana Patráňová, "Muslimové vyjádřili společnou modlitbou na Letné nesouhlas s policejním zásahem." ["Muslims expressed their protest against the police crackdown by a joint prayer at the Letná plain"] *Český Rozhlas*. May 2, 2014, accessed February 20, 2016.
[49] "Muslimové v brněnské mešitě stříleli a fotili se u toho." ["Muslims in the Brno mosque were shooting and takingmaking photographs of themselves doing so"] *iDnes.cz*, November 27, 2012, accessed February 20, 2016.
[50] "Komentáře, střílení v brněnské mešitě" ["Commentaries: shooting in the Brno mosque"]. *Shekel.cz*, January 28, 2012, accessed February 20, 2016.

radical attitudes. For example, the former *imam* of the Prague Mosque and founder of the *Sunna.cz* website,[51] Samer Shehadeh, has published several speeches that relativized terrorist attacks as mere "reactions to provocation" and contained disguised threats.[52] Shehadeh was a notorious personality at the time because of his supportive remarks regarding the 9/11 terrorist attacks and his justification for the killing of innocent children. He was later forced to resign from the position of *imam* in the Prague mosque after his activities became too much even for the Saudi backers of the Prague Muslim Community.

Not even the secular, non-religious migrants from Muslim countries are safe from extremist propaganda. Between 2011 and 2013, some of the Czech Baathist Syrians joined the pro-Assad movement, *European Solidarity Front for Syria* (ESFS). Enthusiastic that somebody seemed to care for the fate of their country, they began to attend rallies and seminars organized by this movement, possibly without even realizing that its founder and leader was Patrik Vondrák – a former leader of the neo-Nazi *National Resistance*. The movement, which was also joined by members of the Communist party and several prominent antisemitic personalities, staged several pro-Assad protests in front of the American embassy and its representatives spoke about the conspiracy theories of a "Greater Israel" as being the cause of the Syrian civil war.[53]

Ultimately, the Czech Muslim community is not seen as being immediately threatened by jihadism or violent Islamism, yet some questions remain unanswered, such as the political influence of foreign fund providers and the incitement caused by several dangerous individuals. The particularism of the community with its division into multiple factions could, paradoxically, slow down the influence of jihadist propaganda to some extent since having access to the two largest Czech mosques does not necessarily lead to the winning of the hearts and minds of Czech Muslims in general.

---

51 *Sunna.cz*, accessed February 20, 2016.
52 "Samer Shehadeh – Vzkaz všem spoluobčanům" ["Samer Shehadeh – Message to all citizens"]. *YouTube*, January 12, 2015, accessed February 20, 2016,
53 Personal observation by the author at the events organized by the ESFS, including a pro-Assad rally on August 1, 2013 and ESFS's seminar "Contemporary situation in Syria", Prague, May 27, 2013. See also Zbyněk Tarant, "Czech Anti-Semitic Movements towards the Muslim World – ISGAP." *YouTube channel of the Jewish Broadcasting Society.* January 28, 2016, accessed February 20, 2016.

Mutual Relationships

When considering the quality of the Muslim-Jewish relationship in the Czech Republic, the conclusion must be drawn that it is limited. Although there is evidence of mutual written correspondence, remarks, expressions of support as well as several incidents, it would be an exaggeration to talk about a "relationship". The term "relationship" should be placed in quotation marks in order to indicate the informal and individualized nature of these contacts and encounters. Most of the contacts and exchanges have been made by individuals without institutional backing, typically by community members who are acting on their own behalf and do not consider their actions to be the actions of the community as a whole. This is a very important characteristic, and in each instance the individual or the institutional nature of such an encounter should be distinguished. The communities can hardly be held responsible for the actions of their rank-and-file members.

The communities have only limited scope for the development of shared interests. The only thing that connects them is the fact that they are small, particularistic religious communities operating within a highly secularized, if not agnostic/atheist society. Several cases of harassment and vandalism have been experienced by both communities.[54] Properties owned by the Jewish community are protected either by the police or by private contractors. The Muslim community has not yet decided to take similar steps, although improvements in video surveillance (CCTV) at the premises of the Brno mosque have been announced, following the latest case of vandalism.[55]

It is possible to imagine joint declarations being issued in cases that affect both communities, such as the issue of ritual slaughter, the circumcision of boys or the wearing of religious symbols in public. So far, the Czech Jewish community has not reacted in any way to debates about the wearing of scarves, nor has it reacted to the proposals to ban the production of *halal* meat.

In 2011, Leo Pavlát described the relationship with the Muslim community as being "excellent", despite the fact that there had been "one controversial sermon

---

[54] For cases of antisemitic attacks, see "Výroční zpráva o projevech antisemitismu v České republice za rok 2014" ["Annual report on antisemitic manifestations in the Czech Republic in 2014"]. Středisko bezpečnosti Židovské obce v Praze. 2015.

[55] "Muslimové mají obavy z útoků, chtějí zvýšit ostrahu mešity v Brně" ["Muslims afraid of attacks. Demand increase of security of the Brno mosque"]. *Deník.cz*. November 16, 2015, accessed February 20, 2016.

made by Lukáš Větrovec in the Brno Mosque".[56] Pavlát was very diplomatic in this proclamation. My own evaluation would be that while current relations between Jews and Muslims in the Czech Republic are not ideal, they can be labeled as being "correct". Both communities are trying to keep a low-profile in the media as both have had negative experiences and hostile reactions that have resulted from even the most positive media appearances. Media coverage in response to conflicts and public exchanges has proven to be harmful to both sides. Both communities have experienced vandalism and harassment. The 2008 case of the anti-Jewish sermons by Lukáš Větrovec, publicised by Lukáš Lhoťan three years later, was unique in that the Federation of the Jewish Communities issued its response and condemnation officially. The chairman of the Centre for Muslim Communities, Muneeb Hassan al-Rawi, responded by sending a formal letter of apology.[57]

From the recent past, several cases of encounters, both positive and negative, between Czech Jews and Czech Muslims can be cited. As for the positive ones, the verbal support provided by Muneeb Hasan al-Rawi to the Prague Jewish community during an attempt by neo-Nazis to stage a march outside the Prague synagogues on the anniversary of *Kristallnacht* is a good example.[58] One can find other examples of positive inter-religious encounters, such as when a Jewish representative participated in the opening ceremony of the Brno Mosque.[59]

However, it is difficult for either community to avoid the political dimension of their particular religion. It has become increasingly difficult in a world where verbal attacks on both Jews and Israel have become an integral part of Muslim identity and where bitter anti-Muslim sentiments accompany support for Israel. The Muslim community actively participated in the protests in front of the Israeli

---

56 Pavlát: "Židé roku 2011 poslouchají Tóru on-line" ["The Jews of 2011 listen to the Torah on-line"]. *Česká Televize*. 12. 29. 2011, accessed February 20, 2016.
57 "Brněnští muslimové se omluvili za nenávistné protižidovské kázání" ["The Muslims of Brno have apologized for the hostile anti-Jewish sermon"]. *Česká Televize*, January 26, 2012, accessed February 20, 2016.
58 Muneeb Hasan al-Rawi: "We are officially against this march. Both the Centre for the Muslim Communities in the Czech Republic and the Islamic Foundation in Brno. Such a thing is anti-human. To celebrate an anniversary of such a deed is wrong." See "Půjdeme i přes zákaz. Jinudy." ["We will march, in spite of the ban. By a different path"]. *Lidovky.cz*, 10. 29. 2007, accessed February 20, 2016.
59 "Historie brněnské mešity" ["History of the Brno Mosque"]. *Islámská nadace v Brně*. September 27, 2008, accessed February 20, 2016.

embassy in 2009, where it staged a public communal prayer,[60] accompanied with slogans such as: "Khaybar, Khaybar, Ya Yahud! Jaish al-Muhammad Sa-Ya'ud!"[61] In 2014, its members participated in a pro-Palestinian protest against the Protective Edge operation. The 2009 protest involved the chanting of religious anti-Jewish slogans and the 2014 protest included calls for BDS, "die-ins"[62], and slogans such as "Down, down Zionism" and "Israel is a terrorist state".[63]

On the other side of the barricade, members of the Jewish community, including its leaders, have participated in pro-Israeli political activism. Some of the public comments, made by community leaders in this context, have not always been diplomatic, to say the least.[64] There have been several major pro-Israeli protests which have involved the participation of Jewish community members, e. g. on October 18, 2015, August 6, 2014, and November 26, 2012, and there have also been annual rallies organized by Christian Zionists to coincide with Yom ha-Sho'a. On April 18, 2010, one of the pro-Israeli rallies was disturbed by a peaceful counter-protest, organized by the local cell of the *International Solidarity Movement*.[65] In most cases, these pro-Israeli rallies have been careful to distinguish between the Israeli-Palestinian conflict and religious issues and speakers have been asked to avoid open Islamophobic rhetoric. The Jewish organizations have been very careful not to participate in openly xenophobic anti-Muslim gatherings as some of these gatherings have recently turned antisemitic as well.[66]

---

60 "Před izraelskou ambasádou v Praze protestovali odpůrci útoků na Gazu" ["Opponents of the Gaza strike were protesting in front of the Israeli embassy in Prague"]. *iDnes.cz*, January 2, 2009, accessed February 20, 2016.
61 "Khaybar, Khaybar, Oh Jews, the army of Muhammad will return". See Lhoťan, Lukáš, *Islám a islamismus v České Republice – druhé vydání* (Prague, 2011), 107.
62 The so-called "die-in" is a Palestinian variant of the "sit-in" protest. It involves protesters lying down on the streets to impersonate dead corpses. It is a common part of BDS protests.
63 Personal observation by the author.
64 As in the case of Rabbi David Peter's remarks that "In Palestine, they are good at arranging good shots for the media", made in the context of the Protective Edge operation. Not that David Peter would be completely wrong. The problem is that inter-communal relations may require the use of more diplomatic language. See "V Gaze umějí narafičit dobré záběry, říká nový vrchní pražský rabín" ["They know how to fake shots well in Gaza, new rabbi of Prague claims"], *Lidovky.cz* August 21, 2014, accessed on February 20, 2016.
65 "Společně proti neonacismu a antisemitismu i okupaci Palestiny." ["Together against neo-Nazism, antisemitism and occupation of Palestine"] *Solidarita* 18 (2010), accessed February 20, 2016.
66 The best example would be the anti-Muslim website *Eurabia.cz*, established by a former pro-Israeli activist, Adam Bartoš, who switched sides and became the most radical antisemite in the Czech Republic. See Zbyněk Tarant, "From Philosemitism to Antisemitism – The Case Study of

The two communities have not participated in these rallies collectively as organizations and it is not clear as to the extent to which the communities can be held responsible for the public actions of their individual members. It is thus difficult to ascertain the extent to which these events have the capacity to influence "inter-communal" relations. It should also be stressed that in spite of the often sharp and exalted rhetoric, both sides have managed to keep their protests orderly and non-violent so far, which is in strong contrast to Western European states, where the 2014 pro-Palestinian protests often turned into violent anti-Jewish riots.

Only a few recommendations can be made about mutual relations given the current circumstances, as the strategy aimed at keeping a low media profile, i.e. of respecting each other and not commenting upon one another, has proven to be effective in most cases. This "respectful silence" is still the better alternative to open hostilities that have the potential to be harmful to both sides. Should there be any further incidents between members of the two communities, time and discretion will be crucial factors to take into account. It would be useful to establish some sort of reliable communication channel, e.g. a "hot line" or a "red phone," and this could be used to resolve any future conflicts before they make it into the media. Such an emergency communication channel could be established either between the chairmen of the main communal organizations, or between any other respectable persons who are regarded as having natural authority among their co-believers and the respect of the other side. Thus, they would be in a position to act as mediators.

Conclusions

Both Jews and Muslims represent small, fragmented communities. There might be about 3,000 Jews in the Czech Republic as well as about 20,000 Muslims. Both communities struggle with particularism and personal grievances among their members, as well as a lack of qualified spiritual leaders. With the Jews, this results in the use of "flying rabbis" or the irregular offering of communal prayers. With the Muslims, it means that radicalized activists are able to take over and open the door to jihadist propaganda. Both communities share the experience of being very small religious societies within a highly secularized, if not agnostic, society. Recent international crises have only had a limited impact on

---

the 'Mladá Pravice' Movement in the Czech Republic", in *Faces of Hatred*, ed. Zbyněk Tarant and Věra Tydlitátová (Pilsen: Nakladatelství ZČU v Plzni, 2013), 114–152.

the communities so far. Most of the issues they are currently facing reach far back beyond the Arab Spring. This is, however, where the similarities end.

While the Jewish community is aging staggeringly, the Czech Muslims are mostly active around student dormitories. While Czech Jews are able to build on the rich heritage of pre-World War II Czechoslovak Jewry, they are made to struggle against the external pressures that force them into the undesirable status of "caretaker communities". Having crossed the demographic "point of no return", the regional communities might easily disappear slowly unless there is a significant influx of fresh members from the outside – with the possible exception of Prague, where the demographic situation seems to be somewhat more stable. It should be noted that the whole of the Czech Republic, with its fertility rate of 1.45 children per woman[67] and a mean age of 41, has also crossed a demographic "point of no return". However, the estimated age profile of the Jewish community is even more serious than is the case with the Czech majority.

The Muslim community finds itself in a completely different situation – with the refugee wave coming to Central Europe, the community may well soon have more members than it can handle. Its representatives will face the difficult task of proving that they are the actual leaders, as many Czech Muslims (e. g. Turks, Shiites etc.) do not recognize the authority of the Centre for Muslim Communities in the first place. Whoever gains the authority to speak on behalf of Czech Muslims will find one's self faced with an even more difficult mission – the need to address the increasing demands of the Czech majority for Muslims to prove their loyalty to the Republic. Some of these demands may become increasingly hostile due to the growing anti-Muslim panic within Czech society. If the situation is not to escalate, then these Muslim leaders will have to "turn the other cheek". They will have to protect their communities from jihadist voices that tend to thrive in hostile environments. Their task will be to prove that there is indeed a niche for moderate Islam in the heart of Europe.

Bibliography

"Adresář modliteben" ["Directory of Prayer rooms"]. *Infomuslim.cz*. http://www.infomuslim.cz/modlitebny/. Accessed February 20, 2016.

"AKTUALIZACE: Brno se bouří proti nové islámské modlitebně" ["UPDATE: Brno rebels against new Islamic prayer room"]. *IVČRN*. February 4, 2016. http://www.ivcrn.cz/brno-se-bouri-proti-nove-islamske-modlitebne/. Accessed February 20, 2016

---

67 *Czech Republic – CIA World Factbook*, accessed February 20, 2016.

Arava-Novotná, L. "Židovská tradice mezi pokračováním a mytologizací. Pražský judaismus a jeho předávání v letech 1945–2005" ["Jewish tradition between continuation and mythologization. Prague Judaism and its tradition between 1945–2005"]. *Mýtus – "Realita" – Identita. Národní metropole v čase "Návratu do Evropy"* [*Myth – "Reality" – Identity. National metropolis at the time of "Returning to Europe"*], edited by Blanka Soukupová and Andrzej Stawarz. Urban studies, volume 9. Prague: Faculty of Humanities [FHS], 2015.

Bečka, Jiří, and Mendel, Miloš. *Islám a České země* [*Islam and the Czech lands*]. Olomouc: Votobia, 1998.

Bezcenzury.org. Available only via WebArchive. https://web.archive.org/web/20110126053836/http://www.bezcenzury.org/rs/index.php. Accessed February 20, 2016.

"Brněnští muslimové se omluvili za nenávistné protižidovské kázání" ["The Muslims of Brno have apologized for the hostile anti-Jewish sermon"]. *Česká Televize*. January 26, 2012. http://www.ceskatelevize.cz/ct24/regiony/1194968-brnensti-muslimove-se-omluvili-za-nenavistne-protizidovske-kazani. Accessed February 20, 2016.

"Cizinci podle státního občanství k 12.31.2014" ["Foreigners according to nationality" up to 12.31.2014]. *Czech Statistical Office*. https://vdb2.czso.cz/vdbvo2/faces/index.jsf?page=vystup-objekt&evo=&str=&vyhltext=&pvo=CIZ08&udIdent=&verze=-1&nahled=N&sp=N&nuid=&zs=&skupId=&pvokc=&filtr=G~F_M~F_Z~F_R~F_P~_S~_null_null_&katalog=all&pvoch=&zo=N&z=T. Accessed February 20, 2016.

Czech Republic – CIA World Factbook. https://www.cia.gov/library/publications/the-world-factbook/rankorder/2127rank.html. Accessed February 20, 2016.

DellaPergola, Sergio. "World Jewish Population." *Berman Jewish DataBank*, 2013. http://www.jewishdatabank.org/studies/downloadFile.cfm?FileID=3113. Accessed February 20, 2016.

"Důvodová zpráva k návrhu zákona, kterým se mění zákon č. 153/1994 Sb., o zpravodajských službách České republiky, ve znění pozdějších předpisů" ["Explanatory report to the proposal of an act that mends the the Act no. 153/1994 sb."]. *Bezpečnostní informační služba*. http://www.ceska-justice.cz/wp-content/uploads/2015/01/důvodová-zpráva-BIS.pdf. Accessed February 20, 2016.

Federace židovských obcí – organizace. [*Federation of the Jewish Communities – About Us*]. https://www.fzo.cz/o-nas/organizace/. Accessed February 20, 2016.

Heitlingerová, Alena. *Ve stínu holocaustu a komunismu – čeští a slovenští židé po roce 1945* [*In the shadow of the Holocaust and Communism – Czech and Slovak Jews after 1945*]. Prague: G Plus G, 2007.

"Historie brněnské mešity" ["History of the Brno Mosque"]. *Islámská nadace v Brně*. September 27, 2008. http://www.mesita.cz/node/3. Accessed February 20, 2016.

"Komentáře, střílení v brněnské mešitě" ["Commentaries – shooting in the Brno mosque"]. *Shekel.cz*. January 28, 2012. http://www.shekel.cz/25666/strileni-v-brnenske-mesite. Accessed February 20, 2016.

Lhoťan, Lukáš. *Islám a islamismus v České Republice* [*Islam and Islamism in the Czech Republic*]. Prague, 2011.

Lhoťan, Lukáš. "Jak čeští islamisté 'nepřijímají' peníze ze Saúdské Arábie" ["How the Czech Islamists are 'not receiving' money from Saudi Arabia"]. *Eurabia.cz*. February 28, 2016. http://eurabia.parlamentnilisty.cz/Articles/30422-jak-cesti-islamiste-neprijimaji-penize-ze-saudske-arabie.aspx. Accessed February 20, 2016.

Lhoťan, Lukáš. "V Praze je mešita, kterou muslimové vydávají jen za jazykovou školu. Podvedli u toho úřady." ["There is a mosque in Prague, which the Muslims present as a mere language school, deceiving the authorities by doing so."] *Reflex.cz*. June 20, 2014. http://www.reflex.cz/clanek/zpravy/57860/v-praze-je-mesita-kterou-muslimove-vydavaji-jen-za-jazykovou-skolu-podvedli-u-toho-urady.html. Accessed February 20, 2016.

Lukáš Lhoťan's YouTube account. *YouTube* 10.11.2011. https://www.youtube.com/watch?v=cRZQSO6hC-I. Accessed February 20, 2016.

Mareš, Miroslav. "Islamismus jako bezpečnostní hrozba pro Českou republiku" ["Islamism as a Security threat for the Czech Republic"]. *Vojenské rozhledy*, 2010/4, 118–128. http://www.vojenskerozhledy.cz/kategorie/islamismus-jako-bezpecnostni-hrozba-pro-ceskou-republiku. Accessed February 20, 2016.

Mareš, Miroslav. "Muslimská politika v ČR na rozcestí." ["Muslim politics in the Czech Republic at the crossroads"]. *Naši politici*. January 20, 2015. http://www.nasipolitici.cz/cs/komentare-tydne/6773-miroslav-mares-muslimska-politika-v-cr-na-rozcesti. Accessed February 20, 2016.

Mendel, Miloš, Bronislav Ostřanský, and Tomáš Rataj. *Islám v srdci Evropy* [*Islam in the heart of Europe*]. Prague: Academia, 2008.

Mrázek, Miloš. "Muslimská unie" ["Muslim Union"]. In *Dingir* 1/2006. http://www.dingir.cz/archiv/Dingir106.pdf. Accessed February 20, 2016.

"Muslimové mají obavy z útoků, chtějí zvýšit ostrahu mešity v Brně" ["Muslims afraid of attacks. Demand increase of security of the Brno mosque"]. *Deník.cz*. November 16, 2015. http://www.denik.cz/z_domova/muslimove-maji-obavy-z-utoku-chteji-zvysit-ostrahu-mesity-v-brne-20151116.html. Accessed February 20, 2016.

"Muslimové v brněnské mešitě stříleli a fotili se u toho." ["Muslims in the Brno mosque were shooting and taking photographs of themselves doing so"]. *iDnes.cz*. November 27, 2012. http://zpravy.idnes.cz/muslimove-v-brnenske-mesite-strileli-a-fotili-se-u-toho-piq-/domaci.aspx?c=A121127_1859592_brno-zpravy_taz. Accessed February 20, 2016.

Národně-vzdělávací institut. Available only through WebArchive. http://www.vzdelavaci-institut.com/_files/index.php?option=com_content&task=view&id=327&Itemid=57. Accessed February 20, 2016.

"Obyvatelstvo podle věku, náboženské víry a podle pohlaví" ["Mean age of the population in regions of the Czech Republic"]. *Czech Statistical Office*. https://vdb.czso.cz/vdbvo2/faces/cs/index.jsf?page=vystup-objekt&str=&evo=&filtr=G~F_M~F_Z~F_R~F_P~_S~_null_null_&pvokc=&katalog=30719&nahled=N&sp=N&nuid=&zs=&skupId=&verze=-1&zo=N&vyhltext=&pvo=SPCR156&pvoch=&udIdent=&z=T. Accessed February 20, 2016.

Patráňová, Zuzana. "Muslimové vyjádřili společnou modlitbou na Letné nesouhlas s policejním zásahem" ["Muslims expressed their protest against the police crackdown by a joint prayer at the Letná plain"]. *Český Rozhlas*. May 2, 2014. http://www.rozhlas.cz/zpravy/politika/_zprava/muslimove-vyjadrili-spolecnou-modlitbou-na-letne-nesouhlas-s-policejnim-zasahem-1345989. Accessed February 20, 2016.

"Pavlát: Židé roku 2011 poslouchají Tóru on-line" ["Jews of 2011 listen to the Torah on-line"]. *Česká Televize*. December 29, 2011. http://www.ceskatelevize.cz/ct24/domaci/1201039-pavlat-zide-roku-2011-poslouchaji-toru-line. Accessed February 20, 2016.

"Před izraelskou ambasádou v Praze protestovali odpůrci útoků na Gazu" ["Opponents of the Gaza strike were protesting in front of the Israeli embassy in Prague"]. *iDnes.cz*. January 2, 2009. http://zpravy.idnes.cz/pred-izraelskou-ambasadou-v-praze-protestovaliodpurci-utoku-na-gazu-1k9-/domaci.aspx?c=A090102_163956_domaci_bar. Accessed February 20, 2016.

"Prohlášení Federace židovských obcí v ČR, Židovské obce v Praze a Židovského muzea v Praze" ["Proclamation of the Federation of the Jewish Communities in the Czech Republic, Jewish Community of Prague and Jewish Museum in Prague"]. *Federace židovských obcí*. December 12, 2011. http://www.fzo.cz/1573/prohlaseni-federace-zidovskych-obci-v-cr-zidovske-obce-v-praze-a-zidovskeho-muzea-v-praze/. Accessed February 20, 2016.

"Průměrný věk obyvatel v krajích České republiky k 12.31.2014" ["Mean age of the population in regions of the Czech Republic"]. *Český statistický úřad*. https://www.czso.cz/csu/xl/130624_vek_kraje. Accessed February 20, 2016.

"Půjdeme i přes zákaz. Jinudy" ["We will march, in spite of the ban. By a different path"]. *Lidovky.cz*. October 29, 2007. http://www.lidovky.cz/ln_domov.asp?c=A071029_073341_ln_praha_svo. Accessed February 20, 2016.

Salner, Peter. *Židia na Slovensku medzi tradíciou a asimiláciou* [*Jews in Slovakia between tradition and assimilation*]. Bratislava: Zing Print, 2002.

"Samer Shehadeh – Vzkaz všem spoluobčanům" ["Samer Shehadeh – Message to all citizens"]. *YouTube*. January 12, 2015. https://www.youtube.com/watch?v=wsH-DxdM9Nk. Accessed February 20, 2016.

"Společně proti neonacismu a antisemitismu i okupaci Palestiny" ["Together against neo-Nazism, antisemitism and even occupation of Palestine"]. *Solidarita*. April 18, 2010. http://solidarita.socsol.cz/2010/domaci/spolen-proti-neonacismu-a-antisemitismu-i-okupaci-palestiny. Accessed February 20, 2016.

Statistika. *Federation of the Jewish Communities*. http://www.fzo.cz/o-nas/statistika/. Accessed February 20, 2016.

Sunna.cz. http://www.sunna.cz. Accessed February 20, 2016.

"T14 Cizinci podle kategorií pobytu, pohlaví a občanství k 12.31.2014" ["We will march, in spite of the ban. By a different path"]. *Czech Statistical Office*. https://www.czso.cz/csu/cizinci/2-ciz_pocet_cizincu-006. Accessed February 20, 2016.

Tarant, Zbyněk. "From Philosemitism to Antisemitism – The Case Study of the 'Mladá Pravice' Movement in the Czech Republic." In *Faces of Hatred*, edited by Zbyněk Tarant and Věra Tydlitátová, 114–152. Pilsen: Nakladatelství ZČU v Plzni, 2013.

Tarant, Zbyněk. "Czech Anti-Semitic Movements towards the Muslim World – ISGAP." *YouTube channel of the Jewish Broadcasting Society*. January 28, 2016. https://www.youtube.com/watch?v=E6xrOn1E_eM. Accessed February 20, 2016.

Taušová, Zuzana. "Kuřim má netušené prvenství. První centrum šíitských muslimů v zemi." ["Unexpected primacy for Kuřim – A first Shiite religious center in the country"]. *iDnes.cz*. August 8, 2012. http://brno.idnes.cz/kurim-ma-netusene-prvenstvi-prvni-centrum-siitskych-muslimu-v-zemi-1if-/brno-zpravy.aspx?c=A120808_1813857_brno-zpravy_taz. Accessed February 20, 2016.

"The Need for a Minyan." In *Teshuvot for the Nineties: Reform Judaism's Answers to Today's Dilemmas*, edited by W. Gunther Plaut and Mark Washofsky, 23–28. New York: Central Conference of American Rabbis, 1997.

Topinka, Daniel, Tomáš Janků, Lenka Linhartová, and Jan Zadina. "Muslimové imigranti v České Republice: Etablování na veřejnosti" ["Muslim immigrants in the Czech Republic: Public emancipation"]. In *Fenomén moci a sociálne nerovnosti – nultý ročník konferencie pre doktorandov a mladých vedeckých pracovníkov*, edited by Lukáš Bomba, Estera Kövérová, and Martin Smrek. Universita Komenského v Bratislave, 2014. http://www2.tf.jcu.cz/~klapetek/janku2.pdf. Accessed February 20, 2016.

Týdeník Politika 96/1992: 2.

"V Gaze umějí narafičit dobré záběry, říká nový vrchní pražský rabín" ["They know how to fake shots well in Gaza, new rabbi of Prague claims"]. *Lidovky.cz*. August 21, 2014. http://www.lidovky.cz/vrchni-prazsky-rabin-david-peter-pro-patek-ln-rodice-rekli-tak-jsem-sel-1z7-/lide.aspx?c=A140821_171856_ln-media_ele. Accessed February 20, 2016.

Valášek, Lukáš. "V Králově Poli vzniká Islámská čtvrť, vyhrožují brněnské letáky" ["An Islamic neighborhood is being created in Královo pole, Brno flyers warn"]. *iDnes.cz*. February 9, 2016. http://brno.idnes.cz/po-utocich-na-brnenskou-mesitu-budi-emoce-i-nove-islamske-kulturni-centrum-v-kralove-poli-gv8-/brno-zpravy.aspx?c=A160209_2224426_brno-zpravy_zde. Accessed February 20, 2016.

Vojtíšek, Zdeněk. "Český boj o mešity" ["The Czech Mosque disputes"]. In *Dingir* 1/2006, 19–21. http://www.dingir.cz/archiv/Dingir106.pdf. Accessed February 20, 2016.

Vojtíšek, Zdeněk. "Posouzení 'Ústředí muslimských obcí' vzhledem k řízení o registraci podle Zákona 3/2002 sb" ["Evaluation of the 'Center of the Muslim Communities' in regard to the process of registration according to the Act no. 3/2002 sb."]. http://www.mkcr.cz/assets/povinne-zverejnovane-informace/Znalecky-posudek-I.pdf. Accessed February 20, 2016.

"Výroční zpráva Bezpečnostní informační služby za rok 2011" ["Annual report of the Security Information Service for 2011"]. *Bezpečnostní informační služba*. https://www.bis.cz/vyrocni-zpravac2ed.html?ArticleID=26. Accessed February 20, 2016.

"Výroční zpráva bezpečnostní informační služby za rok 2014" ["Annual report of the Security Information Service for 2014"]. *Bezpečnostní informační služba*. https://www.bis.cz/vyrocni-zprava6c8d.html?ArticleID=1096. Accessed February 20, 2016.

"Výroční zpráva o projevech antisemitismu v České republice za rok 2014" ["Annual report on antisemitic manifestations in the Czech Republic in 2014"]. Středisko bezpečnosti Židovské obce v Praze. 2015.

"Zátah v mešitě. Policisté chtěli, abychom měli hlavu dole, říká svědek" ["Mosque crackdown: The Police demanded that we keep our heads down, witness recounts"]. *Lidovky.cz*. April 25, 2014. http://www.lidovky.cz/policie-zasahuje-v-praze-kvuli-podezreni-z-sireni-zavadne-knihy-p82-/zpravy-domov.aspx?c=A140425_134327_ln_domov_mct. Accessed February 20, 2016.

Michal Schuster
Jewish-Roma Relations in the former Czechoslovakia: An Alliance Against Racism

Introduction

The Roma and the Jews have a shared experience of persecution and hostility in Christian Europe for centuries. Both groups migrated to escape further persecution and improve their economic situation. They earned a living by certain kinds of work while others were not accessible to them, and for centuries they were not allowed to own land. From the very beginning after their arrival to medieval Europe, the Roma stayed outside of mainstream society most of the time.[1] And naturally they were in contact with another group that was banished from society – the Jews. Various historical sources give evidence for their coexistence on social, economic, and cultural levels until the twentieth century. The shared fates of Jews and Roma were then tragically tangled in the 1930s and 1940s when they were exposed to various forms of persecution on racial grounds, leading up to attempts at the "final solution" in Europe. Jewish and Roma prisoners suffered and died in the same labor, concentration, and extermination camps. Through suffering and genocide during World War II, the long-term oppression culminated.[2]

In this text I will present particular examples of the Jewish-Roma connections on the territory of former Czechoslovakia up from the middle ages, focusing mainly on the twentieth century.

The roots of hostility towards Roma and Jews are very similar: fear of the unknown, hatred of outsiders, distrust of potential alleged spies, nationalism, and chauvinism, but also open forms of racism. The Roma as well as the Jews were also attacked in literature, in the arts, and demonized in popular tradition. Many

---

[1] For more about the history of Roma minority see Ctibor Nečas, *Historický kalendář: Dějiny českých Romů v datech* (Olomouc: Palacký University Press, 1997); Angus Fraser, *The Gypsies: The People of Europe* (Oxford: Blackwell Publishers, 1995); and Ctibor Nečas, *Romové v České republice včera a dnes* (Olomouc: Palacký University Press, 1999).
[2] Petra Lukšíková, "Naučená vzájomnosť (Vzťah majoritnej spoločnosti k Rómom a Židom)", in *Romano džaniben*, 1–2/2000 (Prague: Společnost přátel časopisu Romano džaniben, 2000), 52–57; Gaby Glassmannová, "Židovské a romské dědictví (Několik srovnání)," in *Romano džaniben*, jevend 2002 (Prague: Společnost přátel časopisu Romano džaniben, 2002), 26–29.

stereotypes have survived for centuries, and some remain present in mainstream society until today. Let's mention, for example, "ritual murder" imputed towards the Jews or the myth of Roma kidnapping children from the majority society.

Expulsion and Persecution of Roma (from the Middle Sixteenth Century to the First Half of the Eighteenth Century)

The Roma reached the Byzantine Empire probably in the eleventh century after their long way from India. From the first quarter of the fifteenth century, Roma began appearing in Central and Western Europe. While in the fifteenth century they were quite well received, from the early sixteenth century their position radically changed for various reasons: their exotic look, nomadic way of life, strange language, and suspicion of spying for the Turks. The Church also condemned their magical practices (divination and palmistry). The Roma were ranked among "people outside the society" by the majority in the early modern era – and so began four centuries of cruel expulsion and persecution. European countries issued decrees by which Roma were ordered out of the territories. The greatest persecution in the Czech lands came after the end of the seventeenth century when the Roma were placed outside the law – killing a Roma was not considered a crime in that time.[3]

At the same time there were anti-Jewish pogroms in Europe. Although cultured Jews were better accepted than "Gypsy nomads," the two ethnic groups were seen as foreign elements. The attitude to them was very similar in many ways. Already in 1492, Roma along with Jews were expelled from Spain (after an earlier expulsion of the Arabs). The two groups were bound by social status outside the majority which led to their interaction.[4]

From around the sixteenth century the so-called Zigeunerova synagoga/ Zigeuner Synagogue existed in the old Jewish ghetto in Prague, but it was unfortunately demolished during the sanitation of the ghetto in 1906. The synagogue

---

[3] For more about persecution of the Roma minority, see Eva Procházková, "Perzekuce romských kočovníků v českých zemích v 18. století", in *Sborník archivních prací* 42, no. 2 (1992): 307–409; Ian Hancock, *Země utrpení – Dějiny pronásledování a otroctví Romů* (Prague: Signeta, 2001).
[4] A. G. Alfaro, *Velký proticikánský zátah: Španělsko – všeobecné uvěznění cikánů v roce 1749* (Olomouc: Univerzita Palackého, 1999).

was named after its founder, Moshe Zigeuner/ Cigán, and so was the street where he lived.[5]

The surname of the Czech Jewish rabbi Karol Sidon derives from the term Cikán. He explains its origin as a Jewish surname as follows:

> [The] Jews – at least most of them – lived outside the major society in the Middle Ages. They shared this world on the edge with Gypsies. [...]. Jews and Gypsies were engaged mostly in trade with horses, which was one of the few livelihoods the conditions and laws allowed them. [...] I think the trade could not work without very close cooperation with Gypsies who traded with horses, too. A man who lived in this environment was called by the others Gypsy.[6]

Karol Sidon is known among the Roma in the Czech Republic as a man who understood them and who stood by them whenever they were exposed to acts of violence or discrimination. As a journalist, he wrote about them in the 1960s, his articles being the most daring at the time of Communist assimilation of the Roma.[7]

The First Czechoslovak Republic (1918–1938)

In pre-war Czechoslovakia, the vast majority of Roma lived in Slovakia – most of them were so-called Slovak Roma, and another group were Hungarian Roma in Hungarian-speaking enclaves. In Bohemia and Moravia, the majority belonged to a group called the Czech Roma, and the smallest groups were the German Roma, called Sinti, especially in German-speaking areas in the borderlands. The last group in Czechoslovakia were the nomadic Vlach Roma.

The First Republic made an attempt to resolve "the Gypsy question" in 1927 by issuing the Law on Wandering Gypsies. In practice this meant that they all had to apply for identification cards and for permission to stay overnight. The target was to "civilize" their way of life, but the law restricted and deprived the Roma of their civil liberties. The measures discriminated mainly nomadic

---

[5] Milena Hűbschmannová, "Rozhovor s Karolem Efraimem Sidonem, vrchním rabínem v České republice," *Romano džaniben* 1–2 (Prague: Společnost přátel časopisu Romano džaniben, 2000), 3–4.

[6] Hűbschmannová, "Rozhovor s Karolem Efraimem Sidonem, vrchním rabínem v České republice," 5–6.

[7] Hűbschmannová, "Rozhovor s Karolem Efraimem Sidonem, vrchním rabínem v České republice," 3.

Roma, but could also affect the settled Roma.[8] On the other hand, the integration of settled Roma in Moravia was improving, and for the first time it was possible for the Roma to obtain secondary and university education.[9]

Very interesting information on relations between the Roma and the Slovak Jewish population before World War II is provided by the Czech linguist Milena Hübschmannová. From the 1970s, she was creating a unique structural collection of eye-witness testimonies of the fate of Roma in Slovakia during World War II. The recordings she edited were published posthumously in 2005 in Czech and Romani language in the publication *Po židoch cigáni/ Gypsies after Jews: testimonies of Slovak Roma 1939–1945*.[10]

According to these testimonies, the Jews and Roma were an integral part of the social, economic, and cultural life of Slovak villages, and there were specific relationships between these two minorities. Most of the Roma respondents evaluated the Jews very positively.[11]

Elena Lacková, born 1921, Veľký Šariš, Slovakia: "We [Roma] used to live in Šariš right in the town. Eleven Jewish families lived there, too. We did not call them Gadje [non-Roma], we called them Čhinde. They had shops where they sold a variety of goods, and also taverns. We had very good relationships with them." [12]

Jozef Horváth, born 1916, Železník, Slovakia, on the question about how the Jews treated the Roma: "They were our brothers! Just like we, Roma, held together, the Jews stuck together with us! They were so nice to us, Jews helped to one, and they kept saying that we have to stick together."[13]

Tera Fabiánová, born 1930, Žihárec, Slovakia, stated: "Jews always helped Roma, Catholics did not. Jews had shops, they gave you a job, and when you were hungry, they gave you food."[14]

Jews gave the Roma an important opportunity to earn a living not only in the country, but also in towns. During the Sabbath, Roma women operated heating and served in Jewish families (they usually wore water for ritual bathing). Roma men carried goods for Jewish traders. Sometimes Roma women were servants in

---

8 Ctibor Nečas, *Romové v České republice včera a dnes* [*The Roma in the Czech Republic yesterday and today*] (Olomouc: Vydavatelství UP, 1999), 51–65.
9 Nečas, *Romové v České republice včera a dnes*, 34–50.
10 Milena Hübschmannová, ed., *Po Židoch Cigáni: svědectví Romů ze Slovenska 1939–1945* (Prague: Triáda, 2005).
11 Hübschmannová (ed.), *Po Židoch Cigáni*, 680.
12 Hübschmannová (ed.), *Po Židoch Cigáni*, 685.
13 Ibid., 477.
14 Ibid., 694.

family (even as nannies or wet nurses). Jewish businessmen purchased herbs or blacksmith products from the Roma. The Roma also worked on Jewish estates. Unlike Slovak and Hungarian farmers, the Jews payed the Roma for their work in cash instead with food.[15]

Jolana Kurejová, born 1928, in Humenné – Podskalka, Slovakia: "The Jews were used to getting along with Roma, normally. They gave us work. The men wore their suitcases and bags, because Jews owned mostly shops. [...] Only Roma worked for Jews." [16]

Bartoloměj Daniel, born 1924, Šaštín, Slovakia: "Jews were kind to Roma. They always got along well with Roma. There were many Jews in our village. My older sister used to work for them. Many Roma worked for Jews, mainly old Barkóczi – he worked for Kohn. He was delivering flour and sugar, three bags of flour on a shoulder – and he was paid three crowns a day. But you could buy a shirt for two crowns then."[17]

Elena Lacková, born 1921, Veľký Šariš, Slovakia: "Another Jewish woman lived next to us, she also had a shop. [...] My mother sewed for them, crocheted, mended stockings, they payed forty hellers for one hole. They admired the work of my mother very much." [18]

Almost all of the witnesses testified that Jews treated Roma with more respect than the majority. The majority mocked or otherwise humiliated the "gypsies." The Jews also maintained less social distance against the Roma than the majority. Solidarity with the "other" minority group certainly played a big role here.[19]

Elena Lacková, born 1921, Veľký Šariš, Slovakia: "We got along well with Jews. They never mocked us as Gadje did. We had their support. And then came national guards – and the end! [...] And yet the Jews never did anything wrong to anyone! Before the war, they had respect even among farmers. We, Roma, picked our names after them. We called one boy Cajzler. [...] Other one was called Chajim – and so on. The name is given after someone who is rich or happy, or who has respect among people – in short, who somehow stands out. Or after someone we like."[20]

---

15 Ibid., 677–722.
16 Ibid., 705.
17 Ibid., 597.
18 Hübschmannová (ed.), *Po Židoch Cigáni*, 687.
19 Ibid., 677–680.
20 Ibid., 689.

Maxmilián Špira, Jewish survivor from Slovakia: "Before the war there was a shortage of labor, lack of money and they [the Roma] were glad to be in those Jewish families where they received food and some money." [21]

We could also find some examples of Jewish–Roma relations in testimonies of the Roma from Czech lands.[22]

Anežka Klaudová, born 1925, Strážnice, Moravia: "Sometimes they [Roma] went to buy from Jews who sold cheap. There were many Jewish shops. So they mostly went there."[23]

Emílie Danielová, born 1924, Paškovice, Moravia: "Some of the store keepers were Czech, here from Ořechov. They were good. And Jews also gave you food even though you couldn't pay. 'You will pay later, František, you are a good man, you can work, so you will pay later, we will give you some food for your children.' That's what the Jew told us. Some of them Jews were very, very good."[24]

The Nazi Genocide of Roma during World War II (1939–1945)

The greatest tragedy for the European Jewish and Roma population was World War II. Among all the victims of the Nazi regime, the racial and biological principle was particularly decisive in the persecution of the Jews and the Roma. The Protectorate of Bohemia and Moravia followed the German example, continuously introducing all of its anti-gypsy measures. The public viewed these as a continuation of the former Czechoslovak policy on the "gypsy plague," not realizing that this was a different – racial – policy, aimed at eliminating all "inferior races," including Jews and Roma.[25]

---

21 Ibid., 721.
22 Michal Schuster, "Testimonies of witnesses to the Roma holocaust at the Museum of Roma Culture in Brno", in *The Holocaust Between Memory and History. Papers from the conference held in Bratislava on September 30th – October 1st, 2013* (Bratislava, Slovak National Museum – Museum of Jewish Culture, 2014), 143–150.
23 Collection of audio testimonies, Museum of Roma Culture, Brno, Czech Republic, sig. A 43/2003.
24 Ibid., sig. A 14/99.
25 For more about the Nazi persecution of the Roma during WWII see, for example: *From the "Race Science" to the Camps* (Paris, Hatfield: Centre de recherches tsiganes, University of Hertfordshire, 1999); Donald Kenrick, Grattan Puxon, *Gypsies under the Swastika* (Hatfield: University of Hertfordshire Press, 1995); Michael Zimmermann, *Verfolgt, vertrieben, vernichtet: Die national-*

Roma in the protectorate were persecuted in various ways. These included the ban of nomadism, internment of some Roma men in work camps in Lety u Písku and Hodonín u Kunštátu in what was declared as a prevention of criminality, and even the transportation of so-called anti-social individuals to the Auschwitz I camp.

In 1942, the anti-Gypsy politics in the protectorate very quickly turned into a copy of racial discrimination and persecution which were current in the Nazi Germany. It meant the census of "Gypsies and half-Gypsies." One third of them were sent off to the newly opened so-called gypsy camps in Lety and Hodonín (around 2,700 prisoners passed through these camps). These camps became transfer points on the way from Roma to Auschwitz.

Around 5,500 Roma were transported from Czech territory to the Auschwitz-Birkenau concentration camp during 1943 and 1944. Most of them did not survive; the pre-war Roma population in the Czech lands was almost annihilated.[26]

Vlasta Danielová, born 1925, Moravia: "The policemen from Kunovice (South Moravia) never had any problem with us. We were never punished for transgressing law and order; we were friends with all the locals. However, when they took the Jews, rumors spread that Gypsies would follow as well. But we always had jobs and the policemen would say that this did not concern us, that we led normal lives like all locals, that we were at home in the town, which reassured us."[27]

Božena Růžičková, born 1924, Bohemia: "There was a local policeman, his name was Ulrich, and he came one night to warn us we would be taken to the camp. Three months before that there were Jews, they went to school with us but I don't remember their names. Their dad came to tell us they had to go

---

sozialistische Vernichtungspolitik gegen Sinti und Roma (Essen: Klartext Verlag, 1989); Michael Zimmermann, *Rassenutopie und Genozid. Die nationalsozialistische "Lösung der Zigeunerfrage"* (Hamburg: Christians, 1996); Wacław Długoborski (ed.), *Sinti und Roma im KL Auschwitz-Birkenau 1943–44: vor dem Hintergrund ihrer Verfolgung unter der Naziherrschaft* (Oświęcim: Staatliches Museum Auschwitz-Birkenau, 1998); Gerhard Baumgartner, Florian Freund, Harald Greifeneder, *Nationale Minderheiten im Nationalsozialismus, 2. Vermögensentzug, Restitution und Entschädigung der Roma und Sinti* (Wien: Oldenbourg, 2004).

26 For more about the persecution of the Czech Roma during WWII see, for example: Vlasta Kladivová, *Konečná stanice Auschwitz-Birkenau* (Olomouc: Palacký University Press, 1994); Ctibor Nečas, *Andr' oda taboris. Vězňové protektorátních cikánských táborů 194 – 1943* (Brno: Městský výbor Českého svazu protifašistických bojovníků, 1987); Ctibor Nečas, *Andr' oda taboris. Tragédie cikánských táborů v Letech a v Hodoníně* (Brno: Masaryk University, 1995); Ctibor Nečas, *Holocaust českých Romů* (Prague: Prostor, 1999); Michal Schuster, "Pronásledování protektorátních Romů s přihlédnutím k roku 1944," *Válečný rok 1944 v okupované Evropě a v Protektorátu Čechy a Morava* (Prague: Ústav pro studium totalitních režimů, 2015), 140–151.

27 Collection of audio testimonies, Museum of Roma Culture, Brno, Czech Republic, sig. A 32/95.

and we would follow them, too, but we didn't know we would go to a camp, we thought that maybe they would take us to do some work, you know?"[28]

Jan Ištván, born 1921, Letovice, Moravia: "The Roma didn't expect something like that would ever happen... that they would just take us, the Jews, Roma and so on. We were all together, after all. And then in autumn 1939 one Jew came, Mr. Kohn who had a shop with textile, and he said to my father: 'Mr Ištván, it's getting bad, run away from here.' And I remember my father came home and said 'You know this Kohn told me to leave? But where?' He said he didn't know. And he left with his family..."[29]

An interesting fact is that the Jewish doctor, Dr. Alfred Mílek (born 1899), was sent to the "gypsy camp" in Hodonín as a doctor for Roma prisoners. From July 1943 after his deportation to Birkenau, he was replaced by another Jewish doctor, Dr. Michal Bohin (1895–1956), who had worked in the Lety camp before. Both had limited contact with the rest of the population for racial reasons.[30]

Dr. Alfred Mílek talking about the Hodonín camp: "Upon my arrival to the camp I was treated as a gypsy prisoner, I wore black colored military uniform like other Gypsies, I lived in a small chamber without a stove from where I went to the clinic across the room with lying sick people..."[31]

One Roma witness (František Daniel) remembered Dr. Bohin in Hodonín as follows: "I saw a lot of dead and sick people who were dying in pain. Meanwhile one doctor got there, he was so good and nice and very gentle to people. I saw how he treated the patients regretfully and he knew well what's before them. [...], he said: It awaits us all anyway."[32]

Roma prisoners met Dr. Bohin and Mílek again in Auschwitz where they were transported. Two other Jewish doctors from Prague, Berthold Epstein and Rudolf Weisskopf, worked in the hospital of the "gypsy camp" in Birkenau

---

[28] Ibid., sig. V 80/97, NV 214/2012.
[29] Collection of video testimonies, Museum of Roma Culture, Brno, Czech Republic, sig. NV 225/2012.
[30] Markus Pape (ed.), *A nikdo vám nebude věřit: dokument o koncentračním táboře Lety u Písku* (Prague: G plus G, 1997), 142–147; Ctibor Nečas, "Židovští lékaři v cikánských táborech," *Romano džaniben*, 1–2 (Prague: Společnost přátel časopisu Romano džaniben, 2000): 58–61; Markus Pape, "Michal Bohin, lékař-lidumil," *Romano džaniben* 1–2 (Prague: Společnost přátel časopisu Romano džaniben, 2000): 62–65.
[31] Ctibor Nečas, *Romové na Moravě a ve Slezsku (1740–1945)* (Brno: Matice moravská, 2005), 281.
[32] Collection of video testimonies, Museum of Roma Culture, Brno, Czech Republic, sig. V 163/96, V 164/96.

until the end of this camp in August 1944, when almost 3,000 Roma prisoners were gassed. They both left interesting testimonies.[33]

Another story linking the fates of Jewish and Roma at the time of World War II is the one of a Czech Jewish painter, Dinah Gottlieb. She and her family were deported first to the Theresienstadt ghetto and later to the concentration camp Auschwitz-Birkenau. There she was forced to portrait Roma prisoners from Poland, Germany, and France for camp doctor Josef Mengele. Only seven watercolors have been preserved until today – they are exhibited in the Auschwitz museum.[34]

The Roma in Slovakia during the World War II (1939–1945)

The Roma in fascist Slovak Republic were subject to deprivation of rights, forced labor in the labor camps, and mass killings since March 1939. The situation became even worse when the German army occupied Slovakia in August 1944. A special camp was established there for Roma, and the mass killings continued at a larger scale. There were hundreds of Roma victims during World War II in Slovakia.[35]

The book *After Jews Gypsies* also describes how the genocide of the Jews in Slovakia affected the Roma population. Since the summer of 1940, regulations banned the Jews from running tavern businesses and employing "Aryan" persons under forty years of age in their households. Roma, being "non-Aryan," could still do some services for Jewish families, but lost this option with the pro-

---

[33] Nečas, "Židovští lékaři v cikánských táborech," 58–61; Pape, "Michal Bohin, lékař-lidumil," 62–65.

[34] For more about Dinah Gottlieb see, for example: Ctibor Nečas, "Dininy podobizny Romů," *Romano džaniben* 1–2 (Prague: Společnost přátel časopisu *Romano džaniben*, 2000): 68–76; Ctibor Nečas, "Annemarie Babbitt, roz. Gottliebová," *Bulletin Muzea romské kultury*, 18/2009 (Brno: Muzeum romské kultury, 2010): 214–215; Ctibor Nečas, "Nad portréty neznámých vězňů z osvětimského cikánského tábora," *Romano džaniben*, 16/2010 (Prague: Společnost přátel časopisu *Romano džaniben*, 2011): 97–114.

[35] For more about persecution of the Slovak Roma during WWII see, for example: Ctibor Nečas, *Nad osudem českých a slovenských Cikánů v letech 1939–1945* (Brno: UJEP, 1981); Julius Tancoš, René Lužica, *Zatratení a zabudnutí* (Bratislava: Vydavateľstvo IRIS, 2002); Zuza, Kumanová, Arne B. Man (eds.), *Nepriznaný holocaust: Rómovia v rokoch 1939–1945* (Bratislava: Občianske Združenie In Minorita, Slovenske narodne muzeum, Ustav etnologie SAV, 2007); Karol Janas, *Zabudnuté tábory* (Trenčín: Alexander Dubček University in Trenčín, 2008); Karol Janas, *Perzekúcie Rómov v Slovenskej republike (1939–1945)* (Bratislava: Ústav pamäti národa, 2010).

gressing persecution of Jews. Subsequent deportations of Jews since 1942 had a considerable destructive impact on Roma – economic, social, and psychological. They lost social and human support. More than that, they too were threatened by nationalist Hlinka Guards whose slogan "After Jews Gypsies" was known and used also in other countries at that time.[36]

Jozef Horváth, born 1916, Železník, Slovakia: "We did not know what they will do with them [the Jews], when they took them. [...] We heard only after the war that they burned them. Well we knew that they did with them nothing good, because we were threatened, too, that they will collect us and we will go for [be turned into] soap as Jews. Slovaks were shouting at us that way. And so everybody was afraid, we were just waiting when they will collect us as Jews. Well, but in the end this did not happen to us."[37]

Elena Lacková, born 1921, Veľký Šariš, Slovakia: "Then they started to gather the Jews. All of them, also the old ones, who could not walk. Their children supported them, and someone was carried on a stretcher. I can not even speak when I think of it. We, Roma, we all accompanied them, only the youngest children stayed in the village. [...] Not that we were not afraid, but we suspected that it is already end with us too. Guardsmen kept saying: 'You also take your turn! You will go [turn in]to soap and grease!' They used to scare us like this. Soon after that they forced us to move from the village to Korpáš hill. We were expecting the worst, but there was nowhere to escape."[38]

It is important to say now that after their liberation in 1945, the Roma from Slovakia, influenced by their experience of war, started to migrate to the Czech border regions, which were then empty after the expulsion of Germans, as well as to big cities. They were spread out as labour force throughout the industrial areas of Bohemia and Moravia.[39]

---

[36] Hübschmannová (ed.), *Po Židoch Cigáni*, 677–684.
[37] Hübschmannová (ed.), *Po Židoch Cigáni*, 477.
[38] Ibid., 687.
[39] For more about the position of Roma in the Czechoslovakia after 1945, see for example Eva Davidová, *Cesty Romů 1945–1990 – Romano drom* (Olomouc: Palacký University, 1995); Nečas, *Romové v České republice včera a dnes*, 84–97; *Romové v České republice (1945–1998)* (Prague: Socioklub, 1999); Nina Pavelčíková, *Romové v českých zemích v letech 1945–1989* (Prague: Úřad dokumentace a vyšetřování zločinů komunismu PČR, 2004); Michal Schuster, "Jak slovenští Romové přišli do českých zemí...," in *Khatar san? Jak slovenští Romové přišli do českých zemí za prací a co se dělo potom* (Brno: Masaryk University, 2015), 9–12.

After 1945

The experience of the Holocaust could not remain without consequences. After World War II, both groups of survivors had to go through a similar difficult process associated with return, coping with the traumatic experience, but also with recognition from the majority connected to compensations.[40]

Since the second half of the 1950s, the communist regime in Czechoslovakia came with an open policy of assimilation of the Roma. They were supposed to become ordinary Czechs and Slovaks without any relics of their former "Gypsy origin." This was connected with the process of recognition of their persecution during World War II. Although the state authorities were aware of the number of all Czech Roma interned in Auschwitz and also knew the reason of their internment, there was no acknowledgement of their racial persecution.[41]

In European countries, the Roma are still subject to various political and social measures. Old stereotypes and myths appear again and again, and new ones are created. Solutions of the so-called Jewish problem before World War II correspond to current attempts to solve the "Gypsy problem." Examples include separation in public places, special lists and consequent special approaches, proposals to reduce birth rates, proposals for special forced employment, forced evictions from the homes, expulsion from some countries, police raids, discriminatory legislation, and discriminatory attitudes at all levels of life. These include verbal attacks by populist politicians, parties, and movements that are misusing the radical sentiment in mainstream society. Unfortunately, there are also physical attacks which sometimes have fatal consequences.[42] This development is a challenge for both the Jewish and Roma minority in the Czech Republic and should contribute to interaction and mutual solidarity. Also, sharing the topic

---

[40] For more about the compensation process see, for example: Eva Zdařilová, *Proces odškodňování obětí romského holocaustu v České republice* (Prague: Faculty of Arts at Charles University in Prague, 2007); Eva Zdařilová, *Odškodňovací programy v České republice*, last modified March 1, 2016, http://www.romea.cz/cz/romano-vodi/eva-zdarilova-odskodnovaci-programy-v-ceske-republice; Julia von dem Knesebeck, *The Roma Struggle for Compensation in Post-War Germany* (Hertfordshire: University of Hertfordshire Press, 2011).

[41] Michal Schuster, "Genocida Romů v českých zemích a její reflexe," *Romano voďi*, 25.10.2012 (Prague: ROMEA, 2012),10.

[42] *O Roma – Romové: tradice a současnost* (Brno: Svan, 1999); *Romové v České republice (1945–1998)*; Nečas, *Romové v České republice včera a dnes*; *Černobílý život* (Prague: Gallery Press, 2000); Pavel Navrátil, *Romové v české společnosti: jak se nám spolu žije a jaké má naše soužití vyhlídky* (Prague: Portál, 2003).

of Nazi persecution leads to multilevel cooperation and mutual support and cooperation of Jewish and Roma minorities in the Czech Republic.[43]

I would especially like to mention Jewish-Roma cooperation on Holocaust education, remembrance, and research in the Czech Republic. Roma representatives take part in an annual event, Yom ha-Shoah or the Day of Remembrance, for the victims of the Holocaust in Prague and other Czech towns. The event is organized by the Foundation for Holocaust Victims and the Terezín Initiative Institute. The names of Roma victims from Czech lands are also read on this occasion.[44]

The Museum of Romani Culture in Brno has been cooperating with various Jewish organizations and institutions since its formation in 1992, for example with the Jewish Museum in Prague, Terezín Memorial, Terezín Initiative Institute in Prague, the European Shoah Legacy Institute in Prague, and the Jewish Museum in Bratislava in Slovakia. The Museum of Roma Culture is also a member of official Czech delegation at the International Holocaust Remembrance Alliance (formerly the Task Force for International Cooperation on Holocaust Education, Remembrance, and Research).

Every year, the Museum commemorates mass deportations of Roma from the Protectorate of Bohemia and Moravia. As a long-term tradition, the representatives of the Jewish community are among the important guests and speakers on these commemoration events. Our museum is also intensively cooperating with the Brno municipality and the Jewish community from Brno on the project of the Memorial for the Jewish and Roma victims from Brno, which was placed in Brno on September 2014.[45]

---

[43] Daniel Macmillen, "OPINION: As Jews, We should stand up for Roma rights," *Jewish News*, October 13, 2014, http://www.jewishnews.co.uk/opinion-jews-stand-roma-rights/.

[44] "Vzpomínková akce na židovské a romské oběti holocaustu proběhne dnes v sedmi českých městech," *Romea.cz*, April 16, 2015, http://www.romea.cz/cz/zpravodajstvi/domaci/vzpominkova-akce-na-zidovske-a-romske-obeti-holocaustu-probehne-dnes-v-sedmi-ceskych-mestech.

[45] For more about the activities of the Museum of Roma Culture see www.rommuz.cz; Lucie Křížová and Michal Schuster, "Museum of Romani Culture in Brno, Czech Republic. Muzeum romské kultury Brno, Česká republika," *Situation of the Roma Minority in the Czech Republic, Hungary, Poland and Slovakia, vol. 2*, ed. Jaroslav Balvín, Małgorzata Ewa Kowalczyk, and Łukasz Kwadrans (Wrocław: Foundation of Social Integration Prom, 2011), 123–140.

Bibliography

Alfaro, A.G. *Velký proticikánský zátah: Španělsko – všeobecné uvěznění cikánů v roce 1749* [The Great Gypsy Round-Up: The General Imprisonment of Gypsies in 1749]. Olomouc: Palacký University, 1999.

Baumgartner, Gerhard, Florian Freund, and Harald Greifeneder. *Nationale Minderheiten im Nationalsozialismus, 2. Vermögensentzug, Restitution und Entschädigung der Roma und Sinti*. Wien/ Oldenbourg: Oldenbourg Wissenschaftsverlag, 2004.

Collection of video testimonies, Museum of Roma Culture, Brno, Czech Republic, sig. V 80/97, NV 214/2012.

Collection of video testimonies, Museum of Roma Culture, Brno, Czech Republic, sig. NV 225/2012.

Collection of audio testimonies, Museum of Roma Culture, Brno, Czech Republic, sig. A 43/2003.

Collection of audio testimonies, Museum of Roma Culture, Brno, Czech Republic, sig. A 14/99.

Collection of video testimonies, Museum of Roma Culture, Brno, Czech Republic, sig. V 163/96, V 164/96.

Collection of audio testimonies, Museum of Roma Culture, Brno, Czech Republic, sig. A 32/95.

Davidová, Eva. *Cesty Romů 1945–1990 – Romano drom* [The Journeys of the Roma 1945–1990 – Romano drom]. Olomouc: Palacký University, 1995.

Długoborski, Wacław (ed.). *Sinti und Roma im KL Auschwitz-Birkenau 1943–44: vor dem Hintergrund ihrer Verfolgung unter der Naziherrschaft*. Oświęcim: Staatliches Museum Auschwitz-Birkenau, 1998.

Fraser, Angus. *The Gypsies: The People of Europe*. Oxford: Blackwell Publishers, 1995.

Glassmannová, Gaby. *Židovské a romské dědictví (Několik srovnání)* [Jewish and Roma Heritage (Several Comparisons)]. Romano džaniben, jevend. Prague: Společnost přátel časopisu Romano džaniben, 2002.

Hancock, Ian. *Země utrpení – Dějiny pronásledování a otroctví Romů* [Land of Pain – History of the Persecution and Slavery of the Roma]. Prague: Signeta, 2001.

Heuss, Herbert, Frank Sparing, Karola Fings, and Henriette Asséo. *From the "Race Science" to the Camps*. Paris, Hatfield: Centre de recherches tsiganes, University of Hertfordshire, 1999.

Hübschmannová, Milena (ed.). *Po Židoch Cigáni: svědectví Romů ze Slovenska 1939–1945* [Gypsies After the Jews: Testimonies of the Slovak Roma 1939–1945]. Prague: Triáda, 2005.

Hűbschmannová, Milena. "Rozhovor s Karolem Efraimem Sidonem, vrchním rabínem v České republice" ["Interview with Karol Efraim Sidon, Chief Rabbi in the Czech Republic"]. *Romano džaniben* 1–2. Prague: Společnost přátel časopisu Romano džaniben, 2000.

Janas, Karol. *Zabudnuté tábory* [The Forgotten Camps]. Trenčín: Alexander Dubček University, 2008.

Janas, René. *Perzekúcie Rómov v Slovenskej republike (1939–1945)* [Persecution of Roma in Slovak Republic (1939–1945)]. Bratislava: Ústav pamäti národa, 2010.

Kenrick, Donald, and Grattan Puxon. *Gypsies Under the Swastika*. Hatfield: University of Hertfordshire Press, 1995.

Kladivová, Vlasta. *Konečná stanice Auschwitz-Birkenau* [*The Last Stop Auschwitz-Birkenau*]. Olomouc: Palacký University Press, 1994.

Knesebeck, Julia von. *The Roma Struggle for Compensation in Post-War Germany*. Hertfordshire: University of Hertfordshire Press, 2011.

Křížová, Lucie, and Michal Schuster. "Muzeum romské kultury Brno, Česká republika [Museum of Roma Culture in Brno, Czech Republic]." In *Situation of the Roma Minority in the Czech Republic, Hungary, Poland and Slovakia, vol. 2*, edited by Jaroslav Balvín, Małgorzata Ewa Kowalczyk, and Łukasz Kwadrans. Wrocław: Foundation of Social Integration Prom, 2011.

Kumanová, Zuzana, and Arne B. Mann (eds.). *Nepriznaný holocaust: Rómovia v rokoch 1939–1945* [*The Unapproved Holocaust: Roma in the Years 1939–1945*]. Bratislava: Občianske združenie In Minorita, Slovenské národné múzeum, Ústav etnológie SAV, 2007.

Lisá, Helena. *Romové v České republice (1945–1998)* [*The Roma in Czech Republic (1945–1989)*]. Prague: Socioklub, 1999.

Lukšíková, Petra. "Naučená vzájomnosť (Vzťah majoritnej spoločnosti k Rómom a Židom)" ["Learned Reciprocity (The Relationship of the Majority Society to Roma and Jews)"]. *Romano džaniben* 1–2. Prague: Společnost přátel časopisu *Romano džaniben*, 2000.

Macmillen, Daniel. "OPINION: As Jews, We should stand up for Roma rights." *Jewish News*. October 13, 2014, http://www.jewishnews.co.uk/opinion-jews-stand-roma-rights/.

Navrátil, Pavel. *Romové v české společnosti: jak se nám spolu žije a jaké má naše soužití vyhlídky* [*Roma in Czech Society: How We Live Together and What Are the Prospects of Our Coexistence*]. Prague: Portál, 2003.

Nečas, Ctibor. "Nad portréty neznámých vězňů z osvětimského cikánského tábora" ["Over the Portraits of Unknown Prisoners from the Auschwitz Gypsy Camp"]. *Romano džaniben* 16. Prague: Společnost přátel časopisu *Romano džaniben*, 2011.

Nečas, Ctibor. "Annemarie Babbitt, roz. Gottliebová" ["Annemarie Babbitt, born Gottliebová"]. *Bulletin Muzea romské kultury* 18. Brno: Muzeum romské kultury, 2010.

Nečas, Ctibor. *Romové na Moravě a ve Slezsku (1740–1945)* [*The Roma in Moravia and Silesia (1740–1945)*]. Brno: Matice moravská, 2005.

Nečas, Ctibor. "Židovští lékaři v cikánských táborech" ["Jewish Physicians in the Gypsy Camps"]. *Romano džaniben* 1–2. Prague: Společnost přátel časopisu *Romano džaniben*, 2000.

Nečas, Ctibor. "Dininy podobizny Romů" ["Dina's Portraits of Roma"]. *Romano džaniben* 1–2. Prague: Společnost přátel časopisu *Romano džaniben*, 2000.

Nečas, Ctibor. *Romové v České republice včera a dnes* [*The Roma in the Czech Republic, Yesterday and Today*]. Olomouc: Palacký University Press, 1999.

Nečas, Ctibor. *Holocaust českých Romů* [*The Holocaust of Czech Roma*]. Prague: Prostor, 1999.

Nečas, Ctibor. *Historický kalendář: Dějiny českých Romů v datech* [*Historical Calender: History of Czech Roma in Data*]. Olomouc: Palacký University Press, 1997.

Nečas, Ctibor. *Andr' oda taboris. Tragédie cikánských táborů v Letech a v Hodoníně* [*Andr' oda taboris. Tragedy of the Gypsy Camps in Lety and Hodonín*]. Brno: Masaryk University, 1995.

Nečas, Ctibor. *Andr' oda taboris. Vězňové protektorátních cikánských táborů 1942–1943* [*Andr' oda taboris. Prisoners of the Protectorate Gypsy Camps 1942–1943*]. Brno: Městský výbor Českého svazu protifašistických bojovníků, 1987.

Nečas, Ctibor. *Nad osudem českých a slovenských Cikánů v letech 1939–1945* [*Over the Fate of Czech and Slovak Gypsies in the Years 1939–1945*]. Brno: UJEP, 1981.

Pape, Markus. *Michal Bohin, lékař-lidumil* [*Michal Bohin, Physician – Philanthropist*]. *Romano džaniben* 1–2. Prague: Společnost přátel časopisu *Romano džaniben*, 2000.

Pape, Markus (ed.). *A nikdo vám nebude věřit: dokument o koncentračním táboře Lety u Písku* [*And Nobody Will Believe You: The Document on the Concentration Camp Lety near Písek*]. Prague: G plus G, 1997.

Pavelčíková, Nina. *Romové v českých zemích v letech 1945–1989* [*The Roma in Czech Lands in the Years 1945–1989*]. Prague: Úřad dokumentace a vyšetřování zločinů komunismu PČR, 2004.

Procházková, Eva. "Perzekuce romských kočovníků v českých zemích v 18. století". ["The Persecution of Roma Nomads in Czech Lands in 18th Century"]. *Sborník archivních prací* 42, no. 2 (1992).

Schuster, Michal. "Pronásledování protektorátních Romů s přihlédnutím k roku 1944" ["The Persecution of the Protectorate Roma in 1944"]. *Válečný rok 1944 v okupované Evropě a v Protektorátu Čechy a Morava*. Prague: Ústav pro studium totalitních režimů, 2015.

Schuster, Michal. "Jak slovenští Romové přišli do českých zemí..." ["How Slovak Roma Came to the Czech Lands..."] *Khatar san? Jak slovenští Romové přišli do českých zemí za prací a co se dělo potom*. Brno: Masaryk University, 2015.

Schuster, Michal. "Testimonies of witnesses to the Roma holocaust at the Museum of Roma Culture in Brno." *The Holocaust Between Memory and History*. Paper from the conference held in Bratislava on September 30th–October 1st, 2013. Bratislava, Slovak National Museum – Museum of Jewish Culture, 2014.

Schuster, Michal. "Genocida Romů v českých zemích a její reflexe" ["The Genocide of the Roma in the Czech Lands and Its Reflection"]. *Romano voďi*, 25.10.2012. Prague: ROMEA, 2012.

Suleř, Petr. *O Roma – Romové: tradice a současnost* [*O Roma – the Roma: Tradition and Present*]. Brno: Svan, 1999.

Tancoš, Julius, and René Lužica. *Zatratení a zabudnutí* [*The Damned and Forgotten*]. Bratislava: Vydavateľstvo IRIS, 2002.

"Vzpomínková akce na židovské a romské oběti holocaustu proběhne dnes v sedmi českých městech" ["A Commemorative Event on Jewish and Roma Victims of the Holocaust Will Take Place Today in Seven Czech Cities"]. *Romea.cz*, April 16, 2015. http://www.romea.cz/cz/zpravodajstvi/domaci/vzpominkova-akce-na-zidovske-a-romske-obeti-holocaustu-probehne-dnes-v-sedmi-ceskych-mestech.

Zdařilová, Eva. "Odškodňovací programy v České republice" ["Compensation Programs in the Czech Republic"]. *Romea.cz*, March 1, 2016. http://www.romea.cz/cz/romano-vodi/eva-zdarilova-odskodnovaci-programy-v-ceske-republice.

Zdařilová, Eva. *Proces odškodňování obětí romského holocaustu v České republice* [*The Process of Compensation of the Victims of the Roma Holocaust in the Czech Republic*]. Prague: Faculty of Arts at Charles University in Prague, 2007.

Zimmermann, Michael. *Rassenutopie und Genozid. Die nationalsozialistische "Lösung der Zigeunerfrage."* Hamburg: Christians, 1996.

Zimmermann, Michael. *Verfolgt, vertrieben, vernichtet: Die nationalsozialistische Vernichtungspolitik gegen Sinti und Roma*. Essen: Klartext Verlag, 1989.

Section IV: **An Ongoing Struggle with Judeophobia**

Dina Porat
Holocaust Denial as a Symptom of Unresolved European History

Modern antisemitism has found creative ways to generate old-new stereotypes of Jews as the ultimate others, extrinsic and dangerous. In educated populations of the twenty-first century, images of the Jews as "deicides", "racketeers", or "parasites" are rather disliked due to their being obviously primitive or "politically incorrect".[1] Consequently, present-day antisemites use rather subtle images and means in order to discredit Jews and Jewish communities as disturbing coevals, despite or just because of the Holocaust. Holocaust denial and Holocaust relativization are not only emotional reactions to the most horrendous genocidal crime perpetrated in World War II, but also a means to hold Jews responsible for restlessness in society.[2]

Admittedly, parts of the European population had been shocked when realizing the dimension of the Holocaust, the assassination of six million Jewish civilians. Yet the shock faded away, and the Holocaust obviously didn't cause a break with the stereotypes and the feelings of hatred against Jews, that can be traced back to early European Middle Ages. Hatred of and hostility toward Jews have remained deeply engraved in the collective memory. Over the centuries, the surface has changed, but the core of hateful feelings and stereotypes has remained unaltered.[3] Judeophobia sometimes proves to be resistant to education, to argument, to reasoning, to facts – and so the handling of the Holocaust in some countries of Europe, including Central Europe, is recently rather marked by ignorance and new resentment. In Germany, for instance, the mechanisms of rejection, alleged forgetting and Holocaust Fatigue are even more emotionally

---

[1] This text is originally based on my essay "Holocaust Denial and the Image of the Jew, or: 'They Boycott Auschwitz as an Israeli Product'", in *Resurgent Antisemitism: Global Perspectives*, ed. Alvin Rosenfeld (Indiana University Press: Bloomington 2013), 468–481.
[2] In Leo Katcher's book *Post Mortem: The Jews in Germany – Now* (1968), the German Jewish journalist Hilde Walter is quoted with the famous sentence: "It seems the Germans will never forgive us Auschwitz." This is exactly the point, describing discomfort among Germans (and probably among other Europeans as well), when being reminded about the Jewish tragedy in World War II.
[3] "Destroy Israel: Jews are the Evil of the World!" Speech of Prof. Monika Schwarz-Friesel (TU Berlin) on antisemitism at the ICCA, Bundestag Berlin, March 14, 2016.

based then in other countries, thus enlarging perhaps a new gap between non-Jews and Jews.[4]

Hardcore Holocaust denial, which reached its heyday in the 1980s and the 1990s, created a certain image of the "Jew," as Brian Klug put it when he tried to define the distinction between Jews and a "Jew."[5] He argued that antisemitism "is best defined not by an attitude toward Jews but by a definition of a 'Jew'," and that antisemitism is "the process of turning Jews into a 'Jew'." His distinction is equally relevant to both the "Jew" in the singular and "Jews" in the plural, because in both cases the quotation marks turn the Jew/Jews into an idea, a symbol, a stereotype, in which each individual is meant to represent his people at large as a collectivity, and both cease to be recognized as part of reality. The process of turning individuals and a people into "Jew/Jews" is at the heart of the following discussion.

This stereotypical image created by Holocaust deniers derived from older images that developed in the centuries prior to the Holocaust, yet those who advance the very idea of denial take the former images much further, and develop a picture of the most abominable type: for if the story of the Holocaust, as claimed and disseminated by the Jewish people, never in fact happened, then this fiction obviously points to a people with a rare ability to invent unheard-of horror stories that only a sick mind could produce; a people equipped with outstanding skills of self-organization and mobilization capabilities that help spread these lies through the use of all the public media, which they anyhow control, and thereby convince the world that they are truth incarnated; a powerful egoist people capable of brazen blackmailing in order to secure financial and moral gains.

In every period during the long history of antisemitism, different character traits alleged to belong to Jews were at the forefront of verbal and visual portrayals. The characteristics most fostered in modern times are a Jewish people craving for power and the wish to dominate the world. The idea that the grip acquired by Jews over world public opinion through the use of their story of the Holocaust was in fact called by some as proof of the authenticity of *The Protocols of the Elders of Zion*; this supposition has actually been raised by Holocaust deniers themselves, such as by Germar Rudolf.[6] However, the myth of Jewish power and world control was shattered by the realities of the Holocaust, whose destruc-

---

[4] See also Monika Schwarz-Friesel, *Die Sprache der Juden im 21. Jahrhundert* (Berlin/Boston 2013), 6.

[5] Brian Klug, "The Collective Jew: Israel and the New Antisemitism," *Patterns of Prejudice* 37, no. 2 (2003): 124.

[6] Sarah Rembiszewski, *The Final Lie: Holocaust Denial in Germany – A Second Generation Denier as a Test Case* (Tel Aviv: Stephen Roth Institute, Tel Aviv University, 1996), 57.

tive results became possible precisely because the Jewish people at that time were totally helpless and defenseless. Therefore, if one wants, or rather has the urge, to maintain the myth of Jewish power and world control, which serves as the foundation of the leading antisemitic views and convictions in modern times, one is compelled to deny the Holocaust. Denying, distorting, and especially inverting have been techniques of antisemitism since antiquity, and have aimed at creating images of Jews that are the opposite of what is found in reality. One may say that the essence of antisemitism has always been the discrepancy between such images of the "Jew" and his/her/their real abilities and character traits.[7] This discrepancy reached a peak in the wake of the Holocaust.

Since the year 2000, hardcore denial has somewhat weakened, thanks to a number of events and responses: David Irving's trial; the visits of two Popes – John Paul II and Benedict XVI – to Yad Vashem; the Stockholm Forum; and International Holocaust Memorial Day established by the United Nations (UN), followed by UNESCO.[8] Instead, and perhaps as a result of this weakening, the last decade has evinced the flourishing of new terms defining attitudes toward the Holocaust and the use of its memory. They include Holocaust trivialization, minimization, and relativization alongside Holocaust skepticism and softcore denial, as well as terms relating to new types of deniers themselves, such as demi- or semi deniers, and moderate deniers.

Another notable development of the last decade is the activity of the Task Force for International Cooperation on Holocaust Education, Remembrance, and Research, established in the wake of the Stockholm Forum in 2000 and now numbering thirty-four countries. This organization promotes the launching or enhancing of Holocaust education (the teaching of its history, consequences, and implications) in its member states, by allocating budgets, especially for the training of teachers. In 2013 the organization has been re-named into International Holocaust Remembrance Alliance (IHRA).

In light of these developments, one might ask if the image of the "Jew," and the Jewish State populated by "Jews" has changed, weakened, or become more moderate. In pursuing answers, one should bear in mind that Holocaust denial has been transferred during the last decade from the far right to the agenda of leftists and Islamists, and thus has moved from the former theoretical sphere to the political one. Perhaps, while becoming in some ways more moderate, Hol-

---

[7] Klug, 123; see also Dina Porat's expert testimony in the "Michael Adams and Christopher Mayhew against Maariv" case, Supreme Court Judge Yaakov Bazak's verdict, Maariv (August 10, 1978).
[8] See the UN General Assembly resolution of November 21, 2005, no.60/7, on Holocaust Remembrance.

ocaust denial has taken the shape of mainly "denying the meaning and the consequences of the Holocaust rather than necessarily denying the facts of the Holocaust itself."[9]

Denying, distorting, and inverting the Holocaust, its meaning, and its consequences reflect numerous goals, among which the following stand out. There is firstly the wish to abolish the martyr status claimed by the Jews and bestow it on another group. Even if church teachings have lost much of their former influence, basic Christian ideas regarding suffering and salvation are deeply embedded in the Western world, first and foremost in its culture, especially in music, figurative art, and drama, dating from the early medieval era in Europe, where the Holocaust occurred. According to these ideas, the life and death of Jesus are proof that he who suffers brings salvation. The Christian duty is to identify with the supreme sufferer, Jesus Christ. Since Jews have foresworn such identification, martyrdom can certainly not be attributed to them, as they rejected the salvation offered by Jesus long ago.

The second goal, related to the first, is to cancel the notion of the Holocaust's uniqueness. This notion is allegedly fostered by egocentric Jews who constantly claim to be the sole bearers of ultimate victimhood and do not recognize the sufferings of others; moreover, such Jews are said to inflict suffering on "the victims of the victims," as Orientalist Edward Said expressed it.[10] Within this scheme, he who inflicts suffering must become an anti-Christ, equated in the modern world with Nazis, the ultimate symbol of modern evil. The Jews were killed during the Holocaust by Christians, but Christians cannot, a priori, be the anti-Christ, so it is up to the Jews to go on fulfilling that role in this twisted morality play. "If the European memory of the Holocaust recognizes the Jews only as victims, then the moment they cease being victims [such as is the case since the establishment of the State of Israel] they become the guilty party," claims the French Jewish scholar Shmuel Trigano,[11] and their status outside the Holocaust context, without the victim label, is inevitably identified with that of the Nazis. They are either the Chosen People or the cursed one. "The ancient denouncement of the Jew, because of his origin, uniqueness, exclusiveness,

---

9 Dave Rich, "Holocaust Denial as an Anti-Zionist and Anti-Imperialist Tool for the European Far Left," *Post-Holocaust and Anti-Semitism*, no. 65 (a Jerusalem Center for Public Affairs online publication; February 1, 2008).
10 Edward Said, *The Politics of Dispossession* (New York: Pantheon, 1994), 121.
11 Shmuel Trigano, "Europe's Distortion of the Meaning of the Shoah Memory and Its Consequences for the Jews and Israel", *Post-Holocaust and Anti-Semitism* 42 (March 1, 2006), 5; "The Political Theology of the Memory: Europe Is Morally Ready for a Second Holocaust," *Kivunim Hadashim* 17 (January 2008): 87.

his national egoism, his being a closed cast," writes the French Jewish philosopher Alain Finkielkraut, "has been revived due to the trauma caused by Nazism and is expressed in totally modern ways ... [but it still] originates in the Gospel of Paul or its recycling."[12]

The second goal leads to the third, which is the wish to cancel, or at least abate, the allegedly greedy Jewish claim for the return of property looted by Nazi Germany during the Second World War. This claim seems now even greedier than before, given the recent world economic crisis (exacerbated by the Bernard Madoff affair). It raises again the old image of the rich parasitic Jew, who owned so much that perhaps the Germans and others who plundered him did not commit such a terrible crime. When historian Gitta Sereni asked Franz Stangl, commander of Treblinka, in his prison cell in Dusseldorf, what in fact the Jews were killed for, his spontaneous answer was "They wanted the Jews' money," as if the question were naive and out of place.[13] One can easily notice the proximity in time between the Prague Conference convened in June 2009 in order to deal with the necessity to reclaim looted Jewish property, especially works of art, and the Prague Declaration (issued in June 2008 and ratified by the European Parliament in July 2009) that makes the counterclaim: everyone suffered during the Second World War (and the East European countries later had an additional share of suffering at the hands of the Soviet regime); therefore, every individual and every nation is equally entitled to compensation for and commemoration of their suffering.[14] The Prague Declaration reflects the current *Zeitgeist*, a post-heroic post-modern atmosphere in which everyone is equally entitled to a narrative of suffering, and therefore to equal rights and compensation for that suffering. In this cultural context, suffering and victimhood, both personal and collective, have become an asset of pride. Being a victim means having a moral status, a claim for being just, virtuous, and politically correct, to quote Alain Finkielkraut, who has said that "the Jews have the good fortune to be the kings of misfortune."[15] Another author calls Jews "the stars of sorrow,"[16] to cite

---

[12] Alain Finkielkraut, *Au nom de l'Autre, reflextion sur l'antisemitisme qui vient* (Jerusalem: Shalem, 2004), 34–35.
[13] Gitta Sereny, *Into that Darkness: From Mercy Killing to Mass Murder, a Study of Franz Stangl, the Commandant of Treblinka* (London: Deutsch, 1974 and 1995), 232.
[14] The June Prague Declaration on European Conscience and Communism, which turned into the Prague Process, was ratified by the European Parliament on July 2009, a few weeks before the seventieth anniversary of August 23.
[15] See Finkielkraut.
[16] Heinz Heger, *Les hommes au triangle rose*, preface by Guy Hocquenghem (Paris: Persona, 1981), 9.

one more example among many. We are witnessing a "competition of victimhood," says Bernard-Henri Levy, one that has turned society into a compilation of grievances.[17] And if so, needless to say, many groups that claim victimhood status use the term Holocaust to describe their plight and will not settle for a milder term or description. "*Dueños del dolor, dueños del mundo*" was the title of an anti-Jewish article in a Venezuelan newspaper: "They who own the pain own the world."[18] So much so, that Elie Wiesel suggested that Jews stop using the term Holocaust altogether and find themselves another word.[19] In the 1993 UN Human Rights conference convened in Vienna, no precise decision could be taken at the closing plenary because some dozens of groups from all over the globe claimed to be victimized, and in the 2001 Durban I UN conference it took a difficult struggle to reverse the demand to rewrite Holocaust as holocausts.[20]

The fourth wish is inevitable in light of the former ones: to discredit acts of commemoration by the Jewish people that create and perpetuate feelings of guilt, of eternal debt, and of an incriminating memory among their fellow nations, especially the Europeans, by documenting, recording, and publishing Nazi atrocities through every means of communication as well as in museums and monuments. Such acts of commemoration serve as a constant reminder of who took part in the acts of murder and plundering. In order to undermine the validity of the Jewish commemoration activities, and to get rid of these feelings of guilt, the blame is inverted and an opposite claim is made, namely, that the Jews are a nation whose own cruelty, especially toward children, has been genetically coded since biblical times.

It is becoming increasingly clear that the Holocaust, which was supposed to have been a source of empathy and compassion for the Jews, more and more enhances a negative image of the "Jew" and of the Jewish State and fosters post-Holocaust antisemitism. The pre – Holocaust image of the Jew as an all-powerful, avaricious manipulator of power was a crucial motive for the mass murder of European Jewry. Nowadays, in the post – Holocaust era, the Jew is being portrayed in a no less repulsive way, indicating that the changes in social and political cir-

---

17 Bernard-Henri Levy's keynote address at the opening of the World Jewish Congress annual convention, December 17, 2006, Paris. He repeated his three pillars of new antisemitism theory in his book, *Left in Dark Times: A Stand against the New Barbarism* (New York: Random House, 2008), 155–66.
18 Jose Roberto Duque, "Dueños del dolor, dueños del mundo", *Aporrea* (July 20, 2006).
19 Elie Wiesel, "The Memory of the Holocaust on Israel's 60th Anniversary," lecture at Tel Aviv University on May 20, 2008, recorded but not published.
20 I was a member of the Israeli Foreign Ministry delegation to both Vienna and Durban.

cumstances after the Holocaust have become a new source of antisemitism. Holocaust education has also not yet proved itself to be a barrier against antisemitism, for youngsters, whose ignorance is coupled by naiveté, often raise such questions as these: why the Jews? Why all the Jews? What's wrong with them? Was their murder really initiated without any logical reason, or other good motive? Six million – how indeed did so many Jews, who do not seem to be at all helpless today, allow this to be done to them?[21]

A number of countries – but, to date, no more than twenty-five – have enacted laws and other forms of legislation against Holocaust denial, mostly since the 1990s. Denial could have been included under the umbrella of freedom of speech, because the deniers do not resort to physical violence. Rather, deniers express their opinion, so that one may say there is an "absence of a criminal motive" in their activities, as the legal term goes. Yet denial of the Holocaust is punishable outside the United States and Canada, where freedom of speech has gained sanctity, precisely because of the image of the Jew it creates. This image constitutes incitement against a whole group of people, and in a way that might provoke violence against them. Moreover, denial intentionally falsifies the facts by intentionally misusing the documentation, as established in the judges' verdict in Holocaust denier David Irving's trial in London in 2001. In this respect, it might be understood as a form of violence against truth itself or at least as an intentional subversion or corruption of the historical record.

Since early 2005, a working definition of antisemitism, agreed upon by the twenty-seven EU countries, states clearly that "denying the fact, scope, mechanisms (e. g., gas chambers) or intentionality of the genocide of the Jewish people at the hands of National Socialist Germany and its supporters and accomplices during World War II (the Holocaust), [and that] accusing the Jews as a people, or Israel as a state, of inventing or exaggerating the Holocaust" are considered acts of antisemitism. A more recent working definition of Holocaust denial, reached by the International Task Force member states in 2010, which draws on the EU definition, also defines denial as a form of antisemitism.[22] Antisemitism is by now punishable in all countries that have a law against racism – antisemitism is included. However, only five countries legislatives's mention antisemitism specifically.

However, some might interpret the new definitions as resulting from Jewish behind-the-scenes pressure as one more manifestation of Jewish power. Thus,

---

[21] See Anders Lange, *A Survey on Teachers' Experiences and Perceptions in Relation to Teaching about the Holocaust* (Stockholm: The Living History Forum, 2008).
[22] It was reached during the ITF Haifa plenary in mid-December 2010 by the ITF subcommittee on antisemitism and Holocaust denial.

this very achievement can be used to confirm, still again, the image of a powerful, self-aggrandizing group that cares exclusively about itself, is steeped in its sorrowful past while ignoring the sufferings of others and brings to trial and even jail courageous individuals who are out to expose what they consider to be the truth and do not shy away from confronting the powerful and the influential. Deniers present themselves as researchers adhering to a basic principle of historical research: they doubt and scrutinize the written word, namely the documents and the testimonies, and they do not readily accept the conventional and the agreed upon. Their skeptical attitude is presumed to be part and parcel of freedom of speech, the core of democratic society. Today they call themselves revisionists, not deniers, because revisionism of history is a main trend of contemporary historiography. They claim that those who oppose open or free scientific criticism belong to the narrow-minded dark forces that shun enlightenment. Human rights activity now is largely about diversity, about universal values, not particular ones. NGO idealists have turned tolerance into a religion and equal rights almost into a cult. In such a context, Jews find themselves facing an ironic twist: they, who have always hoisted the flag of universalism and universal human rights, are depicted in the post-Holocaust period as fostering a backward particularity that goes against many of the current convictions of people who regard themselves as progressive.

The issue of the definition of antisemitism and the legislation against it and against Holocaust denial brings us to a second meeting point between counter mirror images, that of Jews *vis-à-vis* the liberal European intellectual left, and not only the radical left. The eminent Yad Vashem scholar Israel Gutman argues that during the 1940s worldwide public opinion regarded antisemitism as the major underlying cause for the Holocaust. However, in the post-Holocaust period the international community could not internalize the implications of a crime of such colossal scale. Thus, a new definition of antisemitism that was called for after the fall of Nazism, and even the very mentioning of antisemitism, were evaded in postwar declarations, treaties, and in other major texts.[23] A striking example is Eleanor Roosevelt's introduction to the first English edition of Anne Frank's Diary. Roosevelt, who took a central part in the formulation of the 1948 Universal Declaration of Human Rights, did not mention the terms anti-

---

23 See Israel Gutman, "Denying the Holocaust" (lecture at the Study Circle on Diaspora Jewry in the home of the president of Israel (Jerusalem: the Hebrew University, 1985)), 12 and 16. In May 2005, Gutman elaborated on this lecture in the Yad Vashem research seminar, recorded but not published.

semitism, Germans, or Nazis at all, nor did she mention the fact that Anne was Jewish, a perilous status that forced her to spend two years in hiding![24]

Evading mention of the term antisemitism, Gutman continues, makes it possible to ignore it as a cause for the Holocaust, and also to blame others for it, first and foremost the Jews themselves, especially the Zionists. The following step was to equate the Zionists with the Nazis and thus to try to do away with the heavy cloud of guilt that has hovered over Europe since 1945. The Left in particular needs this equation to maintain its own sense of self-righteousness. It is not surprising, then, to note that it was the Soviet Union that first came up with this equation after the war, because the Left in Germany, first and foremost the Communist Party, did not have the stamina, courage, and unity to stop the Nazi Party from coming to power before 1933 and thus did not act to prevent the Holocaust. Moreover, it was the Soviet Union that signed the notorious August 1939 Molotov-Ribbentrop Pact, which paved the way for the bloody war during which it became possible to carry out large-scale murder, and not only of Jews. Millions of Soviet citizens and soldiers perished, and the postwar Soviet leaders, who were reluctant to shoulder the heavy responsibility for the pact and for ignoring the signs of an impending German invasion in June 1941, needed a culprit. Therefore, using as always the old tactic of inversion, it was the Soviet Union, joined by extreme leftists in the West, that first accused the Zionist movement of collaborating with the Nazis.[25] It was also the Soviet Union that mustered its followers among UN members to vote for the infamous 1975 decision, according to which Zionism was equated with racism. The racist ideology, needless to say, was the fundamental tenet of the Nazi party and of its aspirations to reorganize the world accordingly.

By equating Israelis and Jews with the Nazis – the most extreme of the rightist movement – today's Left establishes itself at the opposite pole, the reverse image: its followers are to see themselves as righteous and virtuous, an image the Left cannot do without.[26] Thus cruelty, in the forms of contemporary fascism, colonialism, capitalism, and racism is allegedly the true essence of today's Zionist, leaving little doubt that he, the Zionist, and his brother the Jew, could have cooperated with the Nazis, their parallels, during the Holocaust. Therefore, it is from the Left side of the political map, not only from the Arab-Muslim world, that the call for the abolishment of the State of Israel is heard. And since the

---

[24] Anne Frank, *The Diary of a Young Girl* (New York: Doubleday, 1952), introduction by Eleanor Roosevelt.
[25] Jim Allen, *Perdition: A Play in Two Acts* (London: Ithaca Press, 1987).
[26] See Gil Michaeli, an interview in Paris with Finkielkraut, entitled "The Very Existence of Israel Thuned in the Eyes of Many into a Monstrous Phenomenon," *Maariv* (November 24, 2006).

abolishment of a state, moreover, one established by the UN and ever since a full member of that body, is in itself a colossal crime, unprecedented so far, an accusation of an equally unheard-of colossal crime is required in order to justify it.[27] These extreme equations, accusations, and comparisons sometimes exploit specifically designated commemoration dates in the Israeli-Jewish calendar as proof of the alleged Zionism-Holocaust connection, especially when the circumstances that led to choosing these dates are forgotten, unknown, or deliberately disregarded. Indeed, Holocaust Memorial Day is commemorated shortly after Passover (in the Hebrew month of Nissan), and a week later, on the fifth of Iyar, there follows Israel's Independence Day. This sequence of dates could be construed to show that Israel's Independence Day is closely related to Holocaust Memorial Day, because the Holocaust has served Zionism in order to build the Jewish State. Unfortunately, every Israeli leader and speaker on Holocaust Memorial Days hails the fact that the now thriving state "was born out of the ashes" and is a sweet revenge on Nazi Germany.

Attributing this sequence of dates to a deliberate decision by Israeli authorities in order to form a link between these two events is a grave historical mistake for a number of reasons. The Warsaw Ghetto uprising started on Passover eve, April 19, 1943, because the German SS command wished to present the destruction of the ghetto as a birthday gift to Hitler, who was born on April 20. This intention was not fulfilled because the uprising went on for a few weeks. After liberation, survivors in displaced persons' camps wanted to institute a Holocaust Memorial Day that would be connected to the heroism of Warsaw Ghetto uprising, and since Passover eve itself was out of the question for religious reasons, another date had to be found. In 1959, after innumerable debates and controversies that lasted almost a decade, the date of the twenty-seventh of Nissan was chosen. It comes right after the end of the holiday and is a date on which the fighting in the ghetto was exceptionally successful. In the meantime, the State of Israel was born on the fifth of Iyar, 1948, so that there is actually only an incidental connection between the two dates.

Moreover, the Holocaust did not found the State: modern Zionism established its first settlement in 1860 and, over the years, built a thriving self-governing community. Had there not been a 600,000-strong Yishuv (the Zionist Jewish entity that resided in pre-State Israel) the 360,000 survivors would not have found a shelter. And the UN November 1947 partition resolution, voting for the establishment of a Jewish State, came indeed after the Holocaust but not as its direct result. Political considerations, such as the Soviet interest in replacing Britain in

---

27 Levy, 155.

the Middle East and in preventing American future influence in the area, were much more instrumental than belated empathy.[28]

Despite such solid historical arguments, the more the Holocaust becomes an integral part of Israel's public life, the more this alleged Holocaust-State connection is proclaimed, without a recognition that it plays into the hands of anti-Israeli Holocaust deniers, such as those in the Iranian high echelons, who keep using their "no Holocaust-no State" mantra. They allege that there would be no justification for the Jewish state if not for the Holocaust, which even Israelis acknowledge as the moral basis for their state. These claims serve as yet further proof that the Holocaust is a politically inspired invention used as an instrument in Jewish-Zionist hands in order to extort national gains.

A third and final meeting point between counter mirror images concerns intergenerational relations and postwar national self-images. Second and third generation descendants of Nazi and pro-Nazi perpetrators and collaborators understandably welcome any idea or terms that minimize the Holocaust, because such notions assuage tensions within families and communities. The image of their predecessors is thus transformed from cruel murderers and torturers to respectable law-abiding citizens who had to do their share within the framework of their respective regimes during a terrible war and under circumstances beyond their control.[29]

Postwar national self-images, especially in European countries that were Nazi Germany's allies or under its occupation, are seriously challenged by the history and the memory of the Holocaust. Coming to grips with past realities entails a public acknowledgment, linked to educational efforts on a broad scale, of each country's share in the plight of its local Jewish citizens and in its collaboration with Nazi Germany. Such efforts are difficult, for they pinpoint sectors and individuals as the culprits. Seen in these terms, the memory of the Holocaust can be wrenching. It is an obstacle on the road to national reconciliation, complained the French president Francois Mitterrand, when survivor Serge Klarsfeld protested against putting a wreath of flowers on Marshall Philippe Petain's grave. It was his duty, said Mitterrand, "to try to appease the eternal civil wars

---

[28] See Dan Michman, "From Holocaust to Resurrection! From Holocaust to Resurrection?" *Iyunim Bitkumat Israel* (Studies in Israeli and Modern Jewish Society) 10 (2000): 234–58. Yehuda Bauer reacted in "Did the Holocaust Bring about the Establishment of the State of Israel?" *Iyunim Bitkumat Israel* 12 (2002): 653–54. See also Michman, "A Reaction to a Reaction," *Iyunim Bitkumat Israel* 13 (2003): 393–95.

[29] Dan Baron, *Legacy of Silence: Encounters with Children of the Third Reich* (Cambridge, Mass.: Harvard University Press, 1989), with a 2003 edition by the Koerber Foundation, Hamburg.

between the French."[30] The problem is more acute in Eastern Europe, where leaders and parties that had collaborated with the Germans are hailed as anti-communist heroes in today's post-communist era. They are reburied in state funerals and their newly placed statues sometimes adorn the main squares. No wonder, then, that much of the material that denies or minimizes the Holocaust in Eastern Europe and the extensive part taken by the locals in its various stages is being published by *landsmanschaften* members who immigrated mainly to the Americas. Their wish is to maintain a positive image of their fatherlands, for themselves, for their compatriots, and for posterity. As anti-communists, many of them become right-wingers and supporters of the European Right that has recently been gaining strength, paving its way to power by anti-immigration racist arguments, and by traditional antisemitism that is rejustified once the Holocaust may be presented as a Jewish lie, or at least as a gross exaggeration.[31]

In several countries of Central Europe, including for example Hungary and Poland, competing collective narratives have turned into harsh fights over the "sovereignty of interpretation", and the fights escalate over the years with special historical memoirs and meanings. The search for a new collective identity, as it seems, has just begun in some of the countries of the former Eastern bloc. An extremely alarming development was recently seen in Poland, where a law was on its way to be enacted, at the beginning of 2018, determining that "whoever accuses, publicly and against the facts, the Polish nation, or the Polish state, of being responsible or complicit in the Nazi crimes committed by the Third German Reich... shall be subject to a fine or a penalty of imprisonment of up to three years."[32] Even the fact that the Polish government has stepped back in the summer of 2018 and finally moved to change the law to decriminalise the offence, a new social and public "climate" cannot be created, at least in some countries of Europe where considerable parts of the public want to rid themselves of the shadows of the past, either by ignoring former collaboration or even by contesting the crimes.

Beyond these obvious strategies of political and historical self-relief in European and non-European, majoritarian non-Jewish societies, there are yet recent phenomena of Holocaust denial and/or relativization which had already been

---

[30] Justice Georges Kiejman quotes Mitterrand in *Liberation* (October 22, 1992). Klarsfeld's angry answer, in *Le Nouvel Observateur* (October 25, 1992).
[31] Michael Shafir, "Between Denial and 'Comparative Trivialization': Holocaust Negationism in Post-Communist East-Central Europe" (Jerusalem: The Vidal Sassoon International Center for the Study of Anti-Semitism, 2002).
[32] *BBC News*, June 27, 2018, see https://www.bbc.com/news/world-europe-44627129, accessed August 21, 2018.

considered as "extinct". Thus, in 2009, the prominent British Catholic bishop Richard Williamson declared, in an interview with Swedish television, his belief that Nazi Germany did not use gas chambers during the Holocaust. In Germany, Ursula Haverbeck, well-known as the "Nazi grandma", had recently published a series of articles in the right-wing magazine "Stimme des Reiches" ("Voice of the Empire") in which she insisted that the Holocaust and murder of Jews at Auschwitz extermination camp in German occupied Poland would not be historically proven. Moreover, in one occasion she was convicted for calling the Holocaust "the biggest and longest-lasting lie in history." Haverbeck also co-founded a – meanwhile closed – right wing "education center" called Collegium Humanum. In the summer of 2018, she was sentenced to two and a half years in prison by a German court.[33]

In December 2017, the notorious Swiss-French right-wing essayist Alain Soral was sentenced to a 6,000 euro fine for having published and put on sale on his website a poster with denying, defamatory and inciting hatred against Jews. A drawing entitled "Mémorial Pornography", was published in August 2014 depicting a woman dressed in a corset, with stars of David on her chest, standing in front of a representation of the entrance to Auschwitz–Birkenau, holding coins in one hand and a bra in the other. However, the court did not order the withdrawal of the poster from Alain Soral's site.[34]

Holocaust denial, whether hardcore or in the form of minimization, trivialization, skepticism, and so on, is not the deniers' principal goal. The image of the "Jew" depicted by all sorts of deniers as a figure to be scorned, distrusted, derided, and worse, is not a side effect of the denial but rather its target. It enables the denier, individually and collectively, to live in peace with himself and keep his pre-Holocaust world intact. Thus, the battle for the memory of the Holocaust has become a struggle for historical ownership and for the projection of respectable images. What is, maybe, even more alarming than the periodical incidents of Holocaust denial and relativization in Europe and beyond is the public attention attracted by the self-staging of the deniers in the twenty-first century. Denying the Holocaust might mutate to a "cultural event", not taken that seriously by large publics but disdaining the victims and the survivors.

It is of utmost importance to realize that the across-Europe-urge to deny or at least to relativize the Holocaust is not a problem of uneducated social strata, on

---

[33] *The Telegraph*, August 4, 2018, see https://www.telegraph.co.uk/news/2018/08/04/german-court-rejects-nazi-grandma-appeal-rules-holocaust-denial/, accessed August 21, 2018.
[34] See https://www.archyworldys.com/alain-soral-again-sentenced-for-holocaust-denial/, accessed August 21, 2018.

the contrary.[35] Highly educated people join discourses in the social media to assure themselves that the genocide of the Jewish people never happened, or that it had taken place on a much smaller scale, while at the same time joining political and civic forces delegitimizing the State of Israel. Educated authors of anti-Jewish texts and speeches certainly know that the ideas they express are inflammatory, and the dubious attempt to hold Jews responsible for restlessness in society is thus proceeding. It should still be emphasized that democratic western countries that have realized the potential threat antisemitism poses to public order and social values have attempted in recent years to contain it, by allocating budgets for security, education, legislation, and the creation of a balanced public atmosphere.

Bibliography

Allen, Jim. *Perdition: A Play in Two Acts*. London: Ithaca Press, 1987.
Baron, Dan. *Legacy of Silence: Encounters with Children of the Third Reich*. Cambridge/Mass.: Harvard University Press, 1989 (new edition in 2003 by the Koerber Foundation, Hamburg).
Bauer, Yehuda. "Did the Holocaust Bring about the Establishment of the State of Israel?" *Iyunim Bitkumat Israel* 12 (2002): 653–54.
Duque, Jose Roberto. "Dueños del dolor, dueños del mundo." In *Aporrea* (July 20, 2006).
Finkielkraut, Alain. *Au nom de l'Autre, reflexion sur l'antisemitisme qui vient*. Jerusalem: Shalem, 2004.
Finkielkraut, Alain. *L'avenir d'une negation: Reflexion sur la question du genocide*. Paris: Edition du Seuil, 1982.
Frank, Anne. *The Diary of a Young Girl*. Introduction by Eleanor Roosevelt. New York: Doubleday, 1952.
Gutman, Israel. "Denying the Holocaust." Lecture at the Study Circle on Diaspora Jewry in the Home of the President of Israel, the Hebrew University, Jerusalem, 1985. Recorded but not published.
Heger, Heinz. *Les hommes au triangle rose*. Preface by Guy Hocquenghem. Paris: Persona, 1981.
Katcher, Leo. *Postmortem: Jews in Germany Now*. New York: Hamish Hamilton Ltd., 1968.
Klug, Brian. "The Collective Jew: Israel and the New Antisemitism." *Patterns of Prejudice* 37, no. 2 (2003): 117–138.
Lange, Anders. *A Survey on Teachers' Experiences and Perceptions in Relation to Teaching about the Holocaust*. Stockholm: The Living History Forum, 2008.
Levy, Bernard-Henri. *Left in Dark Times: A Stand against the New Barbarism*. New York: Random House, 2008.

---

[35] See in this context also Monika Schwarz-Friesel (ed.), *Gebildeter Antisemitismus: Eine Herausforderung für Politik und Zivilgesellschaft* (Baden-Baden: Nomos, 2015).

Michaeli, Gil. An interview in Paris with Alain Finkielkraut: "The Very Existence of Israel Thuned in the Eyes of Many into a Monstrous Phenomenon." In *Maariv* (November 24, 2006).

Michman, Dan. "From Holocaust to Resurrection! From Holocaust to Resurrection?" *Iyunim Bitkumat Israel* (*Studies in Israeli and Modern Jewish Society*) 10 (2000): 234–58.

Michman, Dan. "A Reaction to a Reaction." *Iyunim Bitkumat Israel* 13 (2003): 393–95.

Rembiszewski, Sarah. *The Final Lie: Holocaust Denial in Germany – A Second Generation Denier as a Test Case*. Tel Aviv: Stephen Roth Institute/Tel Aviv University, 1996.

Said, Edward. *The Politics of Dispossession*. New York: Pantheon, 1994.

Sereny, Gitta. *Into that Darkness: From Mercy Killing to Mass Murder. A Study of Franz Stangl, the Commandant of Treblinka*. London, 1974/1995.

Shafir, Michael. *Between Denial and 'Comparative Trivialization': Holocaust Negationism in Post-Communist East-Central Europe*. Jerusalem: The Vidal Sassoon International Center for the Study of Antisemitism, 2002.

Trigano, Shmuel. "The Political Theology of the Memory: Europe Is Morally Ready for a Second Holocaust." *Kivunim Hadashim* 17 (January 2008).

UN General Assembly resolution of November 21, 2005, no. 60/7, on Holocaust Remembrance.

Wiesel, Elie. "The Memory of the Holocaust on Israel's 60th Anniversary." Lecture at Tel Aviv University on May 20, 2008. Recorded but not published.

Online Sources

"Alain Soral again sentenced for Holocaust denial." https://www.archyworldys.com/alain-soral-again-sentenced-for-holocaust-denial/. Accessed August 21, 2018.

"German court rejects 'Nazi grandma' appeal, as it rules Holocaust denial is not covered by free speech." *The Telegraph*, August 4, 2018. https://www.telegraph.co.uk/news/2018/08/04/german-court-rejects-nazi-grandma-appeal-rules-holocaust-denial/. Accessed August 21, 2018.

"Poland Holocaust law: Government U-turn on jail threat." *BBC News*, June 27, 2018. https://www.bbc.com/news/world-europe-44627129. Accessed August 21, 2018.

Rich, Dave. "Holocaust Denial as an Anti-Zionist and Anti-Imperialist Tool for the European Far Left". *Post-Holocaust and Anti-Semitism*, no. 65. Jerusalem Center for Public Affairs online publication; February 1, 2008. http://jcpa.org/article/holocaust-denial-as-an-anti-zionist-and-anti-imperialist-tool-for-the-european-far-left/. Accessed September 6, 2018.

Trigano, Shmuel. "Europe's Distortion of the Meaning of the Shoah Memory and Its Consequences for the Jews and Israel." *Post-Holocaust and Anti-Semitism* 42 (March 1, 2006). www.jcpa.org/phas/phas-042-trigano.htm. Accessed August 21, 2018.

Haim Fireberg
The Antisemitic Paradox in Europe: Empirical Evidences and Jewish Perceptions. A Comparative Study Between the West and East

Reviewing antisemitic violent incidents worldwide reveals that the aftermath of Jewish hatred is constantly thriving. Whilst studying and comparing contemporary antisemitism in EU Member States, several peculiarities, almost paradoxes, can be identified. One of them hints that the level of violent antisemitism, as shown by the number of violent incidents, does not necessarily indicate the level of antisemitic sentiment. It could be considered as a necessary condition in defining antisemitism, but undoubtedly not a sufficient one. This article analyzes the contradictory factors that influence the understanding of antisemitism by individuals, organizations, and states using the latest empirical evidence available.

In late 2012, the European Union Agency for Fundamental Rights (FRA) conducted a survey on antisemitism in eight EU Member States (United Kingdom, France, Germany, Italy, Belgium, Sweden, Hungary, and Latvia).[1] An extended follow-up survey and research is taking place in 2018, in which thirteen Member States are taking part, including Austria, that did not participate in the original survey.[2] The survey reached out to 5,847 Jews, and the results were published in November 2013. Sixty-six percent of the participants declared that antisemitism in their countries is "a big problem" or "a fairly big problem." The most concerned group to see antisemitism as a huge problem were the Hungarian Jews (90%), and almost neck-and-neck were the French Jews (85%). On the other hand, in the UK (52%) and Latvia (54%), most of the respondents agreed that antisemitism is "not a very big problem" or "not a problem at all."[3] The survey

---

[1] EU Agency for Fundamental Rights. Discrimination and hate crimes against Jews in EU member states: experiences and perceptions of antisemitism [FRA 2013], http://fra.europa.eu/en/publication/2013/discrimination-and-hate-crime-against-jews-eu-member-states-experiences-and.
[2] EU Agency for Fundamental Rights, http://fra.europa.eu/en/press-release/2017/major-eu-anti semitism-survey-planned-2018. The countries covered are Austria, Belgium, Denmark, France, Germany, Hungary, Italy, Latvia, the Netherlands, Poland, Spain, Sweden, and the UK.
[3] FRA 2013, 16.

https://doi.org/10.1515/9783110582369-014

has shown that the diversity of antisemitic perceptions are not divided along geographic lines.

Although country differences exist, the bigger picture reveals that more than three-quarters (76%) of the overall participants consider that antisemitism "has worsened over the past five years in the country where they live."[4] France and Hungary lead this perception – Hungary with 91% of respondents, and France with 88%. But also in the UK, where only a minority believed that antisemitism is a real problem, 66% of participants still agreed that the situation has worsened. In Latvia, only 39% claimed so.[5]

In France and Hungary, where according to the FRA survey antisemitism is a considerable problem, the levels of antisemitic violence based on Tel Aviv University [TAU] publications[6] sharply differ. In France, between the years 2009 and 2015, we recorded a yearly average of one hundred and forty-two violent incidents. In Hungary, on the other hand, the seven-year average was around nine incidents only. It seems that violence alone cannot explains the fact that in both countries the sense of insecurity is a major factor in contemporary Jewish life.

In the UK and Latvia, where the Jewish population estimated almost no problems with antisemitism, TAU data reveal an even more astonishing situation: in sharp contrast to Latvia where the recorded average of violent incidents from 2009 until 2015 was only one, the average yearly amount in the UK was one hundred and forty-four. The UK infamously holds the European record.

Violent antisemitism is not a problem in Latvia and matches the FRA survey results. In Hungary, with a large Jewish population (the third highest in Europe after France and Britain), almost all the targets were Jewish facilities and memorial sites, not human beings. Would it be right to assume that neither the total sum of violent cases in Hungary nor their nature should be considered as an explanation to the outcome of the survey; i.e. antisemitism in Hungary is flourishing and is "a big problem"?[7]

---

4 Ibid.

5 FRA 2013, 17.

6 Data on violent antisemitic events is based mostly on the TAU annual analyses that have been published yearly for the last twenty-six years by the Kantor Center for the Study of Contemporary European Jewry.

7 The Kantor Center, http://www.kantorcenter.tau.ac.il/sites/default/files/Doch2014-2.pdf; http://www.kantorcenter.tau.ac.il/sites/default/files/Doch_2013.pdf. In Hungary, fourteen violent incidents against Jews were recorded in 2013, and fifteeen in 2014. In 2013, one hundred and sixteen of violent manifestations were recorded in France; in forty-seven cases persons have been attacked (forty-one percent of the cases). Of one hundred and sixty-four incidents recorded in 2014, eighty-eight (fifty-four percent) targeted persons. In the UK, ninety-five violent incidents were recorded in 2013, sixty-three of which (two-thirds, sixty-seven percent) targeted human beings. In 2014, the

The year 2015 presented a new notorious record in murderous antisemitic violence in Europe. To mention just three: the attack that occurred in Paris on January 9 where an Islamist killed four Jewish shoppers at a kosher supermarket; the attack on the central synagogue in Copenhagen in February where a security guard was murdered and two others were injured by an Islamist gunman; and in Manchester in September, four Jewish boys were brutally beaten and severely injured in an antisemitic attack. Although the state of violent antisemitism in France and the UK has a lot in common, the survey shows that the people in both countries understand the problem differently.

If it is therefore not the extent of violence that generates the same anxiety towards antisemitism, what could France and Hungary – for instance – have in common that makes the Jews there very concerned with "the problem of antisemitism"?

In January 2015, in the wake of the Paris massacres, the French PM, Manuel Valls, recalled the visions of the French Revolution. "The choice was made by the French Revolution in 1789 to recognize Jews as full citizens. To understand what the idea of the republic is about, you have to understand the central role played by the emancipation of the Jews. It is a founding principle". Valls did not try to defend the current situation in France; on the contrary, he spoke sharply and bitterly: "If ... 100,000 Jews leave, France will no longer be France. The French Republic will be judged a failure."[8]

Although Valls was known for his warm attitudes towards French Jews, his decisive words should also be understood in connection with the continuous emotional erosion that many French Jews have about the being French citizens and practicing open Judaism at the same time. For many, it was only the tip of the iceberg; a process lasting almost a decade in which Jews felt that they are not only under attack by vast groups of radical Muslims, but that the main political groups, primarily from the French Left, abandoned them. A popular Jewish opinion is that there is an unholy – although undeclared publicly – alliance between French radical Left and Muslim extremists against a common enemy – the Jewish community in France for an alleged unequivocal support of Israel.[9] When such a

---

overall number of violent incidents were 141, eight-two (fifty-eight percent) of which were against people.
[8] *The Atlantic*, http://www.theatlantic.com/international/archive/2015/01/french-prime-minister-warns-if-jews-flee-the-republic-will-be-judged-a-failure/384410.
[9] In 2015, one of the prominent Jewish intellectuals in France, Shmuel Trigano, expressed this view in his publication, *A Journey Through French Anti-Semitism* (Spring, 2015). For online summaries on Trigano's attitudes, see https://jewishreviewofbooks.com/articles/1534/a-journey-through-french-anti-semitism/.

connection is being made, even casual criticism of Israeli policy towards a two-state solution becomes antisemitic in nature and involves calls to Boycott, Divestment and Sanctions [BDS] on Israel, while it is illegal according to French law.[10]

Ron Azogui, a member of the Service de Protection de la Communauté Juive [SPCJ] concluded:

> Antisemitism in France cannot be considered anymore as a temporary situation associated with the situation in the Middle East; it is a structural problem that has not been fought as such and has not been halted yet. … Forty percent of racist violence perpetrated in France in 2013 targeted Jews. However, Jews represent less than 1 percent of the French population. … [We believe] that antisemitic violence has settled and is anchored in society. But the aggravating factor is that French Jews feel isolated in their fight against antisemitism. Aren't the values that are attacked by this scourge those of a whole nation?[11]

Unfortunately, his rhetorical question as well as Valls' remarks are still part of the French Jewish community's common experience.

In Hungary, we find that the quest for Hungarian national identity in the twenty-first century, the uncertainty about its common values and the place of minorities, including Jews, in this nation-state are central issues in Hungary's discourse today. Dr. Rafi Vago, the renowned scholar, has thus described the situation there:

> […] deep divisions in Hungarian political life [could be found], between the center-right party *Fidesz* [the ruling party], the extremist right wing party, *Jobbik*, and the liberal-left, over Hungary's past. It … became a test case for the delicate balance and relationship between various parts of Hungarian society, the media and the political spectrum.[12]

Vago stressed that standing in the center of the conflict was

> […] the need to face the fate of its almost 600,000 Jews who perished in the Holocaust and evaluate the inter-war and war time Horthy regime. … The [recent years] … became [also] the focal point of strong differences of opinion [about] the rise of antisemitism, and the government's handling of those issues.[13]

---

10 Jean Yves Camus, *Antisemitism in France, 2014*, http://www.kantorcenter.tau.ac.il/sites/default/files/Doch2014-2.pdf, 63–66.
11 *The Kantor Center*, http://www.kantorcenter.tau.ac.il/sites/default/files/Doch_2013.pdf, 52.
12 *The Kantor Center*, http://www.kantorcenter.tau.ac.il/sites/default/files/Doch2014-2.pdf, 54.
13 Ibid., 54.

Although the number of violent antisemitic incidents in Hungary is relatively small and rare, there are strong feelings among the Jews. "Jews can now feel antisemitism in the streets,"[14] stressed Rabbi Schlomo Koves [Slomo Köves], the executive Rabbi of the Unified Hungarian Congregation. Many consider that the bad atmosphere is the result of the nationalistic discourse, especially the revisionism of Hungary's WWII past:

> The main danger in Hungary is the attempt to 'whitewash' the anti-semitic past, rehabilitate aspects of the Horthy era, emphasize Hungary's alleged loss of sovereignty in March 1944, with the German occupation, thus as attempt to relativize Hungary's role in the destruction of its Jewry.[15]

It is not a debate that takes place behind closed doors and in academic circles only, but in the media and in public demonstrations as well. A few examples in short:
- In 2014, the government, ruled by *Fidesz*, erected a statue at Freedom Square, showing Germany's imperial eagle striking down on archangel Gabriel, Hungary's guardian angel, symbolizing Hungary's innocence versus Nazi aggression, in commemoration of March 19, 1944, the date of the Nazi occupation of Hungary. Continuing opposition to the statue reflects the attitudes of wide segments of the Hungarian public, not only among Jews, that the statue deflects Hungary's responsibility for the Holocaust. The leading historian of the Holocaust in Hungary, and Holocaust survivor, Prof. Randolph L. Braham, returned a prestigious state award to the Hungarian government in protest of rewriting Hungary's history.[16]
- Another source of dispute is the project of the "House of Fates," a planned educational center and a Holocaust museum in the eighth district, now home to many Jews. It became the focus of ongoing debates claiming that the project's aims are not clear, that the voice of the Jewish community has not been taken into consideration.[17] Although the dispute has nothing to do with antisemitism per-se, and many prominent members of the Jewish

---

14 *The Kantor Center*, http://www.kantorcenter.tau.ac.il/sites/default/files/Doch_2013.pdf, 44.
15 Ibid., 44.
16 *The Kantor Center*, http://www.kantorcenter.tau.ac.il/sites/default/files/Doch2014-2.pdf. Randolph L. Braham, "Hungary: The Assault on the Historical Memory of the Holocaust," in *The Holocaust on Hungary: Seventy Years Later*, ed. Randolph L. Braham, Andreas Kovacs (Budapest: CEU press, 2016), 261–309.
17 *Hungarian Free Press*, http://hungarianfreepress.com/2015/05/07/budapest-politicians-tour-new-holocaust-museum-described-as-shocking; *The Kantor Center*, http://www.kantorcenter.tau.ac.il/sites/default/files/Doch2014-2.pdf, 55.

community, although not the official ones, are involved in the project, the contemporary opinion among the Jewish leadership in the Federation of Hungarian Jewish Communities [*Mazsihisz*] – and it is spreading down to the ranks – is of cessation from the government.[18]

The FRA survey has shown many parallels between French Jewry and Hungarian Jewry that support the analysis above. Although France and Hungary differ in their political systems and civic ethos, the willingness of Jewish citizens to emigrate "because of not feeling safe living there as a Jew" in both countries is the highest according to the FRA survey: 48% of Hungarian Jews and 46% of French Jews.[19] The feelings of insecurity led to an increasing disbelief in the future of the community and has weakened the sense belonging to the nation as would be shown later on. While the two countries differ in the overall number of violent antisemitic incidents, the fear of becoming a victim of threats, insults and verbal harassment – important factors in creating an antisemitic environment – is very similar. Sixty-five percent of Hungarian Jews and seventy-six of French Jews were worried of being exposed to these kinds of attacks.

The only actual difference was found when they were asked about their worries of being personally attacked. Seventy-one percent of French Jews answered that they are worried, in comparison to forty-three percent of Hungarian Jews. Indeed, this difference could be explained by their specific experiences.[20] On the other hand, Hungary has one unique factor – as 66% of respondents emphasized – which is the vast exposure of individuals to antisemitism in the public sphere, first and foremost, by the eagerness of mainstream politicians to adopt publicly antisemitic attitudes and antisemitic rhetoric, something that almost does not exist in French politics.[21]

On the other side we have the UK and Latvia. The FRA survey revealed, as was mentioned above, that 52% of UK Jews believe that antisemitism is not, or is almost not, a problem in Britain, and it is the highest rate among the EU members. The Jewish Policy Research (JPR) concluded in the wake of the survey:

---

18 An update on this controversy: *The Hungarian Spectrum*, http://hungarianspectrum.org/tag/mazsihisz/. Another controversy was about the anti-George Soros campaign in Hungary that deepened the gap between the Government and the veteran Jewish establishment headed today by Andras Heisler. *The Times of Israel*, https://www.timesofisrael.com/decrying-netanyahu-betrayal-hungary-jews-say-pm-ignoring-them/.
19 FRA 2013, 37.
20 Ibid., 33.
21 FRA 2013, 26.

"[The British Jewish population has] a strong sense of belonging to the UK."[22] There are several empirical indications that support this assumption, e.g., 77% of the respondents declared that they "have not considered emigrating" because of their fear of antisemitism.[23] Knowing that the UK, according to TAU data, has the highest rate of antisemitic violence in Europe causes one to wonder how it is that only a relatively small proportion of UK respondents to the FRA survey claimed to be worried of being a victim of violence. Thirty-five percent were worried about verbal harassment and twenty-five percent were worried about physical attacks, the smallest rate of all participants in the survey.[24] Even though the number of violent antisemitic incidents in Britain is the highest among EU members, the sense of security of the British Jews, and their belief as shown in the survey of being integrated into the British society, are remarkable.[25]

In comparison to France, a considerably lower rate of respondents claimed to be suffering from antisemitic attitudes in the public sphere or at social events. The survey revealed that UK Jews sense that many in the political establishment – at least when it comes to the Conservative Party and the more moderate representatives of the Labour Party – are standing with them in fighting antisemitism and discrimination.[26]

Concerns and questions have been raised in Britain recently about the future of Jewish existence in the country. The sense of security was diluted in the wake of the murderous antisemitic incidents in 2014 (Brussels), and in 2015 (Paris, Copenhagen and Manchester), and the hate against the Jews shown in the streets of European capitals during the summer of 2014 and "Operation Protective Edge." A prominent Londoner lawyer, Hillary Freeman, summarized the popular feel-

---

22 *The Kantor Center*, http://www.kantorcenter.tau.ac.il/sites/default/files/Doch2014-2.pdf, 71.
23 FRA 2013, 37.
24 Ibid., 33. It is interesting that when asked about their personal experience, the rate of those who have suffered from violence is climbing to nineteen percent. This is above the percentage of Latvia's respondents (sixteen percent) and Italy, with the lowest rate of people who actually suffered from violence (twelve percent).
25 My analysis totally objects to Jonathan Boyd's conclusion that "most European Jewish populations appear to feel a strong sense of belonging to the countries in which they live, and most seem to be able to comfortably manage the relationship between their Jewish and wider national identities. Even in the countries where levels of antisemitism are revealed by these and other data to be highest, Jews feel remarkably attached to the nations in which they live: over 70 percent of respondents in Hungary feel a strong sense of belonging to Hungary, and over 80 percent of respondents in France feel a strong sense of belonging to France". *Institute for Jewish Policy Research*, http://www.jpr.org.uk/documents/Jewish%20life%20in%20Europe%20-%20Impending%20catastrophe%20or%20imminent%20renaissance.pdf, 12.
26 Ibid., 26.

ings: "As a Jew, I find this particularly offensive. It's taking the Holocaust – the greatest tragedy in the history of the Jewish people – and using it as a stick to beat us with... I am horrified that my grandma, now 96, might live to see the country that gave her sanctuary over 70 years ago become a place that is no longer safe for Jews. But the terrifying truth is that once the genie of antisemitism has been released from the bottle, it is almost impossible to put it back."[27]

The Paris massacres have sent shock waves throughout the UK's Jewish society, and its members have been looking for reassurance to their civilian status in the UK. In 2015, during a meeting with Jewish Leaders David Cameron, PM, praised the sense of mutual solidarity and belonging that unifies all Britons:

> I know that everyone will be very concerned about what happened in Paris and the appalling attacks. ... I want to reassure you that we will try and do everything we can to make sure that your organizations are properly engaged with our police and security services right across the board to see if there is anything more we can do to ensure security. ... But I think we should use the momentum of those great demonstrations to emphasize what we are in this country: a very successful multi-ethnic, multi-faith democracy.[28]

On November 2017, while celebrating a centenary to the Balfour declaration, Theresa May stressed that there

> [...] can be no excuses for any kind of hatred towards the Jewish people. Criticizing the actions of Israel is never – and can never be – an excuse for questioning Israel's right to exist, any more than criticizing the actions of Britain could be an excuse for questioning our right to exist[29]

The Jewish leadership has continuously played down a survey from January 2015, called the "Antisemitism Barometer" – blaming it for severe methodological faults – which claimed that almost half of UK Jews are now considering emigrating.[30] By doing it, they also preferred to neglect the consequences from the

---

27 *The Daily Mail*, August 8, 2014, http://www.dailymail.co.uk/debate/article-2720381/Why-British-Jew-I-m-terrified-anti-Semitism-suddenly-sweeping-country.html.
28 *Jewish Leadership Council*, http://www.thejlc.org/2015/01/jewish-community-leaders-meet-with-prime-minister-david-cameron-3.
29 *The Telegraph*, November 2, 2017, http://www.telegraph.co.uk/news/2017/11/02/theresa-may-says-can-no-excuse-anti-semitism-marks-balfour-centenary/.
30 *Jerusalem Post*, http://www.jpost.com/Diaspora/Nearly-half-of-British-Jews-says-they-have-no-future-in-Europe-study-finds-387693; https://antisemitism.uk/barometer/.

changes of the Labour Party's leadership, headed by Jeremy Corbyn, its harsh new policies towards Israel, and the antisemitism in the party's ranks.[31]

Jonathan Arkush, the president of the Board of Deputies, has pushed forward the idea of reaching out to the growing Muslim society, offering to tighten the bond of citizenship and integration of British society, in order to fortify again the sense of security among the UK Jewish population. In his words: "I want to meet Muslims and show them that Jews are actually human beings and you can combine being a good Muslim with being a good British citizen and hopefully take them away from being at risk of flirting with jihadi ideas."[32]

Latvia, like Hungary and almost all post-Soviet States, deals mainly with its past in its quest for the future: the quest for a national identity. Violent antisemitism is almost a non-issue in Latvia. Unfortunately for the small community, the adoration of the Latvian Nazi-era SS units and other Nazi collaborators became part of the national discourse of the new Latvia.[33] Even so, a high proportion of the respondents (sixty-eight percent) in the FRA survey showed a high sense of belonging and answered that they are "not considered emigrating." On one hand, with similarities to Hungary, we see rising nationalism and the eulogizing of their own Nazi past, but on the other hand, with similarities to the UK, we see a declaration of trust in their homeland.[34]

The respondents were asked to point to several statements that are "possible contexts for negative statements about Jews." In almost every possible "context," the Latvian proportions were the lowest. Every statement that involved public attitudes or politicians' attitudes against Jews had gotten a low rating, in contradiction to France, Hungary, and even better than the UK's results.[35] The empirical findings show that what is being considered from the outside as neo-Nazi and nationalistic debate has not yet converted into antisemitism and has not yet given rise to a new generation of extreme antisemites. One more factor is that the Israeli-Arab conflict, which according to the Latvian respondents has the smallest influence on antisemitic discourse or violence in comparison to France,

---

31 *The Times of Israel*, https://www.timesofisrael.com/british-jews-fight-to-regain-the-labour-party-they-once-called-family/. For an extensive analysis of the contemporary relations between the British Left and the British Jews, see David Hirsh, *Contemporary Left Antisemitism* (Routledge, 2018); and also Dave Rich's article from January 2018, *The Kantor Center*, http://kantorcenter.tau.ac.il/sites/default/files/Dave%20Rich%20180128.pdf.
32 *Jerusalem Post*, http://www.jpost.com/Diaspora/New-UK-Jewish-leader-I-want-to-meet-Muslims-407127.
33 *The Kantor Center*, http://www.kantorcenter.tau.ac.il/sites/default/files/Doch_2013.pdf; http://www.kantorcenter.tau.ac.il/sites/default/files/Doch2014-2.pdf, 19–20.
34 FRA 2013, 37.
35 Ibid., 26.

Hungary, and the UK (fourteen, fifty-six, forty-nine, and thirty-five percent respectively).[36]

In Conclusion

A few patterns have been found whilst studying and comparing contemporary antisemitism in four EU Member States, two from Western Europe – France and UK – and two from Central and Eastern Europe – Hungary and Latvia.
- The level of violent antisemitism, as is shown by the number of violent incidents, does not necessarily indicate the state of antisemitic perceptions. Generally, it could be a necessary condition in defining an antisemitic atmosphere, but undoubtedly not the sufficient one. France and Britain have the highest level of recorded incidents, but their Jewish population's self-perception on antisemitism is almost the opposite. The same could have been indicated in Hungary and Latvia. Both countries have a low level of violent antisemitism, but their perceptions of antisemitism in their countries differ from each other.
- Both countries where respondents have indicated that antisemitism is a severe problem (i.e. France and Hungary) show either a high ratio of estrangement from the ethos that have been chosen for national identity (Hungary), or demonstrate ongoing dissatisfaction from the state of civic consolidation and express worries about society's disintegration (France). In both Latvia and Britain there is a higher level of confidence in the civic order, especially in Britain, and a strong belief that the British society is on the right path in dealing with the challenges of new antisemitism. But confidence in government and society, or lack of it, is only one part of the sufficient conditions.
- Frustration from the political establishment, from ruling parties, and from the solutions they supply in order to control violent antisemitism, but much more importantly, to supply a common basis for all fractions of society to unite around, are the major factors in adopting harsh perceptions about antisemitism. Without belief in the future of the country, and without confidence that Jews are an important component of its society, Jews feel abandoned. And this lack of confidence is the main sufficient condition in adopting the hard antisemitic atmosphere. In the UK we have shown a high level of cooperation between the authorities and the Jewish community; in Latvia it was not a declared issue but empirically given that no friction was found

---

36 Ibid., 24.

between the Jews and the government. But in France, though de-facto, the authorities are trying their best to confront antisemitism and to find new paths to civic integration, the level of trust towards the political establishment is still very low.[37] Much of the same could still be said about Hungary.

Gilles Clavreul, who until recently was the French Inter-ministerial Delegate against Racism and antisemitism, summarizes the paradox very well:

> [...] What we have are citizens – Jews, Arabs, Blacks, Christians – they are first and foremost French citizens. Each French citizen has the right to live safely, to practice his own religion, to be protected by the French government. ... The idea that some of our French compatriots, Jews, do not see their place in France, we cannot accept. It is a cut that will hurt the whole body [of France]. Changing consciousness and bringing people back together will take time. There is no reason to give up.[38]

Bibliography

Online sources

Ahren, Raphael. "Decrying 'betrayal,' Hungary Jews say Netanyahu ignoring them." *Times of Israel*, July 20, 2017. https://www.timesofisrael.com/decrying-netanyahu-betrayal-hungary-jews-say-pm-ignoring-them/.
Bandler, Kenneth. "On my Mind: The battle for France." *The Jerusalem Post*, September 29, 2015. http://www.jpost.com/Opinion/On-my-mind-The-battle-for-France-419428.
Borschel-Dan, Amanda. "British Jews fight to regain the Labour party they once called 'family.'" *Times of Israel*, November 9, 2017. https://www.timesofisrael.com/british-jews-fight-to-regain-the-labour-party-they-once-called-family/.
Boyd, Jonathan. "Jewish life in Europe: Impending catastrophe, or imminent renaissance?" *Institute for Jewish Policy Research*, November 2013. http://www.jpr.org.uk/documents/Jewish%20life%20in%20Europe%20-%20Impending%20catastrophe%20or%20imminent%20renaissance.pdf.
Braham, Randolph L. "Hungary: The Assault on the Historical Memory of the Holocaust." In *The Holocaust on Hungary: Seventy Years Later*, edited by Randolph L. Kovacs and Andreas Kovacs. Budapest: CEU press, 2016.
"Budapest politicians tour new holocaust museum described as shocking." *Hungarian Free Press*, May 7, 2015. http://hungarianfreepress.com/2015/05/07/budapest-politicians-tour-new-holocaust-museum-described-as-shocking.
Campaign Against Antisemitism. *The Antisemitism Barometer*. https://antisemitism.uk/barometer/.

---

37 *Jerusalem Post*, http://www.jpost.com/Opinion/On-my-mind-The-battle-for-France-419428.
38 Ibid.

EU Agency for Fundamental Rights (FRA). *Discrimination and hate crime against Jews in EU Member States: experiences and perceptions of antisemitism.* November 2013. http://fra.europa.eu/en/publication/2013/discrimination-and-hate-crime-against-jews-eu-member-states-experiences-and.

EU Agency for Fundamental Rights (FRA). *Major EU antisemitism survey planned for 2018.* December 13, 2017. http://fra.europa.eu/en/press-release/2017/major-eu-antisemitism-survey-planned-2018.

Freeman, Hilary. "Why, as a British Jew, I'm terrified by the anti-Semitism suddenly sweeping my country." *The Daily Mail,* August 9, 2018. http://www.dailymail.co.uk/debate/article-2720381/Why-British-Jew-I-m-terrified-anti-Semitism-suddenly-sweeping-country.html.

General analyses worldwide (different years and multi authors). *The Kantor Center.* http://kantorcenter.tau.ac.il/general-analyses-antisemitism-worldwide.

Goldberg, Jeffrey. "French Prime Minister: If Jews flee, the Republic will be a failure." *The Atlantic,* November 10, 2015. http://www.theatlantic.com/international/archive/2015/01/french-prime-minister-warns-if-jews-flee-the-republic-will-be-judged-a-failure/384410.

Jewish Leadership Council. *Jewish Community Leaders Meet with Prime Minister David Cameron.* January 10, 2013. http://www.thejlc.org/2015/01/jewish-community-leaders-meet-with-prime-minister-david-cameron-3.

Lewis, Jerry. "Nearly half of British Jews say they have no future in Europe, study finds." *The Jerusalem Post,* January 14, 2015. http://www.jpost.com/Diaspora/Nearly-half-of-British-Jews-says-they-have-no-future-in-Europe-study-finds-387693.

Linde, Steve. "New UK Jewish leader: I want to meet Muslims." *The Jerusalem Post,* June 25, 2015. http://www.jpost.com/Diaspora/New-UK-Jewish-leader-I-want-to-meet-Muslims-407127.

Rich, Dave. "Antisemitism in the radical left and the British Labour Party." *The Kantor Center,* January 2018. http://kantorcenter.tau.ac.il/sites/default/files/Dave%20Rich%20180128.pdf.

Sanchez, Raf. "Theresa May says there can be 'no excuse' for anti-Semitism as she marks Balfour centenary with Netanyahu." *The Telegraph,* November 2, 2017. http://www.telegraph.co.uk/news/2017/11/02/theresa-may-says-can-no-excuse-anti-semitism-marks-balfour-centenary/.

Trigano, Shmuel. "A Journey Through French Anti-Semitism." *Jewish Review of Books,* Spring, 2015. https://jewishreviewofbooks.com/articles/1534/a-journey-through-french-anti-semitism/.

Appendix: Memories, Reflections, and Prospects

Konstanty Gebert
What is Jewish about Contemporary Central European Jewish Culture?

Throughout our part of the continent, Jewish culture is undergoing an apparent revival. Jewish-themed cafes with klezmer music and cholent on their menus are springing up from Vilnius to Bucharest, Jewish cultural festivals are all the rave in Kraków and Budapest, and books of Jewish interest have print runs occasionally far exceeding the total national memberships of Jewish communities. Yet by now it is common knowledge that this revival is largely staged by non-Jews for the benefit of other non-Jews, and while it might involve, as e.g. in Kraków, the participation of top-notch Jewish performers, Jewish communities, as a rule, have little in common with these highly visible developments.

The phenomenon is not new; its beginning roughly coincides non-coincidentally with the fall of Communism in 1989. It was first documented by Ruth Gruber in her magisterial book, *Virtually Jewish*,[1] which still remains the indispensable source of knowledge about the topic, more than fifteen years after publication. Yet this phenomenon of a Jewish-less Jewish revival continues to develop and deepen, proving it is no passing fad, but in fact a permanent element of the Central European cultural scene – both in terms of popular as well as high culture. It is therefore legitimate to reflect on what, if anything, is Jewish about it.

My remarks are limited to developments in my native Poland that I know best, even if a walk through Prague's bustling Jewish district proves again that the phenomenon knows no borders. The district is bustling – both with tourists and with businesses catering to their needs, or creating them. While there obviously are Jews among those tourists, they too are targeted as tourists, not as Jews – and if there might be Czech Jews among the owners and operators of tour guide offices, restaurants and shops, their wares are no different from those offered by the non-Jewish majority. What remains unquestionably Jewish are the physical sites themselves, and some of the small staff which operates them. They make the Jewish revival possible, but are hardly its most salient feature.

Let me make it immediately clear that I do not consider this something *a priori* morally or historically wrong: this is simply the way tourism operates worldwide. Nor are Jews obviously in any way excluded from the revival: their absence

---

[1] Ruth Ellen Gruber, *Virtually Jewish. Reinventing Jewish Culture in Europe* (Berkeley: University of California Press, 2002).

is mainly due to their drastically post-Shoah reduced demography. Yet as a Jew, I am concerned by this turn of events. If Jewish culture in Central Europe is largely produced by non-Jews for non-Jews, then do Jews have anything in common with it anymore? Should they? Is there room left for a Jewish culture by Jews and for Jews? And if so, how would it be different?

Relevant developments in Poland are in many respects highly representative of the larger Central European trends. The one country which is different is of course Hungary, with its exceptionally large Jewish community. There, Jewish culture for the non-Jews, similar to that which exists elsewhere, coexists with culture produced by Jews for Jews. Within the context of our part of the continent this juxtaposition is extraordinary and would necessitate a separate study.

At the same time, the Hungarian case highlights a fundamental issue which needs to be addressed: what, if anything, is wrong with the idea of a Jewish culture produced for non-Jews? Culture, after all, is universal and if we do not object to Jews being the consumers, or organizers of non-Jewish cultural production, the reverse must by definition be unobjectionable – and in fact is. A problem, however, arises when we have non-Jews not only organizing or consuming Jewish culture, but actually producing it: writing novels on Jewish themes, for example, or composing and performing klezmer music. At first sight, the approach to such phenomena should be no different to that outlined above; what is good for the goose is good for the gander, and Jews have actively and creatively contributed to the cultures they have assimilated into. And yet there is a difference between a minority contributing to the majority culture, and a majority significantly taking over cultural production for the minority: in the latter case, there is the risk of the very contents of that cultural production changing as a result.

Two counterarguments can be raised to that objection. The first is that any intervention in culture changes it, so there should be no difference between, e.g. Julian Tuwim, a great twentieth century Polish poet of Jewish extraction, writing in Polish, and Kapela (now Orkiestra) Klezmerska Teatru Sejneńskiego,[2] an excellent contemporary music band where all its members happen to be non-Jewish, performing klezmer music. This, of course, is true, but there is a difference of scale: Tuwim was one, if arguably the pre-eminent, Polish poet among many, most of whom were of ethnically Polish extraction; the Kapela is one of a limited number of klezmer bands, most of which are composed of non-Jews.

---

2 A musical ensemble created in the late 1990s in the small town of Sejny, in north-eastern Poland, near the Lithuanian border, by a team of non-Jewish enthusiasts of local cultures, who set up the Borderlands Centre and Foundation there.

The issue, of course, gets exquisitely more complicated once one allows for the fact that in the interwar period, when he produced his most powerful works, Tuwim was accused by Polish antisemitic critics of falsely trying to pass as Polish while "really" being Jewish; while he himself maintained that he was "really" Polish, his Jewish extraction was a mere incident of biography. After the shock of the Shoah (which Tuwim survived overseas), the poet started proudly claiming his Jewishness, only to be rejected by Jewish nationalists, who would then deny him the right of what they considered an affiliation claimed much too late. Not that this prevented Polish antisemites to continue rejecting him; against both, Tuwim would proudly claim in the poem *Zieleń* (Greenery; first published 1936) of his own "*moja ojczyzna – polszczyzna*" – "my homeland the Polish language."[3]

The second argument is much more pernicious. If the artist's extraction matters, as it goes, is this not a course of reasoning which leads straight to the antisemitic Nuremberg Laws, or rather a perverse anti-non-Jewish version of them (with the treatment meted out to Tuwim a perversion squared)? The answer must be an unequivocal yes. And that should end the debate, if not for the caveat that the concern here is not about the rights of any individual artist to produce whatever culture he or she wants, and indeed the right of any audience to enjoy that production. It is, as with the first counterexample, the cumulative effect we are concerned about: a culture for Jews which would no longer be by Jews and therefore would become, at best, a culture about Jews. It seems reasonable to assume that a community, be it ethnic, religious, or otherwise bonded, will attempt to exert control over its own culture, and consider the loss of that control a very grave development indeed.

This is especially significant as it would appear, based on numerous conversations with young Polish Jews today, that culture plays a particular role in their quest for identity. The religious revival which had played such a significant role in the reconstitution of the community in the waning years of the Communist system, and especially after its fall, seems not to have been passed on to the next generation. Neither the Shoah nor Israel – the two main poles of identity-building among the western Diasporas – seem to be very relevant, either; young Polish Jews clearly do not see themselves as part of a Jewish "community of fate." What bonds them is an enthusiasm for different strands of Jewish culture, from the intellectual appeal of textual study to the artistic rapture of Hasidic dancing, and it is through their commitment to at least participating in, and

---

[3] Julian Tuwim, *Zieleń*; *Dzieła*, vol. I, part II *Wiersze* [*Greenery*; *Works*, vol. I, part II *Poems*] (Warsaw: Czytelnik, 1955).

possibly contributing to, those activities that their Jewish identity is shaped and strengthened. Who determines the production of Jewish culture thus becomes not only a conceptual issue but, indeed, an existential one.

Yet while the importance of Jewish culture for the never-dying angst of Jewish survival is clear, the definition of what makes culture Jewish is, as we have seen, clearly less so. As culture for Jews by Jews is largely replaced with a culture about Jews, this might be grounds for concern. This can be seen in popular culture, by definition of course cosmopolitan, but which has endorsed at least one Jewishly very important trope as its own. The Shoah has become a universal matrix for debates about good and evil, with Jews – and Germans – becoming mere stand-ins for whatever concepts the author wants to make salient. This is best seen in popular feature films such as Steven Spielberg's "Schindler's List," Roman Polanski's "The Pianist," or Agnieszka Holland's "In Darkness." The fact that all three directors happen to be Jewish, though, does not seem to have made a difference. All three are about the cruel persecution of human beings who happen to be Jewish, not about the persecution of the Jews.

In none of them do we hear a word of Yiddish or Hebrew spoken, and only exceptionally do we get glimpses of printed or written Hebrew characters. If anything specifically Jewish is seen on the screen, it is only a visual marker to tell the viewer that these are, in fact, Jews – and not a significant element of the plot. Thus in "The Pianist" we briefly see a bearded gentleman with his head covered getting confused by the system of gates alternatively cutting off the two parts of the ghetto. "In Darkness" shows, somewhat incongruously, one of the characters donning – after weeks spent hiding in the sewers – a snow-white tallit. These are stage props, not relevant elements of the plot. The Jews' Jewishness dwindles into insignificance, and what is salient is the humanity they share with all, including their persecutors.

Now obviously there is nothing wrong with that: the films follow the fates of characters belonging to a minority of assimilated Jews who got caught in the deadly trap of the Nuremberg laws, and whose lives, during the war and before, carried precious little specifically Jewish content. The way they are presented on the screen is not distorted or manipulated. The problem is, however, that they are stand-ins for the Jews as a whole – and this, by implication, makes Jewishness irrelevant to the fate of the Jews, since all were destined to be murdered, even if they were assimilated, and their knowledge or experience of Yiddishkeit was no different than that of their non-Jewish neighbors. This obviously makes for good drama – but disregards the fact that Jewishness mattered a great deal to the non-assimilated Jews and, more importantly, to their killers, who believed that "non-Jewish Jews" are just as "Jewish" as the "Jewish" ones – and that denial is proof of either ignorance or deceit.

The dilemma of how to present the Shoah has been powerfully expressed by Polish-Jewish-American writer Henryk Grynberg, in a polemic with Polish writer Zofia Nałkowska. Nałkowska had given in her merciless *Medallions*,[4] a collection of eight short stories first published in 1946 as Polish literature's first reaction to the Shoah, the following motto: "People had prepared that fate for people." By saying so, she sought to counter the Nazis' antisemitic propaganda, which maintained that Jews were not truly human, which allegedly made them deserve their fate. By rejecting the dehumanization of the perpetrators, she also opposed the simplistic vision that this was a crime committed by monsters, for which nobody else is responsible. Her goals were clearly noble. And yet, wrote Grynberg in 1984, the fundamental premise is wrong: "People had prepared that fate for Jews."[5] The victims' Jewishness was not incidental; it was central to their identity and to their fate. To stress that, let us add, is not to dehumanize them: if humanity can include the perpetrators, it can also accommodate the victims. Denying the importance of their Jewishness, for whatever reason, no matter how noble, eliminates it – and thus, paradoxically, aligns with the goals of the killers.

All this notwithstanding, popular culture, in its treatment of the Shoah, does in fact eliminate the victims' relevant Jewishness, or – worse – makes a cartoon out of it, by stressing visual markers (hairstyles, dress) over cultural content. This is in fact no different from the treatment it accords, in Westerns, to cowboys and "Indians" – and in fact Shoah films are fast becoming a cinematographic genre, akin to Westerns, or *noir*. It is obvious that the authors of Westerns are not at all interested in the vicissitudes of maintaining, developing and passing on "Indian" identities – and there is no special reason they should. Similarly, authors of Shoah films are not concerned about Jewish survival angst. But Jews and "Indians" need to be, as their identities become more and more acultural, as opposed to spiritual and ethnic, and their cultural representation is being largely controlled by someone else.

Incidentally, Jews can at least get some satisfaction from the fact that, in popular culture, their presence is not limited to Shoah films exclusively: "Woody Allens" as a tentative name for the genre which concerns itself with the vagaries of assimilated and neurotic Jewish identities is just as frequent, while "Indians" seem to be permanently fixed between "paleface" and "Great Manitou."

In this respect, Central European culture about Jews is much more reassuring. Not only does it mercifully not center on the Shoah alone, and in fact tends

---

4 Zofia Nałkowska, *Medaliony* [*Medallions*] (Wroclaw: Siedmioróg, 2016).
5 Henryk Grynberg, *Prawda nieartystyczna* [*The Non-Artistic Truth*] (Wołowiec: Czarne, 2002).

to steer clear of it (in a possible tacit recognition that this, for once, is something better left to the Jews themselves), but preserves and cherishes many of the cultural aspects of Yiddishkeit which pop culture dismisses. Obviously, it does so because Yiddishkeit is its *raison d'être*, and not necessarily out of a desire to deepen and develop Jewish culture. Yet even that should be considered a blessing, given the fact that culture by Jews for Jews will remain (outside of the few countries with bigger Jewish communities) feeble, while popular culture representations will unavoidably tend towards the reductive or the cartoonish.

Initially, one of the problems with Jewish-themed cultural productions was that they tended to be meticulously reproductive.[6] This was initially true e. g. of Kapela Sejneńska. With the passage of time, they started feeling more comfortable in the musical language they had adopted, and instead of only recreating what they had encountered, they moved forward to creating their own sounds and tunes.[7] By doing so, and thus departing from the received music, they in fact remained true to the spirit of its original creators. Itinerant klezmer bands had travelled from village to village and town to town, crisscrossing a multiethnic cultural landscape, picking up tunes and tropes as they went, eventually recombining them and including them in their own repertoire. Appropriation is a necessary prerequisite for creation – and in both cases it produced the desired result.

To be sure, this development, in the case of the Sejny Orchestra, was not only due to their growing self-confidence. The band had held many musical workshops with such leading figures of contemporary living klezmer music as Michael Alpert, Stuart Brotman, David Krakauer, Deborah Strauss and Jeff Warschauer.[8] Their close cooperation made the distinction between Jewish culture and Jewish-themed culture largely irrelevant and indicates a way out of the "what is Jewish about Jewish culture" conundrum. Still, it needs to be noticed that music, being the mist abstract form of cultural production, lends itself most easily to this kind of blending. It is therefore obviously also not coincidental that the revival we are speaking about expresses itself disproportionately through that medium: klezmer bands abound in Central Europe, but few Jewish newspapers and almost no *kollelim*. Given the relative importance of these insti-

---

6 "Kapela Klezmerska Teatru Sejneńskiego", *YouTube*, July 17, 2007, https://youtu.be/cMPJ5gNDERI, accessed July 29, 2018.
7 "Orkiestra Klezmerska Teatru Sejneńskiego, Basowiszcza 2013", *YouTube*, July 21, 2013, https://youtu.be/Jy6oYCGyqAY. Accessed July 29, 2018.
8 I wish to thank Ruth E. Gruber for bringing up this point in the discussion following the original version of this paper presented at the international joint-conference of the Charles University and Tel Aviv University in Prague in 2015, entitled "Being Jewish in Central Europe Today."

tutions for Jewish culture and identity, this is hardly a desirable proportion – yet it is to be vastly preferred over the alternative, which would make klezmer as rare as Talmud study groups.

Yet the danger remains, in a way replicating in a subtler way, to be sure, the reductionist process described in popular culture. The kind of Jewish-themed culture which will sustain itself will necessarily avoid the more idiosyncratic aspects of Jewish culture, as both more difficult to access and less immediately interesting, to both groups it targets: non-Jews interested in Jewish things and in cultural diversity in general, and – a much smaller group – Jews themselves. Given, to underscore it once again, the irredeemable weakness of Jewish culture by Jews, this cultural editing will shape, to a substantial degree, the overall picture of Jewish culture, and in consequence, of Jewish identity as such.

To be sure, alternatives will also exist parallel to the mainstream. Small groups of Jews will continue to deepen their knowledge of the more intricate aspects of Jewish spirituality and culture, accompanied in their ventures by motivated groups of non-Jews. These cultural developments will occasionally spill over to the mainstream Jewish-themed culture, infusing it with new tropes. At the same time, the mainstream itself will look for this kind of enrichment, knowing fully well that without this, it risks banality. The extraordinary diversity offered yearly by the Kraków Jewish Festival clearly shows that such a serious effort is sustainable, gratifying and productive. Ongoing cooperation with more vibrant centers of Jewish culture elsewhere – Hungary, France and England, the US and first of all Israel – will in many cases produce the cultural impact outstanding *klezmorim* have had on the Sejny Orchestra.

And yet obviously the end result will be unavoidably the same: the creation of a Jewish-themed culture which will, for its producers as well as for its consumers, be only a strain, if important, of a wider array of relevant cultural tropes. If indeed Jewish culture today is the main source of Jewish identity, this identity is destined to become – depending on one's viewpoint – diluted or enriched by so many other cultural contents that this will ultimately make Jewishness just one element of a much more variegated self-hood.

There is no reason to believe, in any individual case, that this would be a loss. To the contrary – this kind of cultural variety is stimulating and makes for satisfying lives and original cultural creation. A community which is wise enough to allow, indeed accept, such developments at its periphery is sure to benefit from it, while at the same time enriching the society around it. But a community whose very identity consists of such cross-cultural variations cannot expect to maintain its identity for long, let alone perpetuate it. If all we have is the periphery, no matter how colorful, variegated and rich, this means by definition that the center has been lost.

But it has, in fact, been lost. The developments described above are only the most recent chapter in the desperate attempts of an all-but-exterminated community to remain alive, in the face of the bleak reality of numbers. That fact that so much survival has been possible is proof of an unexpected vitality of the Jews – and of the no-less unexpected encouragement and support they received from non-Jewish society at large. But it does seem clear that Jewish culture in Central Europe will be approximately as Jewish as Central Europe itself is now.

Bibliography

Gruber, Ruth Ellen. *Virtually Jewish. Reinventing Jewish Culture in Europe.* Berkeley: University of California Press, 2002.
Grynberg, Henryk. *Prawda nieartystyczna* [*The Non-Artistic Truth*]. Wołowiec: Czarne, 2002.
Nałkowska, Zofia. *Medaliony* [*Medallions*]. Wroclaw: Siedmioróg, 2016.
Tuwim, Julian Tuwim. *Zieleń; Dzieła, vol. I, part II Wiersze* [*Greenery; Works, vol. I, part II Poems*]. Warsaw: Czytelnik, 1955.
Online Sources
"Kapela Klezmerska Teatru Sejneńskiego". *YouTube*, July 17, 2007, https://youtu.be/cMPJ5gNDERI. Accessed July 29, 2018.
"Orkiestra Klezmerska Teatru Sejneńskiego, Basowiszcza 2013". *YouTube*, July 21, 2013, https://youtu.be/Jy6oYCGyqAY. Accessed July 29, 2018.

Anna Chipczyńska
Preserving Jewish Cemeteries as an Actual Challenge in Contemporary Poland

Introduction

Pre-World War Poland was home to the largest Jewish community in Europe, numbering about 3 million. Today's Polish Jews are a small group of a few thousand people. The Shoah, subsequent years of anti-Jewish sentiments, politically inspired antisemitic campaigns and emigrations whittled away that community.

Today there is hardly any discourse on Polish identity and its national memory without a reference to what is broadly termed as "Polish-Jewish relations." The consequences of the human loss not only remain a subject of public debate, but together with the material losses of pre-World War II Poland they define and affect the position of the Jewish community and its relations with the State.

By "Jewish community," I mean the institution – a legal person – that finds its legal basis on the Law of February 20, 1997 on the Relations between the State and the Jewish Religious Communities in the Republic of Poland (further called the "1997 Law").[1] The community, therefore, is not only a group of people with a common interest, identity, religion, and who are ethnically or culturally identifiable, but it is an institution having its legitimacy in a state-enacted law and enjoying international recognition. In my paper I will focus on the legal aspect of Jewish cemeteries where the national law and politics often play a pivotal role by affecting the work of the community, its function and development. The aim is to present an overview of basic legal provisions and their application to specific cases.

Contemporary organizational and financial independence of a Jewish community came into existence with the passing of the 1997 Law that, among other things, regulated the rules for the return of the pre-WWII Jewish communal property to the contemporary Jewish communities in Poland. Until 1997, the Jewish communal property, with a few exceptions, was at the disposal and legal discretion of the state or local authorities. Although the 1997 Law gave the Jewish communities the right of internal self-governance, its financial standing is de-

---

[1] The Law of February 20, 1997 on Relations between the Jewish Religious Communities and the Republic of Poland, Dz. U Nr. 41 poz. 251, http://www.abc.com.pl/du-akt/-/akt/dz-u-1997-41-251, accessed January 24, 2016.

pendent on the system of negotiable return of each property or the financial compensation for its loss. The cemeteries are explicitly mentioned in Art. 30 of the 1997 Law as a separate type of property to be returned to the communities. They became a part of the restitution, yet their shape and devastation did not become a subject of financial compensation. Though the restitution of the communal property has over the years been regarded as the primary source of income, it also vested in the community moral, financial and managerial responsibility for the preservation of the Jewish cemeteries.

There are more than 1,200 Jewish cemeteries in today's Poland, with many of them destroyed by constructions after the war. These were overbuilt by residential areas or public utility buildings such as those in Siedlce, Kalisz, Płońsk, Białystok, and Stoczek Węgrowski. Until today, around fifteen percent of the Jewish cemeteries have so far been returned to the Jewish communities or the Foundation for the Preservation of Jewish Heritage in Poland that was created by the Union of Jewish Communities in Poland and the World Jewish Restitution Organization. Among them are cemeteries in Lodz (at Bracka Street) and Warsaw (at Okopowa Street) that belong to the largest Jewish burial grounds in Europe.

Those cemeteries legally owned by the Jewish communities are not endangered by construction, yet many of them are located in remote parts of the country, meaning their protection and preservation cause significant difficulties. Good relations with local authorities and Polish communities are therefore detrimental to their existence. Their return means that the community becomes automatically responsible for the financial and organizational aspect of their preservation and management; the cost of each cemetery varies from 1,000 to 0.5 million euros per year. The latter was the case of the Warsaw Jewish Cemetery in Brodno, where the Jewish Community of Warsaw spent nearly half a million Euros for the renovation of a fence surrounding the burial ground and the construction of an educational pavilion. Undertaking legal, administrative and advocacy steps to preserve other Jewish cemeteries owned by third parties and endangered by desecration is another regular daily activity of today's Polish Jewish communities. This requires additional monitoring and preventive measures.

Four basic legal instruments are relevant in this field: the Law of 1997 defining relations between the state, the Jewish religious communities and the Republic of Poland; the Law of March 27, 2003 on Spatial Planning[2]; the Law of Janu-

---

[2] The Law of March 27, 2003 on Spatial Planning (with further amendments), Dz. U. 2003 nr 80 poz. 717, http://isap.sejm.gov.pl/DetailsServlet?id=WDU20030800717, last accessed January 24, 2016.

ary 31, 1959 on Cemeteries and Burying the Deceased[3]; and the Law of July 23, 2003 on the Protection and Preservation of National Heritage.[4] These tools are both a challenge and a chance for the protection of Jewish cemeteries in Poland. Their effective application could theoretically make the Jewish community a very successful watchdog and heritage preservation organization, but only theoretically. The legal steps are not the only thing that matters. There at least two more factors to be taken into consideration: the first is successful advocacy work in the region with the aim of close cooperation with local political leaders, civil society representatives and inhabitants; the second is an effective institutional structure that would cover research on each given property and provide for legal and advocacy interventions.

Urban development puts a large number of Jewish cemeteries, including those that were not returned to the Jewish communities, within the boundaries of towns and cities. They are often a part of the so-called "spatial planning zones" that define construction and urban development rules. Frequent changes in spatial planning may endanger the very existence of the cemeteries and necessitate legal interventions by the Jewish community. The legal ground for the interventions constitutes Art. 6 of the 1959 Law on Cemeteries and Burying the Deceased: it states that any changes to the purpose of the cemetery are subject to consent by the religious entity to whom the cemetery belonged. This legal ground shall be a safety mechanism for the preservation of the Jewish cemeteries, yet its application has often not been in use as it will be presented below.

The Białystok Jewish Cemetery Case

The Białystok Jewish Cemetery at Bema Street was built in 1831 and closed in the 1890s. After 1945 it served as an open-air market and is now partially built over by a car park and buildings. A few years ago, Polish and international Jewish media discovered that it was endangered by construction.[5] The newly proposed

---

3 The Law of January 31, 1959 on Cemeteries and Burying the Deceased (with further amendments), Dz.U. 1959 nr 11 poz. 62, http://isap.sejm.gov.pl/DetailsServlet?id=WDU19590110062, last accessed January 24, 2016.
4 The Law of July 23, 2003 on the Preservation and Protection of Heritage (with further amendments), Dz.U. 2003 nr 162 poz. 1568, http://isap.sejm.gov.pl/DetailsServlet?id=WDU20031621568, last accessed January 24, 2016.
5 See "Jewish Cemetery at Bema Street in danger." *Kurier Poranny,* December 21, 2015, http://www.poranny.pl/wiadomosci/bialystok/art/9196676,cmentarz-zydowski-na-ul-bema-nadal-za grozony-plan-miejscowy-zakazujacy-rozbudowy-nieuchwalony,id,t.html; and "PiS City Council

spatial planning document secured the historical boundaries of the cemetery, yet allowed for construction on the burial site. A private investor purchased the land from an entity that overbuilt a part of the burial ground with a meat market hall in the 1950s. The 2006 spatial planning document did not overtly specify the purchased land as a cemetery. In the meantime, the city office worked on the new spatial planning document that would specify the purchased land as the Jewish cemetery, yet still allowed construction works on its site. This document was subject to a vote in the city council. To protect the cemetery, the Jewish Community of Warsaw filed a legal objection in April 2014 that would prevent construction. The Białystok local authorities did not consider the objection filed by the Jewish Community of Warsaw. The debate among counselors ensued on the objection raised by the Jewish Community with some arguing that the Jews should not influence the development work. Just as the first city council meeting, some politicians voiced ostensibly antisemitic arguments; the later meetings with the local leaders that took place on November and December 2014 showed the fear that the change to spatial planning could bring compensation claims by the investor. The Białystok Heritage conservator who participated in the city council meeting on September 2015 proved that the planned construction, which would take place on the site of the cemetery, would violate Jewish religious law. Neither the objections by the Jewish Community of Warsaw nor the conservator's opinions convinced the counselors. Following discussions at two city council meetings on September and December 2015, the local politicians kept in force the previous spatial planning of 2006 that allows construction.[6] The next legal mechanism to be applied in the future with the chance for securing the cemetery is the 2003 Law on the Protection and Preservation of Heritage that does not allow construction work on heritage sites.

The Białystok case shows, first of all, that the 1959 Law on Cemeteries might not be recognized by local decision makers (city council members) as a legal tool to protect the sanctity of cemeteries. Protection of the cemeteries is often subject to spatial planning and can pose a financial challenge for the authorities. Whether the objection succeeds in being positively considered depends on the counselors' conviction of the importance of securing the sanctity of the Jewish cemeteries. It is also dependent on their attitude and understanding of the sensitivity of

---

members want to build on the Jewish bones", http://bialystok.wyborcza.pl/bialystok/1,35241, 18899285,radni-pis-chca-budowac-na-kosciach-bialostockich-zydow.html, September 24, 2015.

[6] "Jewish Cemetery at Bema Street in danger", in *Kurier Poranny*, December 21, 2015, http://www.poranny.pl/wiadomosci/bialystok/art/9196676,cmentarz-zydowski-na-ul-bema-nadal-za grozony-plan-miejscowy-zakazujacy-rozbudowy-nieuchwalony,id,t.html, accessed January 24, 2016.

Polish-Jewish relations. The additional protection mechanism in such cases could be including the cemetery on the list of national heritage sites. The 2003 Law on the Protection and Preservation of Heritage which imposes protection measures on the preparation of spatial planning documents in relation to the heritage sites could be an extra safeguard. The Białystok cemetery case is pending with the 2006 spatial zoning plan being in force. The investor is entitled to apply for the construction permit. The effectiveness of further interventions is a question mark.

With regards to the Białystok Jewish Cemetery constituting first of all a legal playground for the protection of the burial site, the Kalisz Jewish cemetery is a case where politics and attitudes towards Jews play a major role in preserving Jewish cemeteries and also relations with the Jewish community.

The Kalisz Jewish Cemetery Case

The Jewish graveyard in Kalisz was set up ca. 1287 when Duke Przemysław II confirmed in writing the purchase of the land to serve that purpose.

Burials were held there until WWII. The graveyard is the oldest known Jewish cemetery in Poland. During WWII, it was wiped away altogether with the Kalisz Jews. In the 1940s, the Nazis plundered hundreds of matzevot. Some were used to pave the river embankment. A sports field, apartment block, intensive care unit and public education buildings were built over the cemetery in the 1950s and 1960s.

The restitution of the Jewish cemetery in Kalisz to the Jewish community started in 2000 when the Regulatory Commission for the Jewish Communities in Poland officially started its works on this subject. After ten years, the negotiations on the return of the property commenced. Upon agreement, parts of the cemetery territory were to be returned to the Jewish community. The town council received financial support of 3.5 million PLN from the Polish government budget to move the educational premises to the new site. The school was moved to a newly renovated building. On September 2014, before the local elections, the situation changed drastically. The town representatives opposed to proceed with the restitution process and the implementation of the reached agreement. On July 2015, the City Council of Kalisz passed a resolution allowing the re-opening of the school on the site of the cemetery.[7]

---

[7] Kalisz Jewish Cemetery files, Union of the Jewish Communities in Poland, October 5, 2015.

At the end of 2014, the newly elected President of Kalisz, backed by nationalistic and right-wing parties, opposed the restitution of the Jewish Cemetery, arguing it would be in the will of local inhabitants and that it was built over with the educational complex and apartment blocks. According to the president, restituting the property would violate the property rights of the third persons. He also questioned the boundaries of the cemetery.[8] In the end, the newly elected president refused to respect prior agreements.

The Kalisz case shows that the prolonged restitution process, based on the negotiations, is dependent on numerous political changes. The changing political scene can overturn past agreements and decisions. The case of the Kalisz Jewish Cemetery is still pending.

The Brodno Jewish Cemetery Case

The Brodno Jewish Cemetery, located in the Praga district of Warsaw, is the oldest existing cemetery in the city. This eighteenth century burial ground was in use until World War II, and today is the site of some 15,000 assembled gravestones. The cemetery constitutes an example of WWII devastation with further demolition and neglect during the communist regime.

On December 2012, the Jewish Community of Warsaw became the owner of the Brodno cemetery upon the restitution agreement reached with the city of Warsaw. The financial compensation for another property was matched with the return of the cemetery to the Jewish community. It obliged itself to carry out renovation works for the amount of 3 million PLN. They included, among others, renovation of the fence, conservations of the matzevot and the creation of a permanent exhibition. In 2013 and 2014, the community applied to the Ministry of Culture and National Heritage for the co-funding of its renovation project at the cemetery. The application was rejected twice. According to the official response, this eighteenth century cemetery did not possess sufficient historical and societal value for the area. What seemed to be, in the opinion of the Ministry, the weak point of the application was the fact that the site was not on the UNESCO list or among so-called historical monuments. Both titles are granted following state nominations. These administrative weaknesses were the reasons for the non-recognition of the historical authenticity of the site. The community filed complaints in subsequent years – the most recent one remained unanswered despite the written support provided by the Polish Ministry of Foreign Af-

---

8 Ibid.

fairs. The contradictory approaches to the value of the Brodno cemetery prove that opinions among politicians and public officers on the historical meaning and importance of the Jewish cemeteries can be diametrically different. This means that more work in terms of raising political awareness in the discussed field is truly needed. The case of the Brodno Jewish cemetery also shows that the Law on the Protection and Preservation of National Heritage does not automatically provide any state-driven strategy in preserving cemeteries as heritage sites. Additionally, both institutions, i.e. Ministry of Culture and National Heritage on one side, and the City of Warsaw on the other side, have shown a different approach to this heritage site. While it was recognized as such by the local authorities, it has not constituted a particularly high value on the national level.[9]

Conclusions

The Jewish community possesses organizational autonomy and a clear mandate to represent Polish Jews vis-à-vis the state. Nevertheless, its financial standing – the guarantee for its existence and development – is dependent on the lengthy restitution process defined by the 1997 Law. The cemeteries constitute a relevant part of it, and ultimately, so far, their ownership is to be vested in the existing Jewish communities. Their preservation necessitates substantial legal interventions and advocacy work. Other legal instruments such as the 1959 Law on Cemeteries and Burying the Deceased, the 2003 Law on Spatial Planning and the 2003 Law on the Preservation and Protection of Heritage grant the community the role of watchdog, secure its moral and religious values and make it an active player in the field of urban development and preservation of the cemeteries – the areas where local and national politics matters. This is, however, often ignored by the state and by local institutions. The preservation of the sites is not only subject to the timely application of various legal instruments; none of the aforementioned laws guarantee success in this area. The changing urban development, varying local attitudes to Jews as much as the lack of an effective management structure that would provide proper research, supervision and advocacy in this field do not ease the situation. Taking into account the economic and organizational costs of this challenge, the question remains what should be the role and responsibility of the Polish state at the beginning of the twenty-first century regarding protecting and preserving Polish Jewish material heritage? Will the State recognize its pivotal role, and in what manner, to assist the community

---

[9] Brodno Jewish Cemetery files, Jewish Community of Warsaw, October 2015.

in preserving the memory of the once thriving Jewish life? This is the question we will not be able to escape from in the near future.

Note: since the presentation of this paper, the state of the Jewish cemeteries has gained recognition on state level. It is the result of advocacy work by the Jewish communities and awareness raising among decision makers in society. In December 2017, the Polish Ministry of Culture and National Heritage assigned 100 mln PLN for the endowment fund for the renovation of the Jewish Cemetery in Warsaw.[10] The same institution, in cooperation with the Jewish community, enacted conservatory guidelines that aim to halt any works that endanger the existence of cemeteries. As much as these developments constitute positive and strong signals that the preservation of material Jewish heritage has gained recognition by the state authorities, they unfortunately do not constitute sufficient legal protection. Further legal solutions will be needed.

Bibliography

Online Sources

"Jewish Cemetery at Bema Street in danger." In *Kurier Poranny*, December 21, 2015. http://www.poranny.pl/wiadomosci/bialystok/art/9196676,cmentarz-zydowski-na-ul-bema-nadal-zagrozony-plan-miejscowy-zakazujacy-rozbudowy-nieuchwalony,id,t.html. Accessed January 24, 2016.

Jewish Community of Warsaw, News from December 22nd, 2017. http://warszawa.jewish.org.pl/2017/12/100-million-pln-for-the-jewish-cemetery-on-okopowa-str/. Accessed January 24, 2016.

"PiS City Council members want to build on the Jewish bones". http://bialystok.wyborcza.pl/bialystok/1,35241,18899285,radni-pis-chca-budowac-na-kosciach-bialostockich-zydow.html. Accessed January 24, 2016.

The Law of July 23, 2003 on the Preservation and Protection of Heritage (with further amendments), Dz.U. 2003 nr 162 poz. 1568. http://isap.sejm.gov.pl/DetailsServlet?id=WDU20031621568. Accessed January 24, 2016.

The Law of March 27, 2003 on Spatial Planning (with further amendments), Dz. U. 2003 nr 80 poz. 717. http://isap.sejm.gov.pl/DetailsServlet?id=WDU20030800717. Accessed January 24, 2016.

The Law of February 20, 1997 on Relations between the Jewish Religious Communities and the Republic of Poland, Dz. U Nr. 41 poz. 251. http://www.abc.com.pl/du-akt/-/akt/dz-u-1997-41-251. Accessed January 24, 2016.

---

10 Jewish Community of Warsaw, http://warszawa.jewish.org.pl/2017/12/100-million-pln-for-the-jewish-cemetery-on-okopowa-str/, last accessed September 26, 2018.

The Law of January 31, 1959 on Cemeteries and Burying the Deceased (with further amendments), Dz.U. 1959 nr 11 poz. 62. http://isap.sejm.gov.pl/DetailsServlet?id=WDU19590110062. Accessed January 24, 2016.

Natalia Sineaeva-Pankowska
Holocaust Memorialization in Poland: A Case Study of Polin Museum

An Educator's Perspective

The Polin museum in Warsaw was inaugurated in 2014. Its mission, as defined on its website, is "to recall and preserve the memory of the history of Polish Jews, contributing to the mutual understanding and respect amongst Poles and Jews as well as other societies of Europe and the world." The museum can serve as an important field for reconciliation: a place for the restoration and reconstruction of memory and identity. It can also be a space where visitors confront the troubled and not always convenient past and negative legacies and have an opportunity to reflect on it. It should integrate society and influence it to be more critical of its own history. In this context museum visitors are not passive recipients of information, but can also be invited to participate in dialogue with the museum. That being said, museums can also create tension within society, as memory is always selective, and museums are inseparable from politics and ideology.

Indeed, most of the history museums (with rare exceptions) that were established in Eastern Europe, including Poland in the recent decades, are curated to present a dominant (often ethno-nationalist) narrative of history. After the collapse of the former Communist paradigm, they had to reconstruct their identity, looking for other identity symbols, and opted for a nationalistic version. On one hand, the Polish museum landscape is diverse, and on the other, its museums, including those representing the twentieth century, were designed to present the selected nation's positive aspects or martyrdom serving a dominant official narrative.

As scholar Joanna Michlic suggests, Polish-Jewish relations constitute the most difficult aspect of the collective memory of Poles, even compared to the difficult Polish-German, Polish-Ukrainian and Polish-Russian relations.[1] In this context she has written about three types of Holocaust memorialization which have appeared in Poland since the turn of the century: memory to commemorate (pamiętanie dla upamiętnienia), memory for benefit (pamiętanie dla korzyści)

---

[1] Joanna Michlic, *Coming to Terms with the "Dark Past": The Polish Debate about the Jedwabne Massacre* (Jerusalem: Hebrew University, 2002), 3.

and memory to forget (pamiętanie, żeby zapomniec).[2] Memory to commemorate means commemoration of emptiness after the Holocaust and the loss of three million Polish Jews. It attempts to assess and discuss the difficult aspects of Polish-Jewish relations. Memory for benefit means using different aspects of Holocaust memory to gain international prestige and support. Memory to forget involves a more ethno-nationalistic discourse oriented to forget the Holocaust and Jewish history as something foreign and non-Polish.[3] All three types are present in contemporary Polish public discourse and are applicable to the museum space.

In the article I look at the issue of Holocaust memorialization in Poland from the perspective of an educator who has presented Holocaust history to museum visitors. I will present the case study of one museum and my own experience there. However, it remains pertinent to the broader situation in Poland.

I worked at the Education department of the Polin Museum of the History of Polish Jews in Warsaw in Poland for several years, from the opening of the core exhibition in 2014. During that time, I had opportunities to interact, guide and conduct workshops as an educator for many individual visitors and over 600 groups, both Jewish and non-Jewish, mostly from Poland.

Elzbieta Janicka critically assesses the Museum's core exhibition as well as the space surrounding the Museum which, as she writes, presents the dominant narrative of Polish-Jewish history and the Holocaust in Poland today:

> The main narrative uniting the MHJP's surroundings, building and core exhibition is the idyllic myth of Polin which dictates the selection and presenting of information. The story of Polish hosts and Jewish guests that is inherent to the Polin myth establishes inequality and dominance/subjugation as framing principles of a story of majority-minority relations. It also constitutes a mental gag and an instance of emotional blackmail which precludes any rational – analytical and critical – conversation based on historical realities. Furthermore, in practice, it is a part of a pattern of culture which produces – and at the same time legitimizes – violence and exclusion.[4]

---

2 Joanna Michlic, ""Remembering to Remember," "Remembering to Benefit," "Remembering to Forget": The Variety of Memories of Jews and the Holocaust in Postcommunist Poland", in *Przeszłość i pamięć* [*Past and Memory*], 4/2011, (Warsaw) 225246. See online at http://www.fzp.net.pl/spoleczenstwo/modele-pamieci-o-zydach-i-zagladzie.

3 Michlic, ""Remembering to Remember," "Remembering to Benefit," "Remembering to Forget"".

4 See more: Elzbieta Janicka, "The Embassy of Poland in Poland. The Polin myth in the Museum of the History of Polish Jews (MHPJ) as narrative pattern and model of minority-majority relations," in Irena Grudzińska-Gross and Iwa Nawrocki (eds.), *Poland and Polin: New Interpretations in Polish-Jewish Studies:* [Eastern European Culture, Politics and Societies 10] (Frankfurt am Main (etc): Peter Lang Verlag, 2016), 121–172.

The Museum is located at the site of the district where Jews used to live before the war and where Zamenhof Street ran before and during the war. It is the same street that Jews were forced to walk along on their way to Umschlagplatz, from which the trains departed for the extermination camps. The Museum is located in front of the Nathan Rappaport Memorial to the Warsaw Ghetto Uprising erected there in 1948. However, Janicka states there are many Holocaust monuments erected in the area, the majority of which were erected after the fall of Communism and which depict the efforts of Poles to rescue Jews during the war without providing any context. She calls this phenomenon as "De-Holocaustisation", when the essence of the Holocaust was removed from the narrative of the Holocaust, which is accompanied by the process of "Holocaustisation" of the Polish majority.[5]

However, I have found the Polin Museum of the History of Polish Jews to be more inclusive than other Polish museums in presenting and discussing the difficult past. In this article I argue that through its exhibition (and site events), the Museum not only promotes the positive historical legacies of the Polish-Jewish past, but also tackles difficult subjects and negative legacies of the recent past, such as the anti-Jewish pogroms in Poland in 1941 and anti-Jewish violence in 1944–1946. It is critical of some aspects of Polish history. In this respect I argue that the Museum contributes to uncovering the truth and educating the public.

What is the initial concept of the Museum? The idea to create the Polin Museum belongs to a Jewish organization, the Association of the Jewish Historical Institute, supported by the government. Therefore, the Museum is a result of a partnership between the Jewish organization and the government and a minority perspective was at the heart of the Museum's establishment.

The idea to create the Museum was born in 1996. It took more than twenty years to establish the Museum, finally opening as a cultural and educational centre in 2013. Its narrative core exhibition, which covers almost one thousand years of Polish-Jewish history in seven historical galleries, was opened to visitors in 2014. The *genius loci* of the place contribute to its uniqueness.

According to the museum creators, "the museum and the story it tells in the core exhibition will be an agent of transformation. Polish visitors will encounter a history of Poland that is not a national history, not a history of a nation."[6] From start to finish the Museum attempts to present and challenge an integrative and

---

5 Ibid.
6 Barbara Kirshenblatt-Gimblett, "Theater of History", in *Polin 1000 Year History of Polish Jews* (Museum Catalogue, 2014), 19.

diverse Polish history instead of an ethno-nationalistic approach and encourages visitors to participate in the dialogue. In the eighteenth century, Poland was home to the largest Jewish community in the world, at which time it was the center of the Ashkenazy Jewish community. Before World War II, Poland was a multicultural state with national minorities comprising thirty-five percent of its population, of which Jews constituted ten percent – more than three million people. In some places such as Warsaw they constituted thirty, forty and even higher percentages of the population. They coexisted with other minorities, including Ukrainians, Germans, Belorussians and others. As a result of the Holocaust, border changes, post-war emigration and assimilation, the ethnic composition of Poland changed, becoming more uniform, mono-religious and mono-ethnic.

Poland has a diverse and sophisticated history, and it seems that minority narratives should organically constitute the core of the Polish national identity. However, following my own observations from my museum experience, many Poles (museum visitors) have very limited, if any, knowledge about this part of their own history. Therefore, the Polin core exhibition aims to introduce them from start to finish to this history, which is a complex *'story of co-existence, competition, conflict, separation and integration'*,[7] to use the words of the exhibition's chief curator and creator, Barbara Kirshenblatt-Gimblett.

There are many difficult aspects in Polish-Jewish history. Amongst them are the anti-Jewish pogroms in Eastern Poland in 1941, when it could happen that neighbors killed their neighbors (to use the words of Jan Tomasz Gross), and post-war anti-Jewish violence occurred, when Jews who survived the war in hiding started to return to their homes, faced violence and hatred from their Polish neighbors. The Polin museum creators found a solution of how to deal with and present these difficult aspects to Poles. In the words of Barbara Kirshenblatt-Gimblett, "As a portal, catalyst, and forum, Polin Museum offers an alternative model of constructive engagement and trusted zone for critical reflection."[8] That means the exhibition shows all aspects of history, showing all the complexity – not only the negative, but also the positive. The creators also applied a multiple voice approach. Each gallery has a different Jewish narrator, who presents a different perspective. Therefore, the visitors are supposed to be more open to ac-

---

7 Barbara Kirshenblatt-Gimblett, "Historical Space and Critical Museologies: POLIN Museum of the History of Polish Jews in From Museum Critique to Critical Museum", in Katarzyna Murawska-Muthesius, Piotr Piotrowski (eds.), *From Museum Critique to the Critical Museum* (London: Routledge, 2015), 147.
8 Ibid.

cepting difficult truths, when the so-called "trusted zone" is created.⁹ How does it work in practice? What are the most sensitive subjects for the Polish public?

It should be mentioned that the decision to come to the museum usually indicates a certain openness and readiness to engage with another points of view. A travel agency survey conducted for the Museum in April 2016 showed that groups of Polish tourists often turned down offers to visit the Polin Museum, seeing it as something foreign and "not Polish". Some maintain that this reluctance towards the Museum might arise from deeply rooted anti-Jewish prejudices on the part of tourists and the tendency to see the Museum as focusing on a heritage that is not their own.

As I have already mentioned, many Polish visitors do not know the basic facts, such as why Jews came to Poland, what kind of relations have existed between Poles and Jews through the centuries, and how many Jews lived in Poland in different times. Their knowledge reflects stereotypical beliefs prevalent amongst their peers. Moreover, there is a widespread opinion and confidence amongst Polish visitors that the Holocaust is the best-known chapter of the history of Polish Jews.

However, on the basis of my own experiences and observations and those of other museum guides and educators, Holocaust knowledge (for example, the causes, circumstances, and numbers of casualties) amongst Polish visitors remain relatively low.[10] Some name this phenomenon as "Forgetting the Holocaust in the Era of Global Remembrance". The main source of knowledge for them comes from news outlets rather than from academia.

Here I would argue that the most contentious part for Polish visitors is shown in the Holocaust and Post-war galleries, specifically when the different attitudes of Poles towards Jews during the war are presented: from help to hatred; to indifference and passivity, as the most common responses to the Jewish tragedy. However, as Rafal Pankowski states:

> We know there was not a lot of organized collaboration in Poland; (...) it is important to stress that the Holocaust was carried out by German Nazis. That said, I think one way the Polish National Radicals contributed to the Holocaust was through their propaganda and their activities in the 1930s. By reinforcing the social distance between the Polish and Jewish communities, they fostered a kind of hostility, mistrust, and isolation. During the war this turned out to be critical, because in places that were more integrated, where the social distance was smaller – such as in the Żoliborz district in Warsaw – the probabil-

---

9 Ibid.
10 Natalia Sineaeva-Pankowska, "Jak zwiedzający odbierają galerię 'Zagłada. Z notatek przewodniczki" ["Visitors' Reactions to the Holocaust Gallery. From a Guide's Notes"], in *Zagłada Żydów. Studia i Materiały* 12 (2017): 659–678.

ity of survival was much greater. When the communities were segregated or separated by social distance, the chances of survival were minimal. In a wider geographic sense, if you analyse the map of places in Poland where anti-Jewish violence happened, it corresponds in many ways to the same places where the Endecja and the National Radical movement had been strong before the war.[11]

Many Poles believe that they selflessly rescued Jews during the war. It is difficult for Polish visitors to accept that it was in fact not a mass phenomenon, and there were also Poles who harmed Jews, blackmailed them, or were indifferent to their plight, which comprised the largest group. At this part of the exhibition visitors might disagree with the guide and express their emotions. Occasionally, it leads to instrumentalization of symbols: the most meritorious examples, like Irena Sendler, who served in the Polish underground during the Second World War, and Children's Division of *Zegota*, a Polish underground group to assist Jewish people, or Jan Karski, a courier of the Polish Government in exile, who was the first to inform the Allies about the extermination of Jews, are trivialized and reduced to patriotic *clichés*. In this way, the unique historical significance of these individuals can easily become lost. Visitors (particularly Polish teachers) often ask at the beginning of a guided tour to include their stories in the guided tour.

However, the Holocaust Gallery[12] offers a more sophisticated understanding of the issue. For instance, it shows, amongst other things, that the assistance to Jews in the Warsaw Ghetto by *Zegota* was made possible through its cooperation with the Jewish National Committee (Jews themselves) and presents the names of lesser known rescuers. It also shows other attitudes and actions of Poles during the war. This is an example of how the Museum attempts to reconcile Polish visitors with own challenging past, by confronting it directly.

In a way the museum experience reflects the broader situation in Poland, when the topic of *Righteous Among the Nations* (Polish rescuers) is dominant and is overused in discussions about the Holocaust. The contemporary *Righteous Among the Nations* discourse serves the purpose of memory for benefit and Polish pride, to use the definition of Joanna Michlic. Polish museums and exhibitions often tend to display this kind of Holocaust memorialization. Amongst the recent examples is also the new Ulma Family Museum of Poles Saving Jews in the Second World War. It is also widespread in the public and media dis-

---

11 Rafal Pankowski, "Poland's Illiberal Challenge", *NER* 37, no. 4 (2016), http://www.nereview.com/polands-illiberal-challenge/. Accessed September 22, 2018.
12 A conceptual framework of the Holocaust gallery at the Polin Museum was developed by two distinguished scholars, Barbara Engelking and Jacek Leociak.

courses, and other museum exhibitions, as well as museum educational programs.

It can be argued that the new so-called memory bill passed by Poland's Senate in January 2018 contributes to this kind of distortion of history. It directly and indirectly impacts the climate surrounding the museums and the educational work conducted there. It has been argued that "what has been even more problematic and troubling than the wording of the legislation itself (...) is the type and tone of discourse that has accompanied it" and that included many antisemitic statements.[13]

The result of this distortion of history is alarming, as museum visitors have very limited Holocaust knowledge. They also fail to understand the distinction between the death camps and concentration camps, and often see their ancestors – ethnic Poles – to be the primary victims of Nazi Germany. Occasionally, Polish visitors want a guide to draw a comparison between the number of Jews murdered in the aftermath of the deportation of the Warsaw Ghetto and the number of people killed during the Warsaw Uprising. They see Jewish and Polish martyrdom as standing in competition with one another (competing victimhood). Similar questions are asked by members of all kinds of visiting groups: teachers, retirees and individual visitors. It can be argued that in the context of the current political climate recently there have been many more such examples.

Another challenging topic is the anti-Jewish pogroms in the summer of 1941 in Eastern Europe and Eastern Poland, including Jedwabne, and the participation of the local population in it. However, the topic is less emotional than the discussion about the Righteous Among the Nations (thanks to the debate which has lasted for many years following the publication of Jan Tomasz Gross's book about the local population's involvement, Neighbours. The Destruction of the Jewish Community in Jedwabne, in 2000).

There were many more pogroms in Eastern Europe, not only in Poland, as illustrated by the map of pogroms which is present in the Holocaust gallery. Other Eastern European countries did not go through the same process as Poland, and it is much more difficult for their visitors to accept these facts when they visit the Museum. Visitors from these countries usually ask many questions. It is more difficult for them to learn about these questions in their own countries and in their own environment, and there is little or no information in the history museums of Ukraine, Moldova and others.

---

[13] See more: Rafal Pankowski, "The Resurgence of Antisemitic Discourse in Poland", *Israel Journal of Foreign Affairs: Jerusalem* (2018): 1–17, http://www.nigdywiecej.org//docstation/com_docstation/63/r._pankowski_the_resurgence_of_antisemitic_discourse_in_poland._israel_journa.pdf.

Therefore, memory and these difficult debates have symbolically been brought to the museum in Warsaw, especially since the average person would not travel to Jedwabne or to other atrocity sites to face the complicated past by standing at the foot of the monument erected there. In this context, the role of the exhibition is important not only for the process of reconciliation, when the descendants can directly confront these difficult aspects of their own history, but also counter so-called soft Holocaust denial (to use the words of Deborah Lipstadt) or selective and deflective denial (to use Michael Shafir's terminology) in which distorters do not deny that the Holocaust happened, but deny or minimise the role of their own nation in some aspects of it, or transfer the guilt to others.

I have also observed that visitors from Lithuania (information about executions in Ponary, the largest killing site in Lithuania, where massacres were perpetrated by Germans with some local involvement, is included in the exhibition narrative) and the places where, for example, anti-Jewish pogroms happened, such as Kielce, often do not want to discuss and learn more about their own locations at the beginning of their visit. It is difficult for them to accept this information and it is subconsciously blocked. It can also be considered as a "denial", in which protective mechanisms are active. Some visitors said at the end of their visit that they did not want to listen to (i.e. rejected) information about the difficult moments during the guided tours, for example, in the Holocaust gallery, but were willing to accept only positive information. At the same time a certain percentage of these visitors return to the exhibition to learn more and recommend it to their friends and family.

Holocaust survivors constitute the very distinctive and important category of museum visitors. Sometimes they accompany groups that participate in the March of the Living. For some, it is the first visit to Poland after many years, when they want to bring their children and grandchildren with them. Such tours usually take longer. Survivors frequently share their experiences with their guide. As the museum's visitors are not free from prejudice and bias, the guides and educators must be prepared to deal with a variety of attitudes.

The foreigners may already be familiar with the story of Jedwabne. They know that in this case the perpetrators were Poles. They know Gross' book *Neighbours* (this mostly applies to Jewish groups, but also to those non-Jewish from Sweden as Jan Tomasz Gross' book has been translated into many languages including Swedish). They do not know, however, how much effort it required to initiate the discussion of this subject that began in Poland several years ago, or the role played by historians affiliated with the Polish Center for Holocaust Research in this regard.

The Polin museum experience shows how important the museum's role can be in dealing with the difficult parts of history and how it can be a site for reconciliation, for both victims and victimizers. It also shows how this process can be challenging, multi-layered, long-lasting and influenced by political changes.

Bibliography

Janicka, Elzbieta. "The Embassy of Poland in Poland The Polin myth in the Museum of the History of Polish Jews (MHPJ) as narrative pattern and model of minority-majority relations". In *Poland and Polin: New Interpretations in Polish-Jewish Studies: [Eastern European Culture, Politics and Societies 10]*, edited by Irena Grudzińska-Gross and Iwa Nawrocki. Frankfurt am Main (etc.): Peter Lang Verlag, 2016.

Kirshenblatt-Gimblett, Barbara. "Historical Space and Critical Museologies: POLIN Museum of the History of Polish Jews in From Museum Critique to Critical Museum". In *From Museum Critique to the Critical Museum*, edited by Katarzyna Murawska-Muthesiu and Piotr Piotrowski. London: Routledge, 2015.

Kirshenblatt-Gimblett, Barbara. "Theater of History". *Polin 1000 Year History of Polish Jews* (Museum Catalogue) 19 (2014).

Michlic, Joanna. ""Remember to Commemoration", "Remember for Benefit" And "Remember to Forget": Different Models of Memory about Jews and the Holocaust in Post-communist Poland". In *Past and Memory*, 4/2011, 225–246. Warsaw. http://www.fzp.net.pl/spoleczenstwo/modele-pamieci-o-zydach-i-zagladzie.

Michlic, Joanna. *Coming to Terms with the "Dark Past": The Polish Debate about the Jedwabne Massacre (Analysis of Current Trends in Antisemitism)*. Jerusalem: Hebrew University of Jerusalem, 2002.

Pankowski, Rafal. "Poland's Illiberal Challenge." *NER* 37, no. 4 (2016). http://www.nereview.com/polands-illiberal-challenge/. Accessed September 22, 2018.

Pankowski, Rafal. "The Resurgence of Antisemitic Discourse in Poland". *Israel Journal of Foreign Affairs: Jerusalem* (2018) 1–17. http://www.nigdywiecej.org//docstation/com_docstation/63/r._pankowski_the_resurgence_of_antisemitic_discourse_in_poland._israel_journa.pdf.

Sineaeva-Pankowska, Natalia. "Jak zwiedzający odbierają galerię 'Zagłada: Z notatek przewodniczki" ["Visitors' Reactions to the Holocaust Gallery. From a Guide's Notes"]. *Zagłada Żydów. Studia i Materiały* 12 (2017).

Tomáš Kraus
Thirty Years After. The Yesterday, Today, and Tomorrow of the Czech Jewish Community

Introduction

In order to understand the situation of the Czech Jewish Community three decades after the Velvet Revolution, one has to dive quite deep into the history of not only the Jews in Bohemia and Moravia, but also the history of the country itself and – indeed –Europe at large.

The Jews have lived in the region of Central Europe for more than one millennium. Scholars will not take seriously wild theories and legends, but as an ancient Czech proverb goes, in every gossip there is a grain of truth. Some of these early legends say that Jews were sailing central-European rivers as merchants and in locations where they stored their goods they settled down and married local women. That is why some of us have blue eyes. Other legends say that those Jews sometimes did not marry only one wife, like *Prince Samo*, the first Czech historical figure from the seventh century CE who was originally also a merchant and therefore probably Jewish. After all, polygamy was banned by the rabbis only three hundred years later.[1]

Here maybe starts the unique relationship between the Jews and the major society living in Bohemia and Moravia. Or does it start with the princes of the Přemysl dynasty who – according to another wild speculation – might have been descendants of Samo, the Jewish merchant? And did the Jews come to Central Europe as suppliers of Roman legions, as historians and archaeologist suggest,[2] or did they come to Bohemia in the times foreseen in one of the prophecies of *Princess Libuše*, the legendary *founding mother* of the Czech nation?

> ...when your grandson will be ruling over my people, an expelled, oppressed and banished little nation which worships only one God will come to our forests to seek protection. May he accept them very hospitable, as they will bring a great blessing to our meadows... When

---

[1] "Decree of Rabbi Gershom ben Yehuda", *Jewish Encyclopedia*, http://www.jewishencyclopedia.com/articles/12260-polygamy#anchor3, accessed March 1, 2019.
[2] "Josephus Suetonius and Tacitus on the Military Service of Roman Jews", *Academia.edu*, http://www.academia.edu/4732342/Josephus_Suetonius_and_Tacitus_on_the_MilitaryService_of_Roman_Jews, accessed March 1, 2019.

https://doi.org/10.1515/9783110582369-018

in the year 850 AC (during the rule of prince Hostivít) the Vends overflowed Latvia and surrounding areas ... one of local Jewish communities was indeed driven to our parts...³

If we exchange in this romantic saga, Latvia for Crimea, the Vends for Russians and postpone the era for one century, we face a situation caused by the destruction of the Khazar Empire by Kievan Rus. In that case the theory of Arthur Koestler that some Ashkenazi Jews have this specific background may be relevant.⁴

But let us go to real historical facts. The very first literal document about Prague was written by a Jewish diplomat, known under his Arabic name *Ibrahim ibn Jakob*, in 965 CE. He was an envoy of the *chalifa* from Cordoba on his way to Constantinople and he kept records about his travel experiences. In his diary he describes Prague as a vivid city built of stone with a vibrant population where Jews play an important role. Or we can mention one of the most important royal rulers of early Czech statehood, Přemysl Otakar II from the thirteenth century CE. It was he who granted the Jews their status, including royal protection, valid for many centuries to come.⁵ Yes, high taxes were the price, but look at the sign of the *Knight Order of the Crossbreds with a Red Star*. This star is six-pointed and the Order was founded by no-one else as his sister, who – because of her acts of mercy and charity (can we say *tsedaka?*) – became the patron of all ill, poor and suffering. Later, canonised as St. Agnes, she became one of the patrons of Bohemia. The star could have been a coincidence. But there is another one – the same building-works which constructed her convent has built just a few hundred meters away the Old-New Synagogue, the oldest functioning synagogue in Europe today.⁶

And we can go on. These past ten centuries are filled with bizarre tales and surprising stories where Jewish figures are playing a major role. Sometimes it is difficult to separate hyperboles from reality. In German, it has a perfect expression – *Dichtung und Wahrheit*.

---

3 V. V. Tomek, *Pražské židovské pověsti a legendy [Prague Jewish fables and legends]* (Prague: K. Končel, 1932).
4 Arthur Koestler, *The Thirteenth Tribe: The Khazar Empire and Its Heritage* (New York: Random House, 1999).
5 "Statuta Judaeorum", *Wikipedia*, https://cs.wikipedia.org/wiki/Statuta_Judaeorum.
6 Helena Soukupová, *The Convent of St. Agnes of Bohemia* (Prague: National Gallery, Prague, 1993).

Myths and Realities

It's not the intention of the author to write another ordinary footnote of history. Rather, I want to point out to the richness and colorful – sometimes tragic, sometimes happy – life of our ancestors. Only then we can understand from where we come and where we are heading to, as the famous gesture of Chris in *Magnificent Seven* clearly demonstrates.

What are the myths in our history and what is the reality? Let us start with the world famous one, with the *Golem*. This clay figure who was supposedly brought to life by *Rabbi Löw* will forever symbolize Prague and its mysticism. The Golem, incorporated into fundamental Czech literature and later one of the most popular Czech movies of all times, is an integral part of Czech culture and its modern identity. The figure serves many purposes, from best-selling merchandise to parallels with artificial intelligence and a memento of misusing modern technologies. His creator, the legendary Jehuda Liva ben Bezalel, appears in these tales as a miraculous sorcerer, close to the image of Merlin or Gandalf. But who was he in reality? A genial thinker, philosopher, mystic and theologist, an author much ahead of his time.[7] Known as the *Maharal of Prague*, his life and his attitudes are studied by scholars and rabbis until today. His complete work has been translated and published in many languages, including English. And, what a surprise – the word *golem* does not appear anywhere! In addition, rabbi Löw wrote his essential pieces not in Prague but in *Mikulov* (Nikolsburg) where he served for more than 20 years as the chief rabbi of Moravia, before he moved – in his declining years – to Prague.

Now fast forward, from the Renaissance Prague of Rudolf II,[8] to the Baroque era. We have passed the national tragedy after the battle on White Mountain in 1620, after which a few Czech protestant noble families, in order to avoid forceful catholic conversion, decided to convert to Judaism. And we are standing in front of the Crucifix on the famous Charles Bridge, arched over the Vltava River. There are not that many statues on the bridge at this time. But this one is decorated with Hebrew inscriptions: *Kadosh, kadosh, kadosh Adonai Elohim Tzevaot – Holy, holy, holy is the Lord of Hosts*. How is that possible? According to a legend, a Jew who was once passing the Cross did not take his hat off and did not salute the cross, as all others did. Another, more dramatic version says that he even spat on it. In every case, he was arrested and sentenced to death. Only the

---

[7] Rabbi Judah Loew ben Bezalel, *Path of Life* (Prague: Academia, 2009).
[8] Eliška Fučíková et al., *Rudolf II and Prague: The Court and the City* (London: Thames and Hudson, 1997).

local Jewish Community was able to rescue him by paying a fine from which the golden letters were manufactured and installed. What is the reality? There was an internal fight within the Jewish Community of Prague. One group denounced the other of blasphemy and the authority in charge, the Jesuits, did not want to resolve this dispute. So they suggested such a shameful solution.[9] Baroque Prague was a show-place of many tragedies, where forceful conversions played an adverse role. Famous is the case of Shimon Abeles, a child who almost became a Christian saint and whose fate is still a mystery.[10] Many Jews of that time converted to Christianity and some even willingly. No wonder that the Czech population today has up to ten percent Jewish genes.[11]

And again, let us skip a few centuries. Now we are standing in front of the Opera house. The time is said to be marked with a unique and glorious coexistence *of three cultures* – Czech, German and Jewish. However, this nice projection does not work. There was a Czech culture and there was a German culture. And the Jews were on either side, or on both, sometimes in the middle of this national struggle which was rattling and shaking the local societies since the mid-nineteenth century. Sometimes they served as a bridge between these two. And sometimes they were denounced by both. But they still played a unique and unmistakable role, like *Angelo Neumann*. We are standing in a park, in front of the Opera, because of him. This park used to be the best address in Prague, something as Central Park in New York City today. But nowadays it is nicknamed Sherwood because of the nearby Central train station and its homeless inhabitants. Neumann has built the Opera house in 1888 and not only that, after opening he managed the stage, designed the set and hired musicians, singers and conductors, such as a certain Gustav Mahler. And he was a personal friend of Richard Wagner whose music he successfully brought to Prague and other places. Wagner, who was clearly an antisemite, later misused by the Nazi propaganda, always denied his antisemitism, pointing out how many Jewish friends he had. Neumann was one of them, a Jew who officially converted to Christianity. The New German Theatre became the center of the Prague German-speaking intellec-

---

9 Alexandr Putík, "The Hebrew Inscription on the Crucifix at Charles Bridge in Prague: The Case of Elias Backoffen and Berl Tabor in the Appellation Court", in *Judaica Bohemiae XXXII* (Jewish Museum of Prague, 1997).
10 Elisheva Carlebach, *The Death of Simon Abeles, Jewish-Christian Tensions in 17th Century Prague* (New York: Centre for Jewish Studies, Queens College, CUNY, 2001).
11 "Jen třetina Čechů má slovanské kořeny, ukázal výzkum" [Only a third of the Czech population has Slavic roots, shows research], *Novinky.cz*, October 28, 2017, https://www.novinky.cz/veda-skoly/453236-jen-tretina-cechu-ma-slovanske-koreny-ukazal-vyzkum.html, accessed March 1, 2019.

tuals, especially Jews, and one of the most important theatres in Europe.[12] The cult of Wagner was so vivid that – as Max Brod remembered – the worst struggle within the Jewish community of that time was not the split between orthodoxy and modernity, nor a dispute over Zionism and assimilation, but the division between the fans of Richard Wagner and those of Guiseppe Verdi.

Another close associate of Angelo Neumann was a certain *Karl Skraup* who happened to be the nephew of Franz Skraup, the organ player in the Spanish Synagogue in Prague's Old Town. Franz changed his name into Czech for František Škroup and composed the song *Kde domov můj* (*Where Is My Home*) which later became the Czech national anthem.

Speaking of music – it is worth comparing the symphonic poem *Vltava* (*the Moldau*) by Bedřich Smetana with the Israeli national anthem *Hatikvah*. Undoubtedly, it is the same melody – who borrowed from whom, Smetana from Samuel Cohen or vice versa? Who knows? The original is a medieval song *La Mantovana* which inspired many folk songs, including the Czech nursery rhyme *Kočka leze dírou* (*A Cat Crawls through a Hole*) and which might have been Smetana's inspiration. Or did he hear the Hatikvah when he was living with a Jewish family in Gothenburg?[13]

All directors of the New German Theatre after Angelo Neumann, until 1938, were Jews. Heinrich Teweles, former editor-in-chief of the most influential newspaper Prager Tagblatt, served here as his follower for eight years since 1911 and staged few plays by his friend, a Viennese journalist and author by the name of Theodor Herzl who always came to Prague for the premiere of his piece. Together with Teweles, they often attended services in the Old-New Synagogue. Herzl, the founder of the Zionist movement, also wrote a novel called The Old-New Land.[14]

Theodor Herzl and his compatriots organized the first World Zionist Congresses in Basel in the late 1890s. But 1921 and in 1923, the congress came to Czechoslovakia, namely to the famous spa town Karlsbad, Karlovy Vary. The foundations of Israel were laid there, which is quite a well-known fact. Overshadowed by this, another congress which took place in 1937 in close-by Marienbad, Mariánské Lázně, also a famous Western-Bohemian spa resort, remains unnoticed. It was the gathering of Agudath Israel, an orthodox organization led by famous

---

12 Jitka Ludvová, *Až k hořkému konci, Pražské německé divadlo 1845–1945 [Until the bitter end, the Prague German theatre 1845–1945]* (Prague: Academia, 2012).
13 "Švédský pobyt Bedřicha Smetany: Umělecké úspěchy i přítelkyně Froejda" [Bedřich Smetana's Swedish residency: Artistic successes and also friend Froejda], *Radio.cz*, December 26, 2017, https://www.radio.cz/cz/rubrika/special/svedsky-pobyt-bedricha-smetany-umelecke-uspechy-i-pritelkyne-froejda, accessed March 1, 2019.
14 Theodor Herzl, *Altneuland* (Leipzig: Verlag Benjamin Harz, 1902).

rabbis, which completely rejected the idea of a Jewish state.¹⁵ It seems that local healing springs had a confusing effect not only on Goethe, Chopin or Beethoven, on tsar Peter the Great, emperor Franz Josef or king Eduard VII, etc., but also on Jews.

When the first Czechoslovak republic was dismantled after the shameful Munich appeasement treaty in 1938, Prague Castle had an unwelcomed guest. Reinhard Heydrich was not only the architect of the Final Solution of the Jewish Question, i.e. the Shoah, but also an executioner of the Czech nation. This is why he met his final destiny from hands of Czech and Slovak patriots, paratroopers sent from London. During Heydrich's cruel terror rule over the country, everything Jewish had to be removed from public space. This concerned also the statue of Felix Mendelssohn-Bartholdy on the roof of the Rudolfinum concert hall. The German soldiers in charge of this were confused, because when they climbed up they saw many statutes and did not know which one was Mendelssohn. So they took down the statue with the biggest nose. It was Richard Wagner.¹⁶

And finally the glorious chapter in the Czech-Israeli history, the supplies of Czechoslovak arms to a newly established state in 1948. Without this Israel would never have survived, as every Israeli child knows from school. No doubt about that. By not respecting the UN arms embargo the Czechoslovak leaders saved the country. Some of them later paid for their courageous acts with their lives in Soviet-controlled show-trials. But the supplies and the training of personnel was also an excellent business-deal in which the Czechoslovak state got rid of old Messerschmitts, manufactured in Prague during the Nazi occupation, which should have ended in scrap anyway.

"Who Do You Think You Are?"

This is the name of a popular BBC format which was franchised in several countries, which shows celebrities in a search for their family roots. Czech TV released this program some time ago and a few popular personalities found out only then that they are Jewish. On the other hand, some of them, who thought that they had some Jewish background, to their obvious disappointment, learned that they don't.

---

15 https://www.jta.org/1937/08/25/archive/agudath-israel-congress-opposes-jewish-state-asks-fulfillment-of-mandate, accessed March 1, 2019.
16 Jiří Weil, *Na střeše je Mendelssohn* [*Mendelssohn is on the Rooftop*] (Prague: Československý spisovatel, 1960).

As it seems, the Czech Jews of today actually do not know who they are. A simple example can be seen from the 90s: the Federation of Jewish Communities was invited to participate in the work of *Governmental Commission for National Minorities*. But the constituency rejected the invitation, based on the argument that we, the Jews, are not a national minority. We do not fall under the criteria of having a national minority mother tongue. Unlike Germans, Poles, Hungarians and others, what would be considered as the *mamelushen* of Czech Jews? Yiddish? No way! Nobody in Prague spoke Yiddish since the expulsion ordered by Empress Maria Theresa in 1745 (though she had to later permit the Jews to come back). Or Hebrew? Absolutely not! Who speaks Hebrew? Only those who studied in an *Ulpan* since the 1990s.

Obviously, there is no self-image as a "national minority", despite the fact that in pre-War Czechoslovakia Jewish nationality was recognised by the Constitution. Unlike in the Soviet Union, this country allowed individuals to identify with Jewish nationality on a voluntary basis, though with the hope that Jews would give up their German culture. It did not happen because most local Jews – especially in Prague and Brno – had German education. For those few who later survived the Shoah, this might have been a problem when they claimed back their property confiscated by the Nazis. The post-war authorities would turn them down, pointing out that they filed in the census of 1930 German nationality, while all Germans should have been expropriated and expelled from Czechoslovakia by 1946.

So what are the Jews in today's Czech Republic, if not a national minority? A religious minority? The Czechs are said to be the most secular nation in Europe, maybe in the world. Some recent opinion polls confirm that.[17] And secular also are the Czech Jews. That is why the ten Jewish Communities today have around 3,000 members, but the estimation is that there might be up to 20,000 Jews who live in the country today and are not affiliated.

And if their identity may be religious, to what stream of Judaism do Czech Jews belong? Orthodox, liberal, reform? It is difficult to say. It is probably a mixture of all of them, something like *Judaism à la carte*. But the overwhelming majority remains secular. In addition, many Shoah survivors who originally lived in Carpathian Ukraine, the eastern part of Czechoslovakia which was in 1945 given to Soviet Union, came to Bohemia and Moravia. They were Czechoslovak citizens, so-called *repatriants*, and they were in many cases replacing Germans in

---

[17] "Češi jsou nejméně věřící, ale nejvíce vyhranění vůči muslimům" ["The Czechs are the least religious, but most against the Muslims"], *Týden*, 5.12.2018, https://www.tyden.cz/rubriky/domaci/cesi-jsou-nejmene-verici-ale-nejvice-vyhraneni-vuci-muslimum_505743_diskuze.html, accessed March 1, 2019.

territories which were called – not accurately – Sudetenland. This is the reason why in the North of Bohemia today we have more Jewish communities than in the whole of Moravia. The descendants of those people are an integral part of our constituency today.

When these *Carpathians* came also to Prague, it created a friction inside the Jewish community. They brought with them an orthodox way of life which was for those genuine Prague Jews who survived the Shoah something absolutely alien. They spoke Yiddish, they sang different tunes in synagogues, they were reciting *Shma* while sitting with covered eyes instead of proudly standing and singing it aloud. Prague was used for synagogue-cantors who were performing operatic arias by *Lewandowski* and *Sulzer*,[18] for male or even mixed choirs and obviously for organ musical accompaniment. None of this was possible with the Carpathians. On the contrary, they were upset at the Prague Jews. The conflict ended by dividing the synagogues: the Prague Jews got the Jubilee synagogue in Jerusalem Street and the orthodox Carpathian group prayed from then on in the Old-New one.

And another complicated issue appears in the quest for identity: who is a Jew? Is the orthodox *halacha* valid and only the mother counts? Or do we have the luxury after the Shoah and after four decades of Communism to reject people who have only the patrilineal descent? In many cases they were the ones who were bearing the Jewish name of their fathers in difficult times. The majority of society sees them because of that as Jews and now some rabbis come and say that in fact they are not Jewish at all? That is a blow. The Nazis organized transport to concentration camps in several waves, with the last one in late 1944 consisting of people of the so-called mixed marriages. At that time, the crematoria in Auschwitz were not working anymore and that is why many survived. Can these survivors and their children be excluded from Jewish communities because only their father was Jewish?

And after the end of the Cold War, the Americans and Israelis arrived to Prague in the 1990s, which was said to be the Paris of the 60s. What did they bring with them? A mixture of modern Jewish identity, consisting of some Lithuanian-Polish-Russian elements, mixed with Morocco and Yemen, spiced by klezmer music and Yiddish literature. *Klezmer* – and *Ladino* – found a fertile ground with Czech musicians who immediately fell in love with it. In fact, it was a similar situation as in New York when local Jewish jazzmen, on their quest for new

---

[18] Read about the aforementioned composers at http://www.jewish-music.huji.ac.il/content/louis-lewandowski and http://www.jewish-music.huji.ac.il/content/salomon-sulzer, accessed March 1, 2019.

expressions of world music, found the heritage of their ancestors. But here in Prague? No, this was not our music, not our heritage and not our tradition. As we have mentioned, we were admiring Richard Wagner, which was better not to say in Israel.

Back to the Future?

So where do we go from here? Who are the Czech Jews of today and how does their future look like? Our Jewish identity is a confusing mixture. Instead of binding on old traditions of modernity and liberal way of life, during these three decades we have created a specific new bubble where Israeli *shlichim* and American Fiddler on the Roof culture, based on roots of Carpathian *shtetls*, play substantial roles. After all, the Carpathian orthodoxy was the one which helped the Jewish community to survive communism, sometimes in the grey zone and sometimes in underground and dissident circles. The roots of our modern identity range to that time. This is why, after the Velvet revolution, to be Jewish became even a fashion, peak was the popularity of rock-group *Shalom*. Unlike medieval times of forced conversions, we have been witnessing a reverse stream of conversions to Judaism, first by dissidents who wanted to demonstrate their resistance against the communist regime which was antisemitic, afterwards by many graduates from Theological faculties who found Judaism in their curricula and identified with its values even though elderly Jews were pointing out that it is not enough just to put a *kippah* on your head and pretend you are a Jew.

Can we survive? Today – at the end of second decade of the twenty-first century – we, Czech Jews, are in a very comfortable situation. We hardly witness substantial antisemitic incidents, we are integrated into the major society, we are respected by political representation, our institutions are working very well, given the fact that three decades ago we started from scratch. Sometimes we do so well that we suffer from operational blindness. What is then our major challenge? We have 3,000 registered members and we know that the potential is at least three times larger. What shall we do to get those non-affiliated Jews who live in our country to be involved? Can we present more sexy programs for the youth? Can we offer more social and health services to middle and elderly generation? And if so, will this blooming economy enable us this forever? If not, are we prepared for crises of various kinds?

Remember the past. Between the two world wars, Czechoslovakia was viewed as an island of democracy in Europe; it is no wonder that democrats and many Jews fled to Prague from Germany when Hitler seized power. Today, we are in a similar situation. The Czech Republic – with all its controversies in the public arena – is a safe-haven for Jews. But is that enough? Can we survive without new programs, without new ideas, without opening-up, instead of closing down because of narrow-minded approaches? This is something what we, the leaders of Jewish communities of today, have to consider. Otherwise we will end up like the legendary *Tevje* and all his seven daughters.

Bibliography

Carlebach, Elisheva. *The Death of Simon Abeles, Jewish-Christian Tensions in 17th Century Prague*. New York: Centre for Jewish Studies, Queens College, CUNY, 2001.
Fučíková, Eliška, et al. *Rudolf II and Prague: The Court and the City*. London: Thames and Hudson, 1997.
Herzl, Theodor. *Altneuland*. Leipzig: Verlag Benjamin Harz, 1902.
Koestler, Arthur. *The Thirteenth Tribe: The Khazar Empire and Its Heritage*. New York: Random House, 1999.
Ludvová, Jitka. *Až k hořkému konci, Pražské německé divadlo 1845–1945* [Until the bitter end, the Prague German theatre 1845–1945]. Prague: Academia, 2012.
Putík, Aleksandr. "The Hebrew Inscription on the Crucifix at Charles Bridge in Prague: The Case of Elias Backoffen and Berl Tabor in the Appellation Court". In *Judaica Bohemiae XXXII*. Jewish Museum of Prague, 1997.
Rabbi Judah Loew ben Bezalel. *Path of Life*. Prague: Academia, 2009.
Soukupová, Helena. *The Convent of St. Agnes of Bohemia*. Prague: National Gallery in Prague, 1993.
Tomek, V. V. *Pražské židovské pověsti a legendy* [Prague Jewish fables and legends]. Prague: K. Končel, 1932.
Weil, Jiří. *Na střeše je Mendelssohn* [Mendelssohn is on the Rooftop]. Prague: Československý spisovatel, 1960.

Online Sources

Agudath Israel Congress Opposes Jewish State; Asks Fulfillment of Mandate, August 25, 1937. https://www.jta.org/1937/08/25/archive/agudath-israel-congress-opposes-jewish-state-asks-fulfillment-of-mandate. Accessed March 1, 2019.
"Češi jsou nejméně věřící, ale nejvíce vyhranění vůči muslimům" ["The Czechs are the least religious, but most against the Muslims"]. *Týden*, December 12, 2018. https://www.tyden.cz/rubriky/domaci/cesi-jsou-nejmene-verici-ale-nejvice-vyhraneni-vuci-muslimum_505743_diskuze.html. Accessed March 1, 2019.
"Decree of Rabbi Gershom ben Yehuda". *Jewish Encyclopedia*. http://www.jewishencyclopedia.com/articles/12260-polygamy#anchor3. Accessed March 1, 2019.
"Jen třetina Čechů má slovanské kořeny, ukázal výzkum" ["Only a third of the Czech population has Slavic roots, shows research"]. *Novinky.cz*, October 28, 2018. https://

www.novinky.cz/veda-skoly/453236-jen-tretina-cechu-ma-slovanske-koreny-ukazal-vyzkum.html. Accessed March 1, 2019.

"Josephus Suetonius and Tacitus on the Military Service of Roman Jews", *Academia.edu*. http://www.academia.edu/4732342/Josephus_Suetonius_and_Tacitus_on_the_MilitaryService_of_Roman_Jews. Accessed March 1, 2019.

"Statuta Judaeorum", *Wikipedia*. https://cs.wikipedia.org/wiki/Statuta_Judaeorum. Accessed March 1, 2019.

"Švédský pobyt Bedřicha Smetany: Umělecké úspěchy i přítelkyně Froejda" ["Bedřich Smetana's Swedish residency: Artistic successes and also friend Froejda"]. *Radio.cz*, December 26, 2017. https://www.radio.cz/cz/rubrika/special/svedsky-pobyt-bedricha-smetany-umelecke-uspechy-i-pritelkyne-froejda. Accessed March 1, 2019.

About the Authors

Barna, Ildikó is Associate Professor of Sociology at ELTE University, Faculty of Social Sciences in Budapest, where she also serves as Head of the Department for Social Research Methodology. Her research topics include antisemitism, xenophobia, post-Holocaust Jewry, and quantitative research on archival sources. Her most recent publications include *Jews and Jewry in Hungary in 2017* (in Hungarian, 2018), co-edited with András Kovács, the chapter about Hungary in *Modern Antisemitism in the Visegrád Countries*, edited with Anikó Félix, and a paper titled "Interdisciplinary Analysis of Hungarian Jewish Displaced Persons and Children Using the ITS Digital Archive."

Chipczyńska, Anna is President and Deputy President of the Jewish Community of Warsaw, a role she has held since 2014. She holds degrees in International Relations from Warsaw University, Central European University in Budapest and a graduate degree in Law from BPP Law School in London. In 2003–2004, she was a Research fellow at the Department of Political Science of Tel Aviv University. She worked, among others, in the humanitarian sector and for the OSCE Office for Democratic Institutions and Human Rights. In 2016, Chipczyńska was awarded an honorary membership by the Polish Society of the Righteous Among the Nations. She is also a member of the Founders' Assembly of the Foundation for the Preservation of Jewish Heritage in Poland and a Council member of the Union of Jewish Communities in Poland.

DellaPergola, Sergio, born in Italy in 1942 and living in Israel since 1966, received his M.A. in Political Sciences from the University of Pavia and his Ph.D. in Social Sciences and Contemporary Jewry from the Hebrew University of Jerusalem. He is Professor Emeritus and former Chairman of the Hebrew University's Avraham Harman Institute of Contemporary Jewry. A specialist on the demography of world Jewry, he has published or edited sixty books and monographs including *Jewish Demographic Policies: Population Trends and Options* (2011) and over three hundred papers on historical demography, family, international migration, Jewish identification, antisemitism, and projections in the Diaspora and in Israel. He has lectured at over one hundred universities and research centers all over the world and was Senior policy consultant to the President of Israel, the Israeli Government, the Jerusalem Municipality, and major national and international organizations. He is also chief editor of *Hagira – Israel Journal of Migration* and member of Yad Vashem's Committee for the Righteous of the Nations. He has won the Marshall Sklare Award for distinguished achievement in the Social Scientific Study of Jewry (1999) and the Michael Landau Prize for Demography and Migration (2013).

Fireberg, Haim is Research associate at the Kantor Center for the Study of Contemporary European Jewry, Tel Aviv University and Head of research programs at the Center. His main foci of research are the social and demographic situations of Jewish communities in Europe, urban history of the Jewish community (H'Yishuv) in Palestine in the twentieth century and during the first decade of the State of Israel, and the study of virtual Jewish communities (maintaining Jewish and Israeli life in cyberspace). Fireberg is also actively monitoring and researching contemporary antisemitism, concentrating on Europe, and is in charge of analyzing violent incidents worldwide. His most recent publication is *Being Jewish in 21st Century Germany* (co-edited with Olaf Glöckner, 2015). At present, Fireberg (with partners) is conducting a vast social and cultural research of the three main Jewish communities in Europe.

Gebert, Konstanty is a journalist in Warsaw, Poland. He co-founded the unofficial Jewish Flying University in 1979 and an independent trade union, NTO, in 1980. In the 1980s, he was an editor and columnist of underground publications. In 1989, he helped launch *Gazeta Wyborcza* and co-founded the Polish Council of Christians and Jews. In 1997, he founded the Polish Jewish monthly *Midrasz*. In 2006, he launched the Warsaw office of the European Council on Foreign Relations and he is a board member of the Dutch Jewish Humanitarian Fund, the Einstein Forum and Paideia. He has taught in Poland, the USA, and Israel. Gebert is the author of eleven books on the Polish transformation, the Yugoslav wars, Torah commentary, Israeli history and the European twentieth century. His essays have also appeared in two dozen collective works in Poland and abroad.

Glöckner, Olaf is Senior Researcher at the Moses Mendelssohn Centre for European-Jewish Studies at University of Potsdam, Germany where he also teaches Modern History and Jewish Studies. His main foci of research are Sociology of Judaism, Jewish Education, European Jewry after the Cold War, Modern Israeli Society, German-Israeli Relations, and Migration to re-unified Germany. Glöckner is also doing research on modern antisemitism in Europe and its prevention options. Among his recent publications are *Deutschland, die Juden und der Staat Israel. Eine politische Bestandsaufnahme [Germany, the Jews and the State of Israel. A political inventory]* (co-edited with Julius H. Schoeps, 2016) and *Das Zeitalter der Genozide: Ursprünge, Formen und Folgen politischer Gewalt [The age of genocide]* (co-edited with Roy Knocke, 2017).

Holý, Jiří is Professor at the Department of Czech Literature and Comparative Literature and Head of the Centre for the Holocaust and Jewish Literature at the Faculty of Arts, Charles University in Prague. He specializes in literary science and research mainly within the area of Czech Literature and Jewish Topics in Literature and Culture. His recent publications include "Nontraditional images of the Holocaust in Czech literature and cinema: comedy and laughter," in *Jews and Gentiles in Central and Eastern Europe during the Holocaust* (2018) and "Jurek Becker: Jakob der Lügner (1969)" (co-written with Hana Nichtburgerová) in *Holocaust. Zeugnis. Literatur. 20 Werke wieder gelesen [Holocaust. Testimony. Literature. 20 works re-read]* (2018).

Khanin, Vladimir (Ze'ev) got his Ph.D. in Political Science from the USSR Academy of Sciences in 1989 and completed his post-doctoral studies at the University of Oxford, UK, in 1991. He is an expert on Russian Jewish community in Israel and the Diaspora and the FSU politics, and currently serves as Chief Scientist of the Israeli Ministry of Aliyah and Integration. He is also Associate Professor in Israeli and Jewish Studies at Ariel University, and lectures Political Studies at Bar-Ilan University, Israel. He publishes extensively on Israeli, East European, and Jewish and African politics and society. Among his recent books are *Joining the Jewish Collective: Formalizing the Jewish Status of Repatriates from the FSU of non-Jewish and Mixed Origin in Israel* (2014) and *"The Third Israel": Russian-speaking Community and Politics in the Contemporary Jewish State* (Moscow, 2014)

Kovács, András is a sociologist with a Ph.D. from the Hungarian Academy of Sciences and Professor at Central European University, Budapest for the Nationalism and Jewish Studies Programs. His past appointments and research stays include Paderborn University (FRG), École des Hautes Études en Sciences Sociales (Paris), New York University (New York), TH Twente (the Netherlands), Salomon Steinheim Institut für Deutsch-Jüdische Geschichte (Duisburg), Institut für die Wissenschaften vom Menschen (Vienna), Moses Mendelssohn Zentrum für europäisch-jüdische Studien (Potsdam), Internationales Forschungszentrum Kulturwissenschaften (Wien), Institut für Soziologie (Universität Wien) and Zentrum für Antisemitismusforschung (TU Berlin). His research subjects include antisemitism, Jews in post-war Hungary, memory and identity, socio-economic attitudes, and political choice. He recently published "Antisemitic Elements in Communist Discourse: a continuity factor in post-communist Hungarian antisemitism" in *Antisemitism in an Era of Transition. Continuities and Impact in Post-Communist Poland and Hungary* (2014) and "Communism's Jewish Question" in *Jewish issues in Communist archives* (2017). He co-edited, with Ildiko Barna, *Jews and Jewry in Hungary, 2017* (in Hungarian, 2018).

Kraus, Tomáš is the Executive Director of the Federation of Jewish Communities in the Czech Republic, a role he has held since 1991. In this position he was rebuilding the infrastructure of Czech Jewish institutions and was in charge of respective legislation. His main task was, however, the negotiations for return of Jewish property and compensation for Holocaust survivors not only on Czech but also on international top political levels.

After the establishment of the Czech-German Future Fund in 1997, he was appointed as member of its Discussion Forum, for several years serving also as the chairman of the Supervisory Board of the Czech Council of Nazi Victims. He was one of those who initiated the Holocaust Era Assets Conference which took place in Prague in 2009 and its direct outcome the European Shoah Legacy Institute, where he was appointed its Supervisory Board Chair. He was elected for several terms the President of B'nai B'rith Renaissance Prague and served also as the President of Czech Society of Christians and Jews. For many years he was active as an executive member of the European Jewish Congress, in 2009 he was elected its Vice-President. He served for four years also as Vice-President of the World Jewish Congress. He regularly publishes articles in various newspapers and magazines, and appears in other media, mainly Czech radio and television. Since 1999 Dr. Kraus has also been lecturing at NYU or Western Michigan University, among other institutions. His subjects are mainly Holocaust and Jewish Studies. Both parents of Dr. Kraus were Holocaust survivors.

Menachem Zoufalá, Marcela is a scientific researcher and lecturer at the Centre for the Study of the Holocaust and Jewish Literature at the Faculty of Arts, Charles University. She received her Ph.D. in Cultural Anthropology and published the outcomes of her doctoral dissertation as the monograph *Judaism and Women in Israel* (in Czech, 2012). She co-authored and edited a collective monograph *Jewish Studies in the 21st Century: Prague – Europe – World* (2014) and co-edited, with Jiří Holý, *The disintegration of Jewish Life: 167 days of the Second* (in Czech, 2016). Her research interests include ethno-religious diversity in Israeli society, contemporary European antisemitism, and Jewish identities. She received the Masaryk Distinguished Chair in 2014 and 2016. She was a Guest Professor in Israel Studies at the Moses Mendelssohn Center for European Jewish Studies, University of Potsdam & Selma Stern Center for Jewish Studies in Berlin-Brandenburg. Together with Jiří Holý, she coordinates an international EU research project among CUNI, TAU, MMZ/UP, and CU entitled *'United in Diversity' – An Interdisciplinary Study of Contemporary European Jewry and Its Reflection*.

Porat, Dina is Professor Emeritus and Head of the Kantor Center for the study of Contemporary European Jewry in Tel Aviv University and also Chief historian of Yad Vashem. She received awarded prizes for some of her many publications (including the 2009 National Jewish Book Award for the biography of Abba Kovner, *The Fall of a Sparrow: The Life and Times of Abba Kovner*), teaching abilities (TAU's Faculty of Humanities best teacher for 2004), and humanitarianism (the Raoul Wallenberg Medal for 2012). She is also on *The Marker Magazine*'s 2013 list of fifty leading Israeli scholars and on *Forbes*'s 2018 list of fifty leading women in Israel. She was a visiting professor in Harvard, Columbia, New York, Venice, and the Hebrew university and has tutored altogether thirty M.A and twenty Ph.D. students. Her main research interests are the History of the Holocaust, Zionism and the Jews of Europe, the "Final Solution" in Lithuania and the Jewish-catholic relations since World War II. She also served as an expert on Israeli Foreign Ministry delegations to UN world conferences as well as the Academic advisor of the International Task Force on Holocaust Education, Remembrance and Research (now IHRA).

Ronen, Avihu is Professor Emeritus of Tel Hai Academic College and Associate lecturer at the University of Haifa. He published several books and papers concerning the Jewish youth movement during the Holocaust and its aftermath. Examples of his publications are *The Battle for Life: The Zionist Underground in Hungary* (Hebrew, 1994; Hungarian, 1998), "Women as leaders" in *Jewish Women: A Comprehensive Historical Encyclopedia* (English, 2006), "Collaborator or Would-Be Rescuer? The Barenblat Trial and the Image of a Judenrat Member in 1960s Israel" (co-authored with Hadas Agmon and Assaf Danziger) in *Yad Vashem Studies* (39, no. 2 (2011)) and *Condemned to Life: The Diaries and Life of Chajka Klinger* (Hebrew, 2011) which won the 2013 Yad Vashem International Book Prize for Holocaust Research. He also edited *I am Writing these Words to You: The original Diaries, Bedzin 1943* by Chajka Klinger (2018).

Schoeps, Julius H. is Professor Emeritus of Modern History and the Founding Director of the Moses Mendelssohn Centre for European-Jewish Studies at the University of Potsdam, Germany. His main research interests are German-Jewish History, Intellectual History, Zionism, German-Jewish Relations, Nazi Looted Art, and antisemitism. Schoeps also did extensive research on Jewish emigration from the former Soviet Union and its impacts on Jewish life in re-unified Germany. His recent publications include *Handbook of Israel. Major Debates* (Co-Editor, 2016), *Begegnungen – Menschen, die meinen Lebensweg kreuzten* [*Encounters – people who crossed my life path*] (2016), and *Düstere Vorahnungen: Deutschlands Juden am Vorabend der Katastrophe (1933–1935)* [*Gloomy premonitions: Germany's Jews on the eve of the disaster*] (2018).

Schuster, Michal graduated in History and Archives at the Faculty of Arts of Masaryk University in Brno, Czech Republic. In the years 2005–2016 he worked as a curator and historian in the Museum of Roma Culture in Brno. Since 2016, he is a historian and researcher in the Terezín Initiative Institute in Prague, working on the Database of Roma Holocaust Victims project. He is also working as an external educator within the Department of Multicultural Education of the Faculty of Education of the Masaryk University in Brno. He specializes in the history of the Roma in the Czech Republic and also occasionally publishes and lectures for the professional and non-professional public. His publications include "Holocaust Romů jako "neznámý" nebo "zapomenutý"" ["Holocaust of the Roma as "Unknown" or "Forgotten""] (co-authored with Helena Sadílková and Milada Závodská) in *Dějiny a současnost 9/2015* and "Pronásledování protektorátních Romů s přihlédnutím k roku 1944" ["The Persecution of the Protectorate Roma in 1944"] in *Válečný rok 1944 v okupované Evropě a v Protektorátu Čechy a Morava* [*War Year 1944 in Occupied Europe and in the Czech and Moravian Protectorate*] (2015).

Sineaeva-Pankowska, Natalia is a Ph.D. candidate in Sociology. Her doctorate dissertation deals with the Holocaust distortion and identity in Moldova. In 2018, she was the Rotary Peace Fellow at Chulalongkorn University in Bangkok and the EHRI Fellow at the Elie Wiesel National Institute for the Study of the Holocaust in Bucharest. She has worked at the POLIN Museum of the History of Polish Jews as an education specialist and as a project coordinator at the 'NEVER AGAIN' Association, which deals with the commemoration of the Holocaust and other genocides as well as contemporary issues of diversity and tolerance with a particular focus on Eastern Europe and South-East Asia. Her main research interests are Holocaust history and memory, genocides in world history, and pedagogy of memorial sites. Among her recent publications is "Jak zwiedzający odbierają galerię Zagłada w Muzeum POLIN" ["How do visitors react to the Holocaust gallery at the POLIN Museum"] in *Zagłada Żydów. Studia i Materiały* [*Holocaust. Studies and Materials*], no. 12 (2016).

Tarant, Zbyněk is Assistant Professor of Near-Eastern and Israel Studies at the Department of Middle-Eastern Studies, University of West Bohemia in Pilsen, Czech Republic. While his main topics of research are the Czech-Israeli relations and the holocaust memorial culture in Israel (the latter being the topic of his dissertation, defended in 2012), he has been actively involved since 2006 in the research of contemporary antisemitism. He focuses mainly on cyber-hate and analysis of emerging threats in contemporary Central European antisemitism. He is the author and co-editor of several monographs in both Czech and English, including the 2013 book *Diaspora paměti* [*The Diaspora of Memory*].

Vago, Raphael, born in Cluj, Romania in 1946 and living in Israel since 1958, earned his academic degrees at Haifa University and Tel Aviv University. Until his retirement in 2014, he served as Senior Lecturer at the Department of History, Tel Aviv University. His teaching and research areas focus on the modern history of East and Central Europe and the Balkans, as well nationality problems, extremist regimes and ideologies, antisemitism, and the Holocaust in the region, especially in Hungary and Romania. Currently he is also Senior Researcher at the Kantor Center for the Study of Contemporary European Jewry and the Cummings Center for Russian and East European Studies at Tel Aviv University. He is a member of the International Commission on the Holocaust in Romania and a frequent lecturer at Yad Vashem's International School for Holocaust Studies. In the past, he also served as a Visiting Professor for an academic year at the University of Calgary and Concordia University in Montreal.

Vincze, Zsofia Kata is currently Associate Professor at Eötvös Loránd University in Budapest. Her latest publication was "Dynamique de la cohésion de groupe, de la promotion et du rejet de la judéité en Hongrie" in *Juifs d'Europe: identités plurielles et mixité* (2017). She graduated from Babes-Bolyai University in ethnology-anthropology, comparative literature, philosophy, and theology. She received her Ph.D. in Budapest. She has continued her academic work in Boston and University of Oxford. She has two books, *Tradition, memory, identity. The Foundation Myth of Exodus*, Budapest 2004, and *Return to the Tradition*, Budapest 2009. She has also published co-authored books and more than fifty-five academic papers on the new nativist Western nationalism, governmental codes referring to antisemitism, Islamophobia, and xenophobia in the "new authoritarian" regimes.

Index of persons

Adday, Abdulrahman 220
Adunka, Evelyn 75 f.
Alpert, Michael 288
Arendt, Hannah 193
Arkush, Jonathan 277
Ash, Timothy Garton 11
Avineri, Shlomo 13
Azogui, Ron 272

Bahbouh, Charif 218
Bányai, František 197
Bar-On, Dan 193
Barzilai, Yisrael 148
Bauman, Zygmunt 18, 115
Bělohradská, Hana 183
Ben-Gurion, David 191 f.
Benshalom, Rafi 149, 152
Bergoglio, Franciscus Jorge Mario 18
Berman, Avraham 148
Berman, Yaacov 148
Bohin, Michal 242 f.
Bondy, François 11
Bondy, Ruth 180
Bor, Josef 183
Bosch, Hieronymus 174
Braham, Randolph L. 273
Brandt, Willy 5
Brotman, Stuart 288
Brynych, Zbyněk 170, 183
Bureš, Toník 165

Cameron, David 276
Čapek, Karel 175
Celan, Paul 13
Chaplin, Charles 172
Clavreul, Gilles 279
Cohen-Weisz, Susanna 74 f.
Corbin, Jeremy 277

Dagan, Avigdor 179
Dalai Lama 172
Daniel, Bartoloměj 5
Danielová, Emílie 6

Danielová, Vlasta 7
Davies, Norman 7
Demény, Pal 147, 149
Diner, Dan 58

Einstein, Albert 200, 202
Eisenberger, Benjámin 149
Epstein, Berthold 242
Evans, Robert John 6

Fabiánová, Tera 4
Finkelkraut, Alain 5
Fischl, Viktor 179 f.
Frank, Anne 260 f.
Frankel, Zacharias 64
Freeman, Hillary 275
Freud, Sigmund 12
Friedl, Rafi 149
Frýd, Norbert 169, 183
Fuks, Ladislav 170–172, 174 f., 183 f.

Gábor, Péter 149, 157
Geiger, Abraham 64
Geminder, Bedřich 149, 157
Gerron, Kurt 180
Goldflam, Arnošt 179 f., 183 f.
Gomulka, Wladyslaw 148, 157
Gorbachev, Mikhail 10
Gottlieb, Dinah 9
Gromyko, Andrei 150
Grosman, Ladislav 176 f., 183 f.
Grossman, Haika 146–148
Grynberg, Henryk 287
Gur, David 147, 156 f.
Gutman, Israel 146, 260 f.

Havel, Václav 11, 175
Haverbeck, Ursula 13
Heisler, András 115, 137, 274
Herz, Juraj 172, 174, 183
Herzl, Theodor 13, 315
Hildesheimer, Esriel 64

Hitler, Adolf 10 f., 57, 60, 165, 176, 180, 183, 196 f., 262, 319
Horváth, Jozef 4, 10
Hübschmannová, Milena 238 f., 244

Ištván, Jan 8

Jaspers, Karl 193
Jélinek, Tomáš 210, 214
Jesus Christ 4
John Paul II 3
Judt, Tony 11

Kádár, Iván 147
Kadár, Ján 176, 183
Kafka, Franz 12, 166, 169, 201 f.
King, Martin Luther 182
Kis, Danilo 12
Klarsfeld, Serge 263 f.
Klos, Elmar 176
Klug, Brian 2
Konrád, György 7, 11
Kopecký, Václav 169
Koves, Schlomo 273
Kovner, Abba 153
Kozlovski, Jan 148
Krakauer, David 288
Kraus, Karl 13
Kraus, Ota 165
Kulbak, Moshe 58
Kulka, Erich 165, 180, 183
Kundera, Milan 4, 7, 10, 12 f., 202 f.
Kurejová, Jolana 5

Lacková, Elena 238 f, 244
Lahola, Leopold 183
Langer, František 198 f.
Langer, Jiří Mordechai 198
Lauder, Ronald 64, 98, 130, 137, 213
Lévy, Bernhard-Henri 67
Lewartowski, Józef 146
Lhoťan, Lukáš 207, 219, 222 f., 226 f.
Lustig, Arnošt 169 f., 174–176, 179, 183

Madoff, Bernard 5
Mahler, Gustav 12
Mareš, Miroslav 218, 222

Masaryk, Tomáš Garrigue 6 f., 10
May, Theresa 276
Mayer-Megyeri, József 147
Meir, Yosef 149
Meir, Yoska 152
Mengele, Josef 9, 182
Meyrink, Gustav 196
Michnik, Adam 11
Mílek, Alfred 242
Mira, Gola 146
Mitterand, Francois 263 f.
Mňačko, Ladislav 183
Moorhouse, Roger 7
Moskalyk, Antonín 174
Moszkowicz, Daniel 146
Muhammad Abbas 218
Muneeb Hasan al-Rawi 218, 226

Nałkowska, Zofia 287
Naumann, Friedrich 5
Němec, Jan 170, 183

Orbán, Viktor 123
Oren, Mordeachi 158
Otčenášek, Jan 169, 183

Palgi, Aryeh 158
Pavlát, Leo 214, 225 f.
Peroutka, Ferdinand 199
Peter, David 197, 214 f., 227
Philips, Bilal 223
Pick, J.R. 183
Pinkus, Benjamin 150
Pinto, Diana 65, 122 f.
Poláček, Jan 172
Poláček, Karel 175
Polanski, Roman 172, 286

Radok, Alfred 165 f., 183 f.
Rapoport, Nathan 151
Reik, Haviva 148, 152
Reinharz, Jehuda 14
Rektor, Joab H. 180
Roosevelt, Eleanor 260
Rosenberg, Moritz 158
Rosenberg, Yaakov (Benito) 149
Rosman, Mordechai 153

Roth, Egon   147, 151
Roth, Josef   12
Roubíček, Josef   167f., 171
Rupnik, Jacques   11
Růžičková, Božena   7

Said, Edward   4
Salner, Peter   214
Schmidt, Helmut   5
Schön, Erich   165
Segev, Tom   191f.
Shapira, Anita   13
Shavit, Yaacov   14
Shehadeh, Samer   224
Sidon, Karol   3, 214
Silbermann, Alphons   67
Škvorecký, Josef   183f.
Smolar, Hirsch   153
Solomon, Yaacov   157
Sólyom, László   147, 149, 157
Soral, Alain   13
Soros, George   121, 274
Špira, Maxmilián   6
Spychalski, Marian   148, 156f.
Stangl, Franz   257
Strauss, Deborah   288
Szmidt, Andrzej   146

Tabori, Georg   180
Tenenbaum, Mordechai   146
Tiso, Jozef   176
Topol, Jáchym   180–184
Tuwim, Julian   284f.

Valls, Manuel   271f.
Větrovec, Lukáš   222f., 226
Vondrák, Patrik   224

Warschauer, Jeff   288
Wasserstein, Bernard   60, 65
Weil, Jiří   166f., 169–171, 183f., 316
Weisskopf, Rudolf   8
Williamson, Richard   13
Wistrich, Robert   13, 21
Wittenberg, Yitzhak   146

Yaari, Aryeh   154
Yavets, Zvi   13
Yelin, Chaim   146

Zachriasz, Szymon   153
Zeman, Miloš   14
Zigeuner, Moshe   237
Zuckerman, Yitzhak   146, 148, 156

www.ingramcontent.com/pod-product-compliance
Lightning Source LLC
Chambersburg PA
CBHW070749230426
43665CB00017B/2308